MW01609412

Student Study Guide/Solutions Manual for use with

BIOCHEMISTRY The Molecular Basis of Life

SIXTH EDITION

Trudy McKee

James R. McKee

Michael G. Sehorn

New York Oxford
OXFORD UNIVERSITY PRESS

Oxford University Press is a department of the University of Oxford.
It furthers the University's objective of excellence in research,
scholarship, and education by publishing worldwide.

Oxford New York
Auckland Cape Town Dar es Salaam Hong Kong Karachi
Kuala Lumpur Madrid Melbourne Mexico City Nairobi
New Delhi Shanghai Taipei Toronto

With offices in
Argentina Austria Brazil Chile Czech Republic France Greece
Guatemala Hungary Italy Japan Poland Portugal Singapore
South Korea Switzerland Thailand Turkey Ukraine Vietnam

For titles covered by Section 112 of the US Higher Education Opportunity Act, please visit www.oup.com/us/he for the latest information about pricing and alternate formats.

Published by Oxford University Press
198 Madison Avenue, New York, New York 10016
http://www.oup.com

Oxford is a registered trademark of Oxford University Press

ISBN: 978-0-19-020991-9

Printing number: 9 8 7 6 5 4 3 2 1

Printed in the United States of America
on acid-free paper

Contents

1 Biochemistry: An Introduction

Brief Outline of Key Terms and Concepts

OVERVIEW

CHEMOSYNTHESIS; PHOTOSYNTHESIS

1.1 WHAT IS LIFE?

All living organisms obey the chemical and physical laws. Life is complex, dynamic, organized, and self-sustaining.

Life is cellular and information-based. Life adapts and evolves.

BIOMOLECULE; MACROMOLECULE; ENZYME; METABOLISM; HOMEOSTASIS GENE; MUTATION

1.2 BIOMOLECULES

In living organisms, most molecules are organic. Organic FUNCTIONAL GROUPS determine the chemical properties of organic molecules. The four classes of small biomolecules are amino acids, sugars, nucleotides, and fatty acids.

HYDROPHOBIC, HYDROPHILIC

BIOPOLYMERS	SMALL BIOMOLECULES
PROTEINS	AMINO ACIDS
POLYSACCHARIDES	SUGARS
NUCLEIC ACIDS	NUCLEOTIDES

Several types of LIPIDS have FATTY ACID components.

AMINO ACIDS AND PROTEINS: POLYPEPTIDE (PEPTIDE, OLIGOPEPTIDE); PEPTIDE BOND; NEUROTRANSMITTER

CARBOHYDRATES: SUGARS, MONOSACCHARIDES

FATTY ACIDS: SATURATED, UNSATURATED

DNA, RNA; PURINE, PYRIMIDINE; GENOME

DNA is an ANTIPARALLEL double helix.

RNA; TRANSCRIPTION; NONCODING RNA

BIOCHEMISTRY IN THE LAB: AN INTRODUCTION

GENOMICS; GENE EXPRESSION; FUNCTIONAL GENOMICS; PROTEOMICS; BIOINFORMATICS

1.3 IS THE LIVING CELL A CHEMICAL FACTORY?

SYSTEM; AUTOPOIESIS

BIOCHEMICAL REACTIONS

- NUCLEOPHILIC SUBSTITUTION: NUCLEOPHILE, ELECTROPHILE, LEAVING GROUP; HYDROLYSIS
- ELIMINATION REACTION: DEHYDRATION
- ADDITION REACTION: HYDRATION
- ISOMERIZATION REACTION
- OXIDATION-REDUCTION (REDOX) REACTIONS
  - OXIDIZE, REDUCE
  - OXIDIZING AGENT, REDUCING AGENT

ENERGY: THE CAPACITY TO MOVE MATTER

In living organisms, energy is usually generated by REDOX REACTIONS.

AUTOTROPH; HETEROTROPH

PHOTOAUTOTROPH; CHEMOAUTOTROPH

CHEMOHETEROTROPH; PHOTOHETEROTROPH

METABOLISM is the sum of all enzyme-catalyzed reactions in a living organism. Classes of biochemical pathway:

- METABOLIC (ANABOLIC and CATABOLIC)
- ENERGY TRANSFER
- SIGNAL TRANSDUCTION

BIOLOGICAL ORDER

In living organisms, a constant input of energy is needed to sustain processes of highly ordered complexity. Examples include synthesis of biomolecules, transport across membranes, active transport, cell movement, and waste removal.

1.4 SYSTEMS BIOLOGY is an attempt to reveal the functional properties of living organisms by developing mathematical models of interactions from available data sets. The systems approach has provided insights into the EMERGENCE, ROBUSTNESS, and MODULARITY of living organisms. Compare the systems approach to REDUCTIONISM.

EMERGENT PROPERTY; DEGENERACY; FEEDBACK CONTROL; NEGATIVE FEEDBACK; POSITIVE FEEDBACK; SYSTEM, NETWORK, AND MODULE

1.1 WHAT IS LIFE?

LIFE IS:

1. Complex and dynamic: biomolecules, biochemical reactions
2. Organized and self-sustaining, characterized by
   —hierarchical order, from atoms to multicellular organisms, that requires a constant influx of energy and matter
   —enzyme-catalyzed reactions; metabolic pathways that can be regulated; *(homeostasis)*
3. Cellular: cell membranes control transport into and out of the cell
4. Information-based
5. Able to adapt and evolve (*mutations*)

1.2 BIOMOLECULES

ORGANIC REVIEW: FAMILIES (FUNCTIONAL GROUPS)

alkene (C=C)	aldehyde (carbonyl)	amine (amino)
alcohol (hydroxyl)	ketone (carbonyl)	amide (amido)
thiol (thiol, –SH)	ester (ester)	carboxylic acid (carboxyl)

NOTE: The α–carbon in carbonyl, carboxyl, and amido groups is the carbon *adjacent to* a carbonyl carbon.

SMALL BIOMOLECULES: MAJOR CLASSES

SMALL BIOMOLECULES ARE BUILDING BLOCKS FOR LARGE BIOMOLECULES:

amino acids	⟶	peptides, polypeptides, proteins
monosaccharides	⟶	carbohydrates; polyglycans
fatty acids	⟶	(components of lipids)
nucleotides	⟶	nucleic acids (RNA, DNA)

Small biomolecules also carry out special biological functions (e.g., as neurotransmitters or hormones), serve as energy sources, and/or take part in complex reaction pathways.

AMINO ACIDS AND PROTEINS

- Amino acids contain an amino group, a carboxylic acid group, and a side chain (R group). In α-amino acids, the amino group is attached to the α-carbon.
- Amino acids are linked together by peptide (amide) bonds, which have double-bond character that impacts the overall structure with its rigidity.
- Amino acid *residues* in proteins
- Polypeptides: peptides (up to 50 amino acids), proteins (longer)

SUGARS AND CARBOHYDRATES: MONOSACCHARIDES, POLYSACCHARIDES

- Sugars contain alcohol groups and either aldehydes (in aldoses) or ketones (in ketoses).

- Names may also indicate number of carbons, such as "aldohexose."
- Polysaccharides (polyglycans) include starch and cellulose (plants), glycogen (animals).
- Nucleotides contain either ribose or deoxyribose.
- Glycoproteins and glycolipids are proteins and lipids that contain carbohydrates.
- Functions: important source of energy, structural support, and participation in intracellular and intercellular communication.

FATTY ACIDS

- Fatty acids contain one carboxylic acid with a long hydrocarbon chain. Fatty acids are usually unbranched with an even number of carbon atoms.
- Unsaturated fatty acids have at least one double bond in the hydrocarbon chain. Saturated fatty acids have only C–C single bonds (they're "saturated" with hydrogen atoms).[1]
- Fatty acids are components of lipid molecules. Lipids are not water soluble.
- Triacylglycerols (three fatty acids + glycerol) store energy.
- Phosphoglycerides (two fatty acids + glycerol + phosphate + a polar compound) are major structural components of cell membranes.

NUCLEOTIDES AND NUCLEIC ACIDS (DNA AND RNA)

- Nucleotides contain a five-carbon sugar (ribose or deoxyribose), a purine or pyrimidine base, and one or more phosphate groups.

  —Purine bases: adenine and guanine

  —Pyrimidine bases: thymine, cytosine (DNA only), and uracil (RNA only)
- Nucleotides are essential in many biosynthetic and energy-generating reactions. ATP is a nucleotide.
- Nucleotides join together via phosphodiester linkages to form nucleic acids—DNA or RNA. The nucleic acid sugar-phosphate backbones alternate . . . sugar-phosphate-sugar-phosphate . . . . A purine or pyrimidine base is connected to each sugar.
- The specific and unique order of the bases—the base sequence—holds genetic information. Genetic information flows from DNA to RNA to proteins.
- Hydrogen bonding occurs between specific base pairs: DNA: AT, GC
RNA: AU, GC
- This hydrogen bonding stabilizes the DNA double helix and allows **TRANSCRIPTION**—the synthesis of RNA from DNA.
- Types of RNA: **mRNA** (messenger RNA), **rRNA**(ribosomal RNA), and **tRNA** (transfer RNA). All three types work together to synthesize polypeptides. Other types of RNA molecules include ncRNA, siRNA, miRNA, snRNA, and snoRNA (noncoding, short interfering, micro, small nuclear, and small nucleolar RNA, respectively).

---

[1] Note the difference between the biochemistry definition of a saturated fatty acid and the organic chemistry definition of a saturated molecule: a saturated fatty acid contains no alkenes, but still has a C=O in its carboxyl group.

EXAMPLES OF EACH CLASS OF SMALL BIOMOLECULE

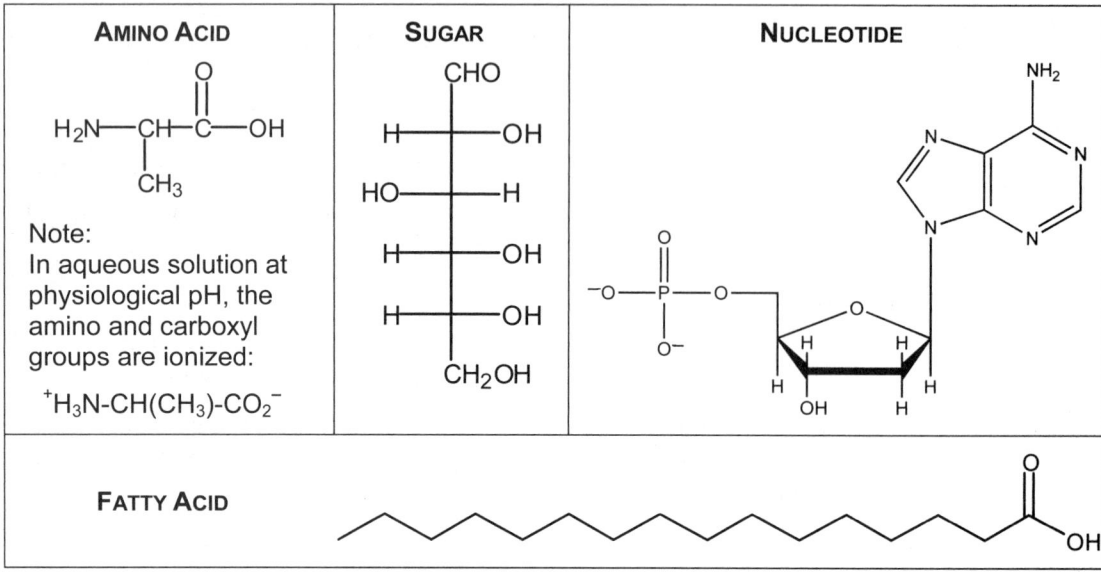

AMINO ACID	SUGAR	NUCLEOTIDE
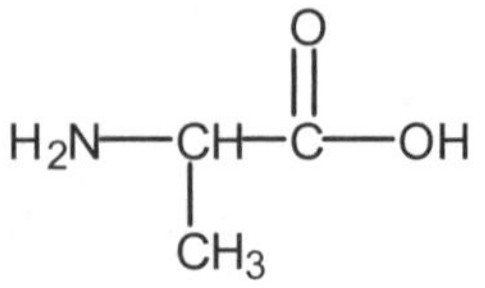 Note: In aqueous solution at physiological pH, the amino and carboxyl groups are ionized: $^{+}H_3N\text{-}CH(CH_3)\text{-}CO_2^{-}$	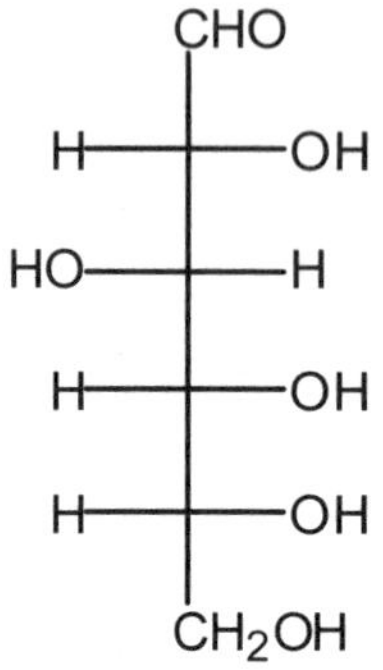	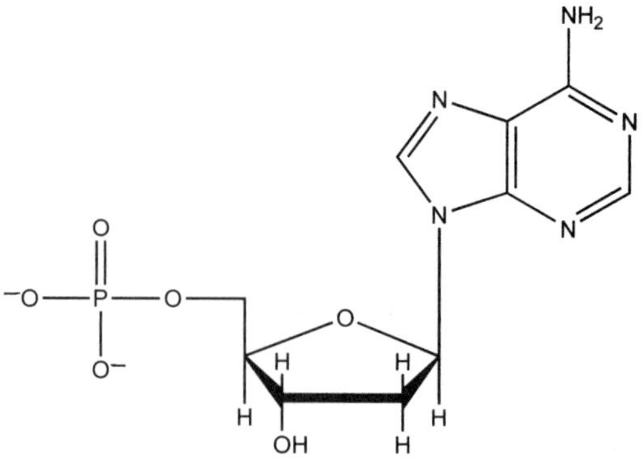
FATTY ACID	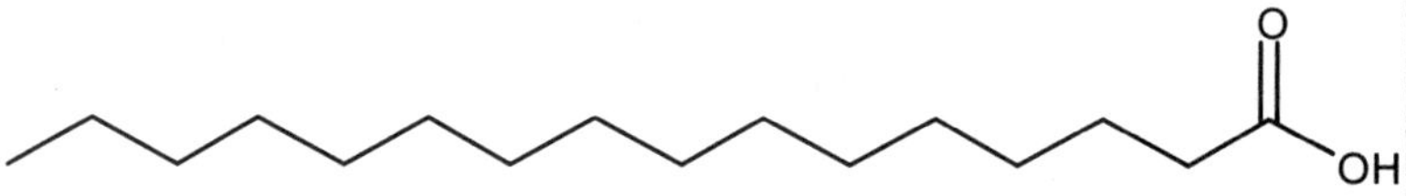	

ORGANIC FUNCTIONAL GROUPS THAT MAY BE NEW TO YOU:

PHOSPHOESTER	PHOSPHODIESTER	PHOSPHOANHYDRIDE	Compare with an ANHYDRIDE:
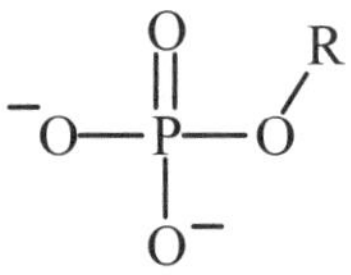	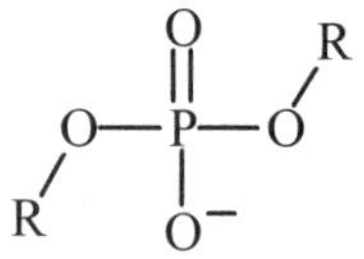	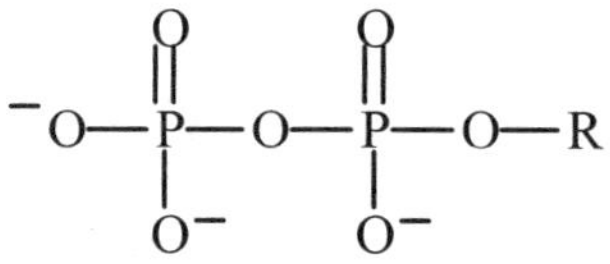	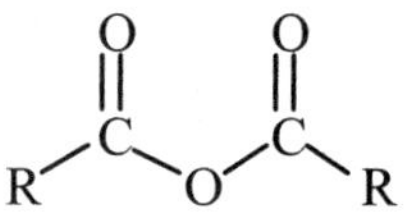

1.3 IS THE LIVING CELL A CHEMICAL FACTORY?

BIOCHEMICAL REACTIONS, ENERGY, METABOLISM, AND BIOLOGICAL ORDER

AUTOPOIESIS and **AUTOPOIETIC SYSTEMS**:
autonomous, self-organizing, and self-maintaining

BIOCHEMICAL REACTIONS ARE CATALYZED BY ENZYMES *(BUT THE REACTIONS THEMSELVES ARE THE SAME ONES LEARNED IN ORGANIC CHEMISTRY)*

MOST COMMON REACTION TYPES (AND SOME EXAMPLES)

- Nucleophilic substitution (hydrolysis of peptides to form amino acids); **NUCLEOPHILE, ELECTROPHILE, LEAVING GROUP**
- Elimination (removal of water to form an alkene)
- Addition (hydration adds H_2O to an alkene; hydrogenation adds H_2)
- Isomerization (converting glucose-6-phosphate to fructose-6-phosphate; both have the same molecular formula but a different arrangement of atoms)
- Oxidation-reduction (redox) reactions (oxidizing an alcohol to an aldehyde)

BIOCHEMICAL OXIDATION AND REDUCTION: REMEMBER THE ORGANIC DEFINITIONS

In general chemistry, you learned that oxidation is the removal of electrons and that reduction adds electrons (resulting in a lower, or *reduced*, charge or oxidation number). In organic chemistry, you learned that oxidation is the addition of oxygen (and/or the removal of hydrogen) and that reduction is the removal of oxygen (and/or the addition of hydrogen). In biochemistry, we'll tap into both of these definitions.

Redox reactions always involve a *transfer* of electrons, so when anything is oxidized, something else must be reduced. An **OXIDIZING AGENT** does the oxidizing (and becomes reduced in the process); a **REDUCING AGENT** does the reducing (and becomes oxidized).

Biochemical redox reactions involve transferring one or two electrons at a time. Usually an H^+ rides along, so what appears to be transferred is an H atom (H•) or a hydride ion (H:–). (Think of the H atom as an H^+ with one electron, and the hydride ion as an H^+ with two electrons.)

Let's look at an essential example: the reduction of the coenzyme NAD^+ to produce NADH. With its additional H, we can identify NADH as the reduced form. With its extra + charge, NAD^+ is the oxidized form. When NAD^+ is reduced, something else must be oxidized. NAD^+ is the oxidizing agent, because NAD^+ oxidizes another molecule. Here's a specific example: NAD^+ oxidizes ethanol to form acetaldehyde:

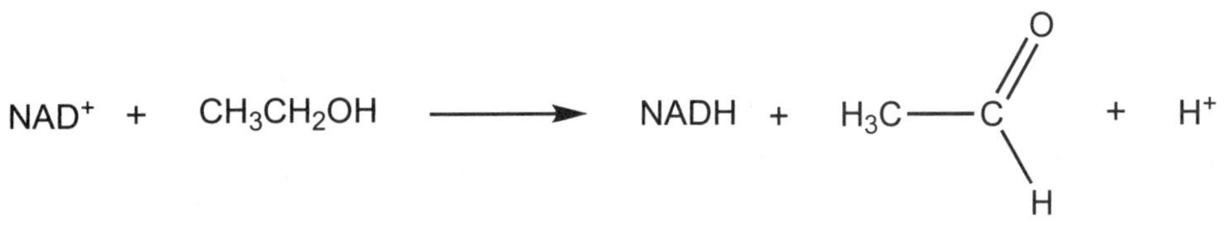

oxidized form	reduced form	reduced form	oxidized form

We could also say that ethanol reduces NAD^+ to form NADH. No matter how we look at it, two electrons (and an H^+) were transferred from ethanol to NAD^+.

ENERGY

Cells obtain energy by oxidizing biomolecules (or certain minerals, or by photosynthesis). Energy is in the form of electrons. ***The more reduced the molecule, the more energy it contains.*** Take another look at the classes of biomolecules. Can you see why fatty acids, with all of those $–CH_2–$ groups, produce more energy (ATP) when they're metabolized than the oxygen-rich sugars do?

Organisms can be classified by how they obtain energy: autotrophs (photoautotrophs or chemoautotrophs) or heterotrophs (chemoheterotrophs or photoheterotrophs).

METABOLISM = SUM OF ALL ENZYME-CATALYZED REACTIONS IN A LIVING ORGANISM

METABOLIC PATHWAYS synthesize and degrade biomolecules.

<u>ANABOLIC PATHWAYS</u>	<u>CATABOLIC PATHWAYS</u>
BIOSYNTHESIS	**DEGRADATION**
(*Anna* builds biomolecules . . .	. . . and the *cat* claws them apart.)
ANABOLISM REQUIRES ENERGY ATP→ADP	**CATABOLISM STORES ENERGY** (as ATP) ADP→ATP
Uses electrons (*typically*) from	Electrons are captured by

NADH, $FADH_2$, or NADPH to form NAD^+, FAD, or $NADP^+$ and larger biomolecules	NAD^+, FAD, or $NADP^+$ to form NADH, $FADH_2$, or NADPH and smaller, more oxidized molecules

ENERGY TRANSFER PATHWAYS: CAPTURE AND TRANSFORM ENERGY

Example: Photosynthesis captures light energy and converts it into chemical bond energy (in sugar molecules).

SIGNAL TRANSDUCTION PATHWAYS: RECEIVE AND RESPOND TO SIGNALS

Steps:
1. Reception
2. Transduction
3. Response

BIOLOGICAL ORDER IS ACCOMPLISHED BY:

1. Synthesis and degradation of biomolecules
2. Transport of ions and molecules across cell membranes (**ACTIVE TRANSPORT** requires energy provided by ATP hydrolysis)
3. Production of motion
4. Removal of metabolic waste products and other toxic substances

LIVING CELLS NEED A CONSTANT FLOW OF ENERGY TO MAINTAIN THIS HIGH DEGREE OF ORDER.

1.4 SYSTEMS BIOLOGY: LIVING SYSTEMS AS INTEGRATED SYSTEMS

EMERGENCE

Certain characteristics result, or *emerge,* from interactions between the components of a system. These **EMERGENT PROPERTIES** cannot be explained by studying the individual parts separately.[2]

ROBUSTNESS: WHAT MAKES A SYSTEM ROBUST?

DEGENERACY: the same or similar functions can be performed by different components

COMPLEX CONTROL MECHANISMS such as the regulation of enzymes via **POSITIVE FEEDBACK** or **NEGATIVE FEEDBACK**

SYSTEMS BIOLOGY MODEL CONCEPTS

SYSTEM: an interconnected and interacting assembly of biomolecules

NETWORK: a group of interconnected biomolecules that perform one or more functions

MODULE: a component or subsystem that performs a specific function

BIOCHEMISTRY IN THE LAB: 1.1 AN INTRODUCTION

THE HUMAN GENOME PROJECT; GENOMICS; FUNCTIONAL GENOMICS; PROTEOMICS; BIOINFORMATICS

---

[2] The opposite of emergence is reductionism (to understand the whole, understand the parts), an incomplete approach to understanding living systems.

HINT: *NOW* is the best time to be *SURE* that you have a SOLID understanding of these concepts from organic:

Polar vs.versus nonpolar	Hydrogen bonding
Hydrophilic vs.versus hydrophobic	Dipole-dipole interactions
Acidic vs.versus basic	van der Waals forces

All of these are *essential* in developing a deep sense of why biomolecules "behave" the way they do and will help you to move from memorization to *learning* much more readily.

Whether a biomolecule is polar or nonpolar has a HUGE impact on its properties, including how it interacts and reacts with other molecules. For example, how a protein's amino acid R groups interact determines its overall shape, and its specific shape determines whether that protein will be fibrous and give structural support or a globular enzyme that catalyzes a very specific reaction.

If you feel the least bit rusty, invest the time *now* to help your understanding throughout the semester.

AFTER STUDYING THIS CHAPTER, YOU SHOULD BE ABLE TO:

- Identify functional groups in a given molecule.
- Recognize the structures of the four classes of small biomolecules. Classify a given compound as an amino acid, sugar, fatty acid, or nucleotide and describe its function in living organisms.
- Recognize the four types of reactions: nucleophilic substitution, elimination, oxidation-reduction, and addition. In oxidation-reduction reactions, recognize which reactant was oxidized and which was reduced.
- Describe the differences between Archaea, Bacteria, and Eukarya; prokaryotic cells and eukaryotic cells; prokaryotes and eukaryotes.
- Demonstrate an understanding of the general characteristics of life, living organisms, and the study of biochemistry, including recent discoveries and the scientific insights that have resulted.

Use this space to note any additional objectives provided by your instructor.

CHAPTER 1: SOLUTIONS TO REVIEW QUESTIONS

1.1 a. biomolecule—the molecules synthesized by living organisms

b. macromolecule—polymers of certain biomolecules

c. enzyme—biomolecular catalyst

d. metabolism—the sum total of all reactions in a living organism

e. homeostasis—the ability of living organisms to regulate their metabolism regardless of their internal and external environments

1.2 a. gene—linear sequence of nucleotides in DNA that ultimately encodes the linear sequence of amino acids in a specific protein

b. mutation—a change in the linear sequence of nucleotides in DNA

c. hydrocarbon—hydrophobic molecules composed of carbon and hydrogen atoms

d. hydrophobic—incapable of hydrogen bonding with polar molecules

e. hydrophilic—capable of hydrogen bonding with polar molecules

1.4 a. polypeptide—polymer of amino acids

b. peptide—a polymer of up to 50 amino acids

c. protein—a molecule consisting of one or more polypeptides

d. peptide bond—amide linkage between adjacent amino acids

e. standard amino acids—20 amino acids commonly found in polypeptides; consist of a specific R-group, an amino group, and a carboxyl group attached to the same α-carbon atom

1.5 a. sugar—basic unit of a carbohydrate

b. monosaccharide—simple sugar consisting of a single sugar molecule

c. polysaccharide—polymer of sugar molecules containing more than 20 monosaccharide units

d. glucose—6-carbon aldohexose sugar

e. cellulose—polymer of glucose with (β1, 4) glycosidic bonds

1.7 a. nucleotide—composed of a 5-carbon sugar, a nitrogenous base, and one or more phosphate groups

b. purine—bicyclic nitrogenous base

c. pyrimidine—monocyclic nitrogenous base

d. nucleic acid—polymer of nucleotides linked together by phosphodiester bonds

e. ribose—5-carbon sugar

1.8 a. DNA—deoxyribonucleic acid

b. RNA—ribonucleic acid

c. double helix—two antiparallel polynucleotide strands wound around each other

d. genome—an organism's entire set of DNA sequences

e. transcription—process where RNA molecules are synthesized from a DNA template

1.10 a. transcription factor—a class of proteins that bind to specific regulatory DNA sequences to regulate gene expression

b. response element—specific regulatory DNA sequence that transcription factors bind

c. signal molecule—a molecule that binds to a receptor protein

d. RNA interference—mediated by siRNA to function as an antiviral defense

e. ribosome—ribonucleoprotein complex that synthesizes polypeptides

1.11 a. nucleophile—an atom or group with an unshared pair of electrons that is involved in a displacement (nucleophilic substitution) reaction

b. electrophile—an electron-deficient species

c. leaving group—the outgoing nucleophile that leaves with its electron pair

d. adenosine triphosphate—the energy carrier molecule

e. anhydride—a molecule containing two carbonyl groups linked through an oxygen atom

1.13 a. redox reaction—a transfer of electron from a donor to an electron acceptor

b. oxidizing agent—atom or group reduced during an oxidation/reduction reaction

c. reducing agent—atom or group oxidized during an oxidation/reduction reaction

d. NADH—reduced form of nicotinamide adenine dinucleotide

e. oxidized molecule—a molecule that donated electrons

1.14 a. FAD—oxidized form of flavin adenine dinucleotide

b. hydride ion—$H^{:-}$

c. energy—the ability to do work

d. electron transport pathway—a series of linked membrane-embedded electron carrier molecules

e. coenzyme—small molecules that function in association with enzymes as carriers of small molecular groups

1.16 a. metabolic pathway—series of chemical reactions occurring in a cell

b. anabolic pathway—small precursors are used to generate large complex molecules

c. catabolic pathway—large complex molecules are degraded into simple products

d. glycolysis—a 10-reaction pathway that degrades glucose to two pyruvates to generate energy

e. signal transduction pathway—a pathway that permits a cell to receive and respond to signals from its environment

1.17 a. system—an interconnected and interacting assembly of biomolecules

b. network—a group of interconnected molecules that perform one or more functions; a metabolic network consists of interconnected biochemical pathways that synthesize and degrade biomolecules

c. emergent property—properties that emerge from interactions among parts; emergent properties have characteristics that cannot be predicted by analyzing their component parts

d. degeneracy—the capacity of structurally different parts to perform the same or similar functions

e. feedback control—the process by which a product of a pathway serves to regulate that same pathway

1.19 The six major elements present in living organisms are carbon, hydrogen, nitrogen, oxygen, phosphorus, and sulfur.

1.20 The functional group(s) in each molecule are:

a. aldehyde group
b. carboxylic acid and amino groups
c. sulfhydryl group
d. ester group
e. alkene
f. amide group
g. ketone group
h. alcohol group

1.22 a. The functions of fatty acids include energy storage and membrane components. Fatty acids are components of larger lipid molecules within cell membranes. Some fatty acids are also precursors to hormones.

b. Sugars function as energy sources and as components of polysaccharides (such as starch, cellulose, glycogen, and chitin), which function as energy sources and/or as structural elements. Nucleotides contain the sugars ribose or deoxyribose. Glycoproteins and glycolipids located on cell membranes also contain sugars and play critical roles in many cellular interactions.

c. Nucleotides are involved in energy transformations. They are also components of DNA and RNA.

d. Most amino acids are the building blocks of proteins. Some have special functions as neurotransmitters or as precursors to other molecules (such as vitamins). Peptides and proteins have a variety of functions, including structural support or catalytic activity.

1.23 DNA is the repository of each organism's genetic information. RNA is the nucleic acid that is involved in the expression of genetic information, primarily in various aspects of protein synthesis.

1.25 Hydrolysis

1.26 NADH is the reducing agent and propanoic acid is the oxidizing agent.

1.28 Glucose converted to cellulose is an example of an anabolic reaction. Conversion of glucose, ADP, and Pi to CO_2, ATP, and NADH is an example of a catabolic reaction.

1.29 CO_2 and CH_4

1.31 The first reaction in the utilization of glucose as an energy course is the phosphorylation of glucose by ATP to yield glucose-6-phosphate and ADP. In this reaction the hydroxyl oxygen is the nucleophile. The phosphorous atom is the electrophile, and ADP is the leaving group.

1.32 Plants dispose of waste products either by degradation or by storage in vacuoles or cell walls.

1.34 The primary functions of metabolism are acquisition and utilization of energy, synthesis of biomolecules, and removal of waste products.

1.35 Examples of the following reactions include:

a. nucleophilic substitution—the reaction of glucose with ATP to produce glucose 6-phosphate and ADP

b. elimination—the dehydration of 2-phosphoglycerate to form phosphoenolpyruvate

c. oxidation-reduction—the oxidation of ethyl alcohol to acetaldehyde

d. addition—the conversion of fumarate to malate

1.37 Both an airplane autopilot system and a biological system are robust (i.e., they have the ability to maintain stability despite changes in the environment or other events that threaten the continuation of system functions). They both have feedback control mechanisms, in which information regarding internal processes is used to adjust functions to maximize performance. The actual fail-safe mechanisms differ in that human-made systems have redundancy (duplicate parts) while biological systems have degeneracy, in which duplicate (or similar) functions may be carried out by different parts of the system.

1.38 In addition to being an important energy source, carbohydrates are important structural molecules in organisms and have a role in intracellular and intercellular communication.

1.40 The largest biomolecules in living organisms are the nucleic acids and the proteins. Nucleic acids store genetic information (DNA) and mediate the synthesis of proteins (RNA). The proteins are the tools that perform all of the tasks required to sustain living processes. Polysaccharides are also large biomolecules with structural and energy storage functions.

1.41 Nucleotides participate in energy-forming and energy-generating reactions. Much of the energy available to drive biochemical reactions is stored in ATP molecules.

1.43 In a hierarchical system, such as that found in living organisms, emergent properties have characteristics that cannot be predicted from the analysis of its component parts. For example, the structural and functional properties of the protein hemoglobin, which is composed of carbon, nitrogen, hydrogen, oxygen, and iron atoms, cannot be determined from analysis of those individual components.

1.44 The nucleotide base sequence of each type of mRNA molecule codes for the amino acid sequence of a specific polypeptide. Each tRNA molecule carries a specific amino acid, which it subsequently delivers to the ribosome for incorporation into a polypeptide during protein synthesis. Ribosomal RNA molecules contribute to the structural and functional properties of ribosomes. Each polypeptide is manufactured as the base sequence information in the mRNA is translated by a ribosome. As base pairing occurs between the codon sequence of mRNA and the anticodon sequence of tRNA molecules, the amino acids are brought into close proximity and a peptide bond is formed.

1.46 Both human-designed complex systems (such as machines or factories) and living systems require raw materials (nutrients) and energy to manufacture components; systems of both types also produce waste products and heat. Machines can be designed to self-regulate upon receiving feedback from the environment (e.g., by monitoring temperature to determine heating or cooling needs). In contrast, living systems are self-sustaining; that is, they produce and repair all of their own structural and functional components and, via nucleic acids, they build the machines (enzymes) that make the components. Even the nucleic acids themselves are reproduced by living systems. This level of self-sustainability is not present in human-designed complex systems.

1.47 Both autopoietic systems and factories take in raw materials (nutrients or parts), use energy, manufacture products, and remove waste. Both maintain an inventory and have mechanisms in place to preserve system function and product integrity via biological feedback and regulation mechanisms or factory quality control. In contrast to factories, autopoietic systems synthesize structural components, energy storage molecules, and molecules that regulate and/or modify system functions. This would be analogous to factories using raw materials to create their own employees, buildings, machines, computers, and fuel. Autopoietic systems are autonomous, self-organizing, and self-maintaining.

CHAPTER 1: SOLUTIONS TO FILL-IN-THE-BLANK QUESTIONS

1.49 1 million

1.50 Degeneracy

1.52 Dehydration

1.53 Mutations

1.55 Autopoiesis

1.56 Photoautotrophs

CHAPTER 1: SOLUTIONS TO SHORT-ANSWER QUESTIONS

1.58 When a species becomes truly extinct, the genetic line dies out. When a species evolves to meet changing environmental conditions, its genetic line continues and the organism is not truly extinct but adapted to its new environment.

1.59 The conversion of the atmosphere to one dominated by oxygen took millions of years. During much of this time, oxygen levels were below the concentration that would react explosively with methane. During this long time period methane levels decreased as it reacted with oxygen to form carbon dioxide. Eventually, all atmospheric methane was consumed before the oxygen levels became high enough for a catastrophic explosion to occur.

1.61 All organisms need energy and nutrients. If a species is able to adapt by increasing its capacity to compete for a known energy source or utilize a new unexploited energy source, it increases its chances for survival.

1.62 The growth of cells is regulated by elaborate control and feedback mechanisms designed to carefully regulate cell growth. Mutations in a just a few of the control genes can disable this elaborate process and result in out-of-control cell growth.

CHAPTER 1: SOLUTIONS TO THOUGHT QUESTIONS

1.64 The C–H bonds of fatty acids are the most reduced form of carbon found in organic molecules. Oxidation of these molecules to form carbon dioxide—the most oxidized form of carbon—has the highest energy yield. Also, fatty acids are stored without the need for water. They are therefore stored in smaller areas and with less mass than polysaccharides.

1.65

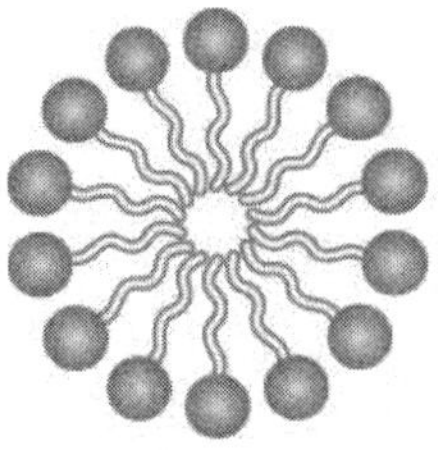

Micelle

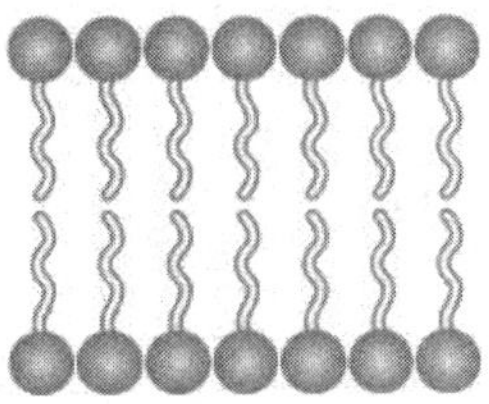

Bilayer segment

The structure of a micelle is illustrated in Figure 3.14 in your text. The phospholipid bilayer is a prominent feature of Figure 2.3 in your text.

1.67 If the bonds between carbon, hydrogen, and oxygen were either more or less stable, the ability to take on partial charge distribution would be disrupted.

1.68 The capacity of healthy bodies to adapt to high-cholesterol diets by inhibiting cholesterol synthesis is an example of the means by which living organisms regulate their metabolic processes.

1.70 The statement "the whole is more than the sum of its parts" refers to how many small parts collaborate to function as a whole. Understanding how the heart contracts to pump blood does not explain how all the cardiac cells work together to produce the contraction.

1.71 The ability of biological systems to have multiple functions is a strength of biological systems. In the case of the genetic code, 61 codons (triple base sequence) code for 20 different amino acids. Most amino acids have more than 1 codon. This example of degeneracy helps to minimize the effect of base substitution mutations in DNA.

1.73 There are 20 standard amino acids, so $X = 20$. The chain length is 10, so $n = 10$. The number of possible decapeptides is 20^{10} or 1.024×10^{13}. To draw all of the possibilities at the rate of 1 every 5 minutes would require about 97,300,000 years.

2 Living Cells

Brief Outline of Key Terms and Concepts

2.1 BASIC THEMES

WATER; **HYDROPHILIC VERSUS HYDROPHOBIC**

BIOLOGICAL MEMBRANES are lipid bilayers. **INTEGRAL AND PERIPHERAL PROTEINS, CHANNEL AND CARRIER PROTEINS, RECEPTORS, ANCHOR PROTEINS**

SELF-ASSEMBLY In living organisms, biomolecules in supramolecular structures are able to assemble spontaneously because of the steric information they contain.

MOLECULAR MACHINES
Molecular machines are complexes in living organisms that function like mechanical devices, with moving parts that perform work. **MOTOR PROTEIN**

MACROMOLECULAR CROWDING
Cells are densely crowded with macromolecules of diverse types. Macromolecular crowding is a significant factor in a wide variety of cellular processes. **EXCLUDED VOLUME**

SIGNAL TRANSDUCTION
Organisms receive, interpret, and respond to environmental information by means of the process of signal transduction, which has three phases: reception, transduction, response, and termination. **SIGNALS, NEUROTRANSMITTERS, HORMONES, CYTOKINES; LIGAND**

PROTEOSTASIS
Cells constantly monitor and restore proteostasis, a protein quality control mechanism. **PROTEOSTASIS NETWORK**

2.2 STRUCTURE OF PROKARYOTIC CELLS

Prokaryotic cells are small and structurally simple, have a **CELL WALL** and a **PLASMA MEMBRANE**, lack a nucleus and other **ORGANELLES**, and have circular DNA in the **NUCLEOID**.

CELL WALL; PEPTIDOGLYCAN; GRAM-POSITIVE, GRAM-NEGATIVE; LIPOPOLYSACCHARIDES (OUTER MEMBRANE); ENDOTOXINS; PORINS, TRANSMEMBRANE PROTEIN COMPLEXES; PERIPLASMIC SPACE; GLYCOCALYX, SLIME LAYERS, BIOFILMS

PLASMA MEMBRANE; PHOTOSYNTHESIS, RESPIRATION
CYTOPLASM; NUCLEOID, CHROMOSOME, PLASMIDS
PILI AND FLAGELLA: CELL MOVEMENT

2.3 STRUCTURE OF EUKARYOTIC CELLS

ENDOMEMBRANE SYSTEM, VESICLES

PLASMA MEMBRANE
Provides strength and shape to the cell, controls transport in and out of the cell, and contains **RECEPTORS** to allow the cell to respond to external stimuli. **GLYCOCALYX, EXTRACELLULAR MATRIX, CELL CORTEX**

ENDOPLASMIC RETICULUM (ER)
CISTERNAL SPACE, LUMEN; ROUGH ER (RER), SMOOTH ER (SER); ER STRESS, UNFOLDED PROTEIN RESPONSE (UPR), ER - ASSOCIATED PROTEIN DEGRADATION (ERAD) APOPTOSIS; BIOTRANSFORMATION REACTIONS; SARCOPLASMIC RETICULUM (SR)

GOLGI APPARATUS (GOLGI COMPLEX)
Formed from relatively large, flattened, sac-like membranous **VESICLES**, the Golgi apparatus packages and secretes cell products. **CISTERNA; *CIS* FACE, *TRANS* FACE; SECRETORY VESICLES (SECRETORY GRANULES); EXOCYTOSIS**

VESICULAR ORGANELLES AND LYSOSOMES: THE ENDOCYTIC PATHWAY
The process of endocytosis is a mechanism whereby exogenous substances are taken into cells. **LYSOSOMES; CLATHRIN-DEPENDENT ENDOCYTOSIS; CAVEOLAR ENDOCYTOSIS; CAVEOLAE; ENDOCYTIC CYCLE**

NUCLEUS
The nucleus contains the cell's genetic information and the machinery for converting that information into protein molecules. **NUCLEOPLASM; CHROMATIN FIBERS; HISTONES; NUCLEAR MATRIX; NUCLEOLUS** (ribosomal RNA synthesis), **SPECKLES; NUCLEAR ENVELOPE, NUCLEAR PORE COMPLEXES, PERINUCLEAR SPACE, INNER MEMBRANE, OUTER MEMBRANE**

MITOCHONDRIA: LOCATION OF AEROBIC RESPIRATION
AEROBIC RESPIRATION generates most of the energy needed by eukaryotes. ATP synthesis occurs in **MITOCHONDRIAL RESPIRATORY CHAINS** embedded in the mitochondrion's **INNER MEMBRANE. OUTER MEMBRANE, INTERMEMBRANE SPACE, CRISTAE, MATRIX, MITOCHONDRIAL FISSION AND FUSION**

PEROXISOMES generate and break down peroxides.
GLYOXYSOMES (in plants)

CHLOROPLASTS; PHOTOSYNTHESIS; THYLAKOID MEMBRANE, GRANA, THYLAKOID LUMEN, STROMA, STROMA LAMELLAE

CYTOSKELETON
This highly structured network of proteinaceous filaments is responsible for cell shape, large- and small-scale cell movement, solid state biochemistry, and signal transduction.
MICROTUBULES, MICROFILAMENTS, INTERMEDIATE FILAMENTS

BIOCHEMISTRY IN THE LAB: CELL TECHNOLOGY

Biochemistry in Perspective: Primary Cilia and Human Disease

OVERVIEW

Understanding more about ***where*** biomolecules react—within specific organelles, in the cytoplasm, or on the outside surface of the cell membrane, for example—places biochemical reactions in the context of their impact within the organism as a whole. Realizing that a specific reaction (and its pathway) occurs in a specific location, for a specific reason, and under a specific set of circumstances gives greater meaning to the functions of the biomolecules, their pathways, and even the overall systems.

STUDY HINTS AND STRATEGIES

Creating your own diagrams, tables, and/or images is a powerful learning tool. Make your learning active by organizing the material in a way that will help you the most, whether as pictures or diagrams of the organelles and cells, as lists or summaries that include functions, or as tables. Small sticky notes with the name of an organelle on one side and its function on the reverse may be helpful. Notes of different colors can be used to keep prokaryotic- or plant-specific features separate.

Create a table in which each row is devoted to a specific organelle. Check out the following example. You will want to tailor the column headers to the emphasis given in your specific class or to challenges that you're facing in learning more about these great little packages.

Item*	Structure	Location	Function	Notes
What is it?	What is it made of?	Where is it?	What does it do?	Why is it important? How is it unique?

* Suggested items for the first column: plasma membrane, endoplasmic reticulum, rer, ser, golgi apparatus, nucleus, mitochondria, vesicular organelles, lysosomes, peroxisomes, plastids, cytoskeleton, ribosomes

2.1 BASIC THEMES

WATER

Whether a biomolecule is **HYDROPHILIC** (polar and/or charged) or **HYDROPHOBIC** (nonpolar)—or both—affects its overall shape. Hydrophilic molecules readily interact with water molecules via dipole-dipole interactions or hydrogen bonds,[1] and hydrophobic molecules avoid water. Large biomolecules that have both hydrophilic and hydrophobic sections arrange themselves accordingly. Example: Proteins fold so that the hydrophobic

[1] Remember that hydrogen bonds are relatively strong *interactions* between an unshared pair of electrons on one molecule and a polarized H on another molecule. That polarized H needs to be attached to an O or N atom that has a lone pair (an O atom in a biomolecule always will; N^+ won't). Hydrogen bonds are very weak, though, compared to a covalent bond.

sections are tucked inside, away from the aqueous environment, and their hydrophilic areas (with polar or charged amino acid R groups) face outward to interact with the surrounding water.

BIOLOGICAL MEMBRANES: SELECTIVE PHYSICAL BARRIERS

Importance of the control of transport across membranes
Basic structure: **LIPID BILAYER** (mostly phospholipids) with proteins
PHOSPHOLIPIDS: hydrophilic head, hydrophobic tails

MEMBRANE PROTEINS:

INTEGRAL PROTEINS (embedded within) vs. versus **PERIPHERAL PROTEINS** (on the surface)
CHANNEL PROTEINS: transport specific ions across the membrane
CARRIER PROTEINS: transport specific molecules across the membrane
RECEPTORS: signal transduction
ANCHOR PROTEINS: attach the membrane to macromolecules

SELF-ASSEMBLY INTO SUPRAMOLECULAR STRUCTURES

Why does self-assembly occur? **STERIC INFORMATION** in the macromolecules: complementary shapes fit together to optimize hydrophilic interactions, hydrophobic interactions, and many, many weak interactions.

MOLECULAR CHAPERONES or templates may provide assistance with assembly (folding).

MOLECULAR MACHINES

- consist of proteins (and protein complexes) that perform work and have moving parts
- convert energy into directed motion via **ENERGY-TRANSDUCING MECHANISMS**:
  1. Nucleotide (e.g., ATP or GTP) binds to **MOTOR PROTEINS** (protein subunits).
  2. Nucleotide hydrolyzes and releases energy.
  3. This energy causes a precisely targeted change in the protein subunit's shape.
  4. This change is transmitted to nearby subunits.

Examples of molecular machines:
RIBOSOMES—rapidly and accurately incorporate amino acids into polypeptides
SARCOMERES—contractile units of skeletal muscle; actin and myosin are proteins

MACROMOLECULAR CROWDING

- Describes cell conditions better than "concentrated"
- **EXCLUDED VOLUME** = volume occupied by macromolecules
- Impacts many intracellular processes

SIGNAL TRANSDUCTION

- How cells receive, interpret, and respond to signals such as molecules or light
- Examples of eukaryotic signal molecules: neurotransmitters, hormones, cytokines

PHASES:
1. **RECEPTION**: signal molecule (**LIGAND**) binds to and activates a **RECEPTOR** on the membrane surface, causing transduction.
2. **TRANSDUCTION**: a change in the receptor's 3D-shape, which triggers:
3. **RESPONSE**: inside the cell, a cascade of events that involves covalent modification of proteins and results in changes such as enzyme activity, gene expression, and motion.

4. **TERMINATION:** The efficiency and effectiveness of cell signal mechanisms requires that they must be terminated in a timely manner.

Calcium ions are used as a signaling device. Cells respond to external stimuli by discrete increases in cytoplasmic Ca^{2+} concentrations that are normally kept quite low.

2.2 Structure of Prokaryotic Cells

Features: relatively small, able to move using pili or flagella, able to retain specific dyes

Identify prokaryotes based upon nutritional requirements, energy sources, chemical composition, and biochemical capacities.

Common features:	Cell wall	Circular DNA molecules
	Plasma membrane	No internal membrane-enclosed organelles

Cell Wall

GRAM-POSITIVE cells have a thick **PEPTIDOGLYCAN** layer outside the plasma membrane

GRAM-NEGATIVE cell walls are more complex. Layers from the outside in:

GLYCOCALYX (slime layer, biofilm, or capsule)

OUTER MEMBRANE consists of **LPS** = **LIPOPOLYSACCHARIDES** (lipid A + polysaccharide)
—contains porins (channel proteins for transport through the membrane)

PERIPLASMIC SPACE with peptidoglycans and proteins

INNER (PLASMA) MEMBRANE

Archaea vary: some are Gram-positive, some Gram-negative, some have no cell wall

Plasma Membrane (cytoplasmic membrane)

selectively permeable barrier
contains **RECEPTOR PROTEINS**; may contain proteins for **PHOTOSYNTHESIS** and **RESPIRATION**

Cytoplasm

NUCLEOID—contains a chromosome (circular DNA molecule)

PLASMIDS—additional small circular DNA molecules

RIBOSOMES (molecular machines composed of RNA and proteins that synthesis polypeptides), **INCLUSION BODIES**

PILI AND FLAGELLA FOR MOTION AND CONJUGATION (singular: pilus and flagellum)

2.3 Structure of Eukaryotic Cells

Endomembrane System; Vesicles

Plasma Membrane

Controls transport of molecules into and out of the cell
Transport is facilitated by carrier and channel proteins
Glycocalyx, receptors, extracellular matrix; cell cortex

ENDOPLASMIC RETICULUM (ER)

LUMEN (OR CISTERNAL SPACE)
ROUGH ER (RER): synthesis of membrane proteins and proteins for export from the cell; contains ribosomes
SMOOTH ER (SER): lipid synthesis, biotransformation
SARCOPLASMIC RETICULUM (SR)—SER in striated muscle

GOLGI APPARATUS (OR GOLGI COMPLEX)

Packages and distributes cell products to internal and external compartments
VESICLES: secretory vesicles (or secretory granules)
Cisterna (plate); *cis* and *trans* faces
TRANSITIONAL ER; ER-GOLGI INTERMEDIATE COMPARTMENT
EXOCYTOSIS (secretion, see Figure 2.17); **CONSTITUTIVE EXOCYTOSIS**

VESICULAR ORGANELLES AND LYSOSOMES: THE ENDOCYTIC PATHWAY

The process of endocytosis is a mechanism whereby exogenous substances are taken into cells, **LYSOSOMES; CLATHRIN-DEPENDENT ENDOCYTOSIS; CAVEOLAR ENDOCYTOSIS; CAVEOLAE; ENDOCYTIC CYCLE**

NUCLEUS

Contains cell''s hereditary information, , which regulates metabolism by directing the synthesis of protein cell components
NUCLEOPLASM; LAMINS, CHROMATIN FIBERS, HISTONES; EUCHROMATIN; HETEROCHROMATIN; NUCLEAR MATRIX (or **NUCLEOSKELETON)**
NUCLEAR ENVELOPE; PERINUCLEAR SPACE; INNER MEMBRANE
NUCLEAR PORE; NUCLEAR PORE COMPLEXES
NUCLEOLUS: synthesis of ribosomal RNA

MITOCHONDRIA

Aerobic metabolism—oxygen-dependent synthesis of ATP
OUTER MEMBRANE/INTERMEMBRANE SPACE/INNER MEMBRANE with **CRISTAE** (folds)/**MATRIX**
MITOCHONDRIAL RESPIRATORY CHAIN: ATP synthesis
Regulation of **APOPTOSIS** (genetically programmed events that lead to cell death)
FISSION; FUSION

PEROXISOMES: CONTAIN OXIDATIVE ENZYMES

Most important function: generate and break down peroxides (R–O–O–R)

CHLOROPLASTS

CHLOROPLASTS are a type of chromoplast (an organelle that accumulate pigments) in which photosynthesis occurs.
PHOTOSYNTHEIS: a process in which light energy is converted into chemical energy.
THYLAKOID MEMBRANE, GRANA, THYLAKOID LUMEN, STROMA, STOMA LAMELLA

CYTOSKELETON

MICROTUBULES: structural support for long, thin cells; protein = tubulin
MICROFILAMENTS: cytoplasmic streaming and amoeboid movement; protein = actin
INTERMEDIATE FILAMENTS: maintain cell shape under mechanical stress; various proteins (Example: keratin filaments in outer skin cells)

FUNCTIONS:

1. maintains cell shape
2. facilitates coherent cellular movement, both large and small scale; provides supporting structure to guide organelle movement within the cell
3. solid solid-state biochemistry: provides a platform for enzyme complexes, greatly increasing reaction rates
4. signal transduction: provides structural continuity for signal cascade protein

BIOCHEMISTRY IN THE LAB: CELL TECHNOLOGY

CELL FRACTIONATION

DIFFERENTIAL CENTRIFUGATION, MICROSOMES, DENSITY-GRADIENT CENTRIFUGATION; MARKER ENZYMES

ELECTRON MICROSCOPY; limit of resolution

AUTORADIOGRAPHY

LIVE CELL IMAGING: Phase contrast microscopy (uses variations in light refraction through substances with different densities); fluorescence microscopy (uses fluorophors to investigate cell function)

AFTER STUDYING THIS CHAPTER, YOU SHOULD BE ABLE TO:

- Draw a diagram of a prokaryotic cell. Label and describe the function(s) of each component.
- Draw a diagram of a eukaryotic cell, with each organelle labeled.
- Describe the functions of eukaryotic organelles.
- Compare the storage of genetic information (DNA) in a prokaryotic cell with that of a eukaryotic cell.
- Given a diagram of a cell, identify the various components and describe their functions. Identify the cell as prokaryotic or eukaryotic.
- Describe the basic structure of chloroplasts.
- Compare the features and functions of rough endoplasmic reticulum, smooth endoplasmic reticulum, sarcoplasmic reticulum, and SER in hepatocytes.
- Draw a diagram of a eukaryotic plasma membrane. Include (and label) the various types of proteins that occur in plasma membranes.
- Describe the features of a prokaryotic cell wall and plasma membrane. Compare these to a eukaryotic plasma membrane.
- Other biological membrane structures include those of the nucleus and of mitochondria. Compare their similarities and differences with the plasma membrane.
- Describe the structure and function(s) of the cytoskeleton.
- Demonstrate an understanding of important intermolecular interactions and their impact on cellular structure and function. Examples include hydrophilicity vs.versus hydrophobicity and biomembrane self-assembly; cell surface receptors and signal transduction; and endocytosis and exocytosis.

Use this space to note any additional objectives provided by your instructor.

CHAPTER 2: SOLUTIONS TO REVIEW QUESTIONS

2.1 a. prokaryote—single-celled organism lacking a nucleus

b. eukaryote—composed of cells that have a nucleus and membrane-bound compartments

c. organelle—large subcellular compartment specialized for a specific task

d. hydrophilic—molecules that possess positive or negative charges that interact with water

e. hydrophobic—molecules that possess few electronegative atoms and do not interact with water

2.2 a. lipid bilayer—biological membrane composed primarily of two layers of phospholipids

b. polar head group—a hydrophilic charged or uncharged polar group in a phospholipid

c. hydrocarbon tail—hydrophobic group such as the fatty acids in a phospholipid

d. integral protein—a protein that is embedded within the membrane

e. peripheral protein—a protein that is not embedded within the membrane but is attached to a membrane protein or lipid

2.4 a. ligand—a molecule that binds to a protein or receptor

b. motor protein—proteins that utilize nucleotide hydrolysis to do work

c. GTP—guanosine triphosphate

d. macromolecular crowding—large numbers of different proteins at low concentrations confined to a small space

e. excluded volume—the volume occupied by macromolecules

2.5 a. signal transduction—the process that organisms use to receive and interpret information

b. neurotransmitter—signaling molecule produced by neurons

c. hormone—signaling molecule produced by glandular cells

d. cytokine—signaling molecule produced by white blood cells

e. LPS—lipid component of bacterial outer membrane lipopolysaccharide

2.7 a. plasma membrane—a phospholipid bilayer that is inside the cell wall of bacteria and the phospholipid bilayer that surrounds eukaryotic cells

b. photosynthesis—process of conversion of light energy into chemical energy

c. respiration—process of oxidation of nutrient molecules to generate energy

d. nucleoid—spacious, irregularly shaped, centrally located region that contains the bacterial chromosome

e. chromosome—organized structure of a single piece of DNA and protein found in cells

2.8 a. pilus—fine, hair-like structures that allow cells to attach to food sources and host tissue

b. conjugation—process that some bacteria use to transfer genetic information from donor to recipient cells

c. flagellum—flexible corkscrew-shaped protein filament used for locomotion

d. endomembrane system—extensive set of interconnecting internal membranes that divide the cell into function compartments

e. vesicle—membranous sacs that bud off from a donor membrane to fuse with the membrane of another compartment of plasma membrane

2.10 a. rough ER—endoplasmic reticulum that has ribosomes on its cytoplasmic surface

b. smooth ER—endoplasmic reticulum that lacks ribosomes

c. clathrin—a protein complex that binds to membrane adaptor proteins to form a basketlike latticework that forces membrane into the shape of a bud

d. unfolded protein response—a cellular process that involves negative regulation of new protein synthesis

e. caveolae—small invaginations formed from a specialized type of plasma membrane microdomains enriched in cholesterol, certain membrane lipids, signaling molecules, ion channel proteins and the membrane protein caveolin.

2.11 a. biotransformation reaction—converts water-insoluble metabolites and xenobiotics into more soluble products that can be excreted

b. sarcoplasmic reticulum—the smooth ER in striated muscle

c. Golgi apparatus—stacks of membranous vesicles involved in the processing, packaging, and delivery of cell products to internal and external destinations

d. Golgi cisterna—the surface of the Golgi apparatus closest to the ER

e. exocytosis—secretory process involving the fusion of membrane-bound granules with the plasma membrane

2.13 a. nuclear pore complex—a 120-MDa structure that consist of ~100 nucleoporins that mediates traffic into and out of the nucleus

b. endocytosis—process of cellular internalization of exogenous substances

c. proteome—the characteristic set of proteins produced by cell

d. endocytic cycle—the continuous recycling of the membrane via endocytosis and exocytosis

e. lysosome—vesicles that contain granules or aggregates of digestive enzymes

2.14 a. acid hydrolase—digestive enzymes located in lysosomes

b. autophagy—process by which lysosomes degrade debris in the cell

c. clathrin-dependent endocytosis—a mechanism for internalizing substances bound to plasma membrane receptors that involves vesicles coated with clathrin triskelia

d. caveolar endocytosis—a form of clathrin-independent endocytosis in which small invaginations form in specialized membrane microdomains called caveolae

e. proteostasis—protein homeostasis; the maintenance of protein folding quality control made possible by a vast array of molecular chaperones and protein degradation mechanisms

2.16 a. intermembrane space—space between the outer and inner membranes of the mitochondrion

b. mitochondrial matrix—the region inside the inner membrane of the mitochondrion

c. peroxisome—small spherical membranous organelles that contain oxidative enzymes

d. peroxin—proteins involved in the assembly of peroxisomes

e. plastid—structures found only in plants, algae, and some protists

2.17 a. outer nuclear membrane—the outermost of two concentric membranes that surround the nucleus; continuous with the rough endoplasmic reticulum

b. inner nuclear membrane—the innermost membrane in the nuclear envelope; contains unique proteins that stabilize nuclear envelope structure, and are involved in chromatin binding, chromatin remodeling protein recruitment and various enzymatic activities

c. chloroplast—a type of chromoplast that is the site of photosynthesis

d. photosynthesis—process involved with the conversion of light energy into chemical energy

e. thylakoid membrane—highly folded internal membrane in chloroplasts that contains the photosystems

2.19 a. microfilament—small fibers composed of polymers of G-actin

b. F-actin—filamentous or polymeric actin

c. G-actin—globular actin

d. ameboid movement—locomotion created by formation of temporary cytoplasmic protrusion

e. intermediate filament—flexible, strong, and stable polymers

f. keratin—an intermediate filament found in skin and hair cells

2.20 a. ciliopathy—defect in primary cilia

b. retinitis pigmentosum—30 different progressive eye disorders that lead to blindness

c. polycystic kidney disease—results from defects in either of two genes that cause cyst formation in kidneys and other organs

d. Bardet-Biedl syndrome—results from the mutation of any of 12 genes that cause retinal degeration or kidney and liver cysts along with other clinical symptoms

e. anterograde transport—the transport of newly synthesized molecules from the ER to the Golgi apparatus and then to cell destinations or to plasma membrane for secretion

2.22 The colon is surrounded by such a large percentage of the body's immune system cells because of the large and diverse microbiota within this organ. The immune system must protect against pathogenic organisms while tolerating nonpathogenic organisms, many of which are beneficial to the body.

2.23 Multiple uses of antibiotics results in the loss of colonization resistance, the capacity of the digestive tract to resist being colonized by pathogenic organisms. Colonization resistance is difficult to reestablish because dysbiosis has been established, i.e. some beneficial organisms have been replaced by pathogens in a process that creates a new less healthy form of colonization resistance.

2.25 Refer to Figure 2.7. The functions of the components of prokaryotic cells are as follows:

a. Nucleoid contains the bacterial chromosome.

b. Plasmid is the site of extrachromosomal DNA, often coding for special functions.

c. Cell wall provides protection and support.

d. Pili allow attachment to other cells.

e. Flagella allow locomotion.

2.26 a. Nucleus—eukaryotes

b. Plasma membrane—eukaryotes and prokaryotes

c. Endoplasmic reticulum—eukaryotes

d. Mitochondria—eukaryotes

e. Nucleolus—eukaryotes

f. Cytoskeleton—eukaryotes (*Note*: Prokaryotes do contain proteins that resemble eukaryotic cytoskeleton proteins in both structure and function, but they lack the network of microtubules, microfilaments, and intermediate fibers that comprise a cytoskeleton.)

2.28 On the surfaces of biomolecules, functional groups that can form noncovalent interactions will facilitate the formation of supramolecular structures with biomolecules that have properties that are similar (e.g., hydrogen bonding) or complementary (e.g., oppositely charged ions). As these noncovalent interactions form, more of the molecules' surfaces are drawn closer to each other, making further interactions possible. Large numbers of these interactions stabilize the complexes formed from these molecules. These interactions are augmented when the biomolecules (such as proteins and nucleic acids) have intricate shapes that are complementary to each other.

2.29 The three phases of signal transduction in living organisms are reception, transduction, and response. Reception is the binding of an external signal molecule to a receptor on the cell surface. Transduction is the conversion of this extracellular message (i.e., the new signal molecule-receptor binding) to an intracellular message via a conformational change in the receptor. The response, the result of this internal message, is a cascade of events that involves covalent modifications of intracellular proteins and may include changes in enzyme activities, cell movement, and other cellular processes.

2.31 The cytoskeleton provides structural continuity for intracellular signal transduction by providing a solid support for signal cascade proteins. Protein-protein interactions trigger sequential protein structure changes, resulting in the flow of information within the cell.

2.32 Lysosomes digest biomolecules of all types. In addition to the normal processing of cellular molecules, lysosomes destroy the components of foreign cells and other exogenous extracellular materials.

2.34 Examples of diseases linked to organelles and their underlying causes are as follows:

(1) Cystic fibrosis is caused by the misfolding of the regulator protein CFTR, which functions as a plasma membrane chloride channel. The misfolded CFTR becomes trapped within the ER and is degraded. Without this Cl^- channel, thick mucus accumulates and compromises the lungs, pancreas, and other organs.

(2) Congenital disorders of glycosylation are caused by mutations in genes that code for glycosylation enzymes or glycosylation-linked transport proteins in the Golgi apparatus.

(3) Progeria is caused by a specific mutation in the lamin A gene that codes for lamin, a component of the lamina of the nuclear envelope.

(4) Emery-Dreifuss muscular dystrophy is caused by the absence or mutation of the gene that codes for emerin, a protein of the inner membrane of the nuclear envelope.

(5) Tay-Sachs disease and Gaucher's disease are lipid storage diseases that are caused by the absence of a lysosomal enzyme.

(6) Pompe's disease (glycogen storage disease type II) is also caused by the absence of a lysosomal enzyme.

(7) I-cell disease is caused by the defective import of enzymes into lysosomes. In addition, ER stress is an important feature of Alzheimer's, Huntington's, and Parkinson's diseases, as well as heart disease and diabetes.

2.35 The highly developed framework of the cytoskeleton performs the following functions in eukaryotic cells: (1) maintenance of overall cell shape, (2) facilitation of coherent cellular movement, (3) provision of a supporting structure that guides the movement of organelles within the cell, and (4) service as a platform for enzyme and signal cascade complexes.

2.37 Among the roles of plasma membrane proteins are transport, response to stimuli, cell-cell contact, and catalytic functions.

2.38 The Golgi apparatus processes, sorts, and packages protein and lipid molecules for distribution to other regions of the cell or for export.

2.39 The principal function of the rough endoplasmic reticulum is the synthesis of membrane proteins and protein for export from the cell. Smooth endoplasmic reticulum, so named because it lacks attached ribosomes, is involved in lipid synthesis and biotransformation processes.

2.40 ER stress results from the accumulation of misfolded proteins. The ER-associated protein degradation pathway is a mechanism that allows the cell to degrade misfolded proteins. The unfolded protein response is a signaling pathway that results in inhibition of new non-heat shock protein synthesis. If the level of unfolded proteins is overwhelming, the ER overload response pathway induces apoptosis.

2.41 Urine flow, wound healing, and sight.

2.43 The kinesins are motor proteins that move particles along the outer pair of micotubules of cilia and flagella toward the cell periphery. Dyneins move molecules in the opposite direction.

2.44 Specific examples of vesicular organelles include lysosomes, which contain digestive enzymes; the glyoxysomes of fat and oil storing cells in seeds that are involved in gluconeogenesis from these fats and oils; the melanin-containing melanosomes, which migrate from the basal layer to the epithelial layer of the skin; and plant vacuoles, which contain numerous enzymes and biomolecules needed for plant growth and development.

2.46 COPII is a type of vesicle coat protein on the surface of vesicles that transport proteins from the rough ER to the Golgi apparatus.

2.47 The intracellular and extracellular structures that protect the eukaryotic plasma membrane are the cell cortex and the extracellular matrix, respectively. The cell cortex, located on the inner surface of the plasma membrane, consists of microfilaments and other proteins in a three-dimensional meshwork that provides mechanical strength. The extracellular matrix is a gelatinous material of structural proteins and complex carbohydrates.

2.49 The nuclear envelope surrounds the nucleus and consists of two membranes strengthened by a network of nuclear lamins. The outer nuclear membrane is continuous with the rough endoplasmic reticulum (RER), the space between the membranes is continuous with the RER lumen, and the two membranes fuse at nuclear pores. By controlling transport into and out of the nucleus, the nuclear envelope separates DNA replication and transcription processes from the cytoplasm and allows for more sophisticated regulation of gene expression than would be possible otherwise.

2.50 Proteostasis is a concept that describes the process whereby cells control the folded conformation of their proteins. Proteostasis is critical to the maintaining healthy cells as evidenced by the size of the proteostasis network (about 2000 genes coding for a diverse array of proteins), a highly interconnected set of pathways that regulate protein folding and degradation.

CHAPTER 2: SOLUTIONS TO FILL-IN-THE-BLANK QUESTIONS

2.52 Macrophages

2.53 Eukaryotic cells

2.55 The universal use of DNA as the genetic material

2.56 Hydrophobic

2.58 Integral

2.59 Energy

CHAPTER 2: SOLUTIONS TO SHORT-ANSWER QUESTIONS

2.61 Eukaryotic cells can generate significantly more energy than can prokaryotic cells; hence, they have a much more complex structure, as indicated by the compartmentation made possible by organelles. This vast increase in complexity requires a larger cell volume.

2.62 The soap anions resemble the lipid bilayer of the cell membrane. They insert into and disrupt the membrane, which inevitably causes leakage of the cell contents. As a result, the cell dies.

2.64 Carboxylic acids are weak acids and are ionized only in a narrow pH range. Carboxylate anions would become protonated and the micelle effect would cease under conditions where protons might build up on the membrane surface. Under these conditions the membrane would break down. Anions of phosphoric acid are derived from a much stronger acid and are not subject to pH changes.

2.65 Cell walls are rigid structures that serve as support for the cell membrane of organisms such as prokaryotes and plant cells. They also restrict the shape of the cell contents. The function of many types of eukaryotic cells, such as many animal cells, requires moment-by-moment changes in their three-dimensional structures, which would be impossible with a cell wall. For example, the immune system cells that move through the body seeking out foreign or damaged cells do so by amoeboid motion.

CHAPTER 2: SOLUTIONS TO THOUGHT QUESTIONS

2.67 Primary cilia contain the sensory components of the motile cilia e.g., but lack several of the motility component (i.e., the central microtubule pair, the dynein arms and the radial spokes within the axoneme). Also a large number of receptors are present in the cilia membrane.

2.68 Because the thick mucoid coat prevents antibodies from binding to surface cellular structures used by the immune system for recognition, it interferes with the immune response.

2.70 This defect could lie anywhere in the synthesis scheme for the LDL receptor. The high cholesterol levels result because cholesterol is not being transferred into the cells and is building up in the blood, thus causing atherosclerosis. The most common defects are improper insertion of the receptor into the plasma membrane or a defective receptor.

2.71 Since the diameter of a spherical mycoplasma cell is 0.3 μm, the radius is 0.15 μm. The volume of a spherical mycoplasma cell: $V = 4\pi r^3/3 = (4)(3.14)(0.15\ \mu m)^3/3 = 0.014\ \mu m^3$.

Assuming that *E. coli* is cylindrical, with dimensions of 1 μm × 2 μm (i.e., a typical rod-shaped bacterium), its volume is $V = \pi r^2 h = (3.14)(0.5\ \mu m)^2(2\ \mu m) = 1.6\ \mu m^3$.

At 0.014 μm^3, the volume of a typical mycoplasma is significantly smaller than that of the cylindrical *E. coli* cell. To be more specific, the mycoplasma is 0.9 % of the size of an *E. coli.* Alternatively, the *E. coli* is about 114 times larger than the mycoplasma.

2.73 The volume of the *E. coli* cell is given by $V = \pi r^2 h = (3.14)(0.5\ \mu m)^2(2\ \mu m) = 1.57\ \mu m^3$.

The surface area is: $A = 2\pi r^2 + 2\pi rh$

$$= (2)(3.14)(0.5\ \mu m)^2 + (2)(3.14)(0.5\ \mu m)(2\ \mu m) = 1.57\ \mu m^2 + 6.28\ \mu m^2 = 7.85\ \mu m^2.$$

The *E. coli* surface-to-volume ratio = $7.85\ \mu m^2/1.57\ \mu m^3 = 5.0\ \mu m^{-1}$.

The volume of the eukaryotic cell is $V = (4/3)(3.14)(10)^3 = 4189\ \mu m^3$.

The surface area is $4\pi r^2 = 4(3.14)(10)^2 = 1256\ \mu m^2$.

The eukaryotic cell surface-to-volume ratio = $1256\ \mu m^2/4189\ \mu m^3 = 3.0\ \mu m^{-1}$.

The eukaryotic cell has a much smaller surface-to-volume ratio than does the *E. coli.* In order to import enough material to sustain the functions of the cell, the membrane must become more efficient. Eukaryotes have significantly greater membrane transport capacity because of membrane transport proteins that are more sophisticated and present in exceptionally large numbers and extensive membrane folding, which increases the surface-to-volume ratio. Note that the loss of the prokaryote cell wall preceded the membrane folding.

Use this space to draw the structure of a prokaryotic or a eukaryotic cell. Label all of the organelles and the various membrane components.

3 Water: The Matrix of Life

Brief Outline of Key Terms and Concepts

3.1 MOLECULAR STRUCTURE OF WATER

POLAR BOND; POLAR MOLECULE; DIPOLE
ELECTROSTATIC INTERACTIONS

3.2 NONCOVALENT BONDING

Noncovalent bonds are important in determining the physical and chemical properties of living systems.

IONIC INTERACTIONS (SALT BRIDGES)

HYDROGEN BONDS,
with both dipole-dipole and covalent character, play a critical role in the properties of water and its place in the structure and function of cells.

VAN DER WAALS FORCES
dipole-dipole interactions; dipole-induced dipole; induced dipole-induced dipole

3.3 THERMAL PROPERTIES OF WATER

Hydrogen bonding is responsible for water's unusually high freezing and boiling points. Because water has a high HEAT CAPACITY, it can absorb and release heat slowly. Water plays an important role in regulating heat in living organisms.

3.4 SOLVENT PROPERTIES OF WATER

Water's dipolar structure and its capacity to form hydrogen bonds enable water to dissolve many ionic and polar substances. SOLVATION SPHERES

HYDROPHILIC MOLECULES, CELL WATER STRUCTURING, and SOL-GEL TRANSITIONS

HYDROPHOBIC MOLECULES AND THE HYDROPHOBIC EFFECT (HYDROPHOBIC INTERACTIONS)
Nonpolar molecules cannot form hydrogen bonds with water and are excluded via clathrate formation.

AMPHIPATHIC MOLECULES, such as fatty acid salts, spontaneously rearrange themselves in water to form MICELLES.

OSMOSIS
Osmosis is the movement of water across a semipermeable membrane from a dilute solution to a more concentrated solution. HYPERTONIC, HYPOTONIC, AND ISOTONIC SOLUTIONS; DIALYSIS
HEMOLYSIS, CRENATION
OSMOTIC PRESSURE is the pressure exerted by water on a semipermeable membrane as a result of a difference in the concentration of solutes on either side of the membrane.
OSMOLARITY = iM M = molarity
i = degree of ionization (van't Hoff factor)
OSMOTIC PRESSURE = $\pi = i$MRT
R = 0.082 L·atm/K·mol T = Kelvin
CELL MEMBRANE POTENTIAL

3.5 IONIZATION OF WATER

ACIDS, BASES, AND pH
STRONG VS. WEAK ACIDS AND BASES; CONJUGATE BASE
Liquid water molecules have a limited capacity to ionize to form H^+ and OH^- ions. **pH SCALE**
Hydrogen ion concentration is a crucial feature of biological systems primarily because of their effects on biochemical reaction rates and protein structure. ACIDOSIS, ALKALOSIS

BUFFERS
A buffer is a mixture of a weak acid and its conjugate base. Buffers prevent changes in pH.
LE CHATELIER'S PRINCIPLE, BUFFERING CAPACITY
HENDERSON-HASSELBALCH EQUATION

$$\text{pH} = \text{p}K_a + \log\frac{[\text{A}^-]}{[\text{HA}]}$$

Buffer ranges = p$K_a \pm 1$

PHYSIOLOGICAL BUFFERS
BICARBONATE BUFFER: CO_2/HCO_3^-
– regulation by lungs and kidneys
– carbonic anhydrase
PHOSPHATE BUFFER $H_2PO_4^-/HPO_4^{2-}$
PROTEIN BUFFERS: Buffering capacity provided by proteins' amino acid side chains.

Biochemistry in Perspective: Cell volume Regulation and Metabolism

OVERVIEW

Without the unique properties of water, life as we know it could not exist. This chapter serves as a refresher for a number of concepts that you have seen in general and organic chemistry (electronegativity, polarity, hydrogen bonds, heat capacity, osmotic pressure, acid-base chemistry, buffers) and places them solidly in the context of biological systems. If some of these concepts happened to slip by you in previous courses, chances are that you'll find them more interesting and accessible here. You will be seeing much of this material applied to amino acids, proteins, and enzymes in the chapters beyond, so be sure to practice (and learn) this material early. This chapter includes examples of buffer problems with detailed solutions and extra practice problems.

3.1 MOLECULAR STRUCTURE OF WATER: H_2O

Water is **POLAR.**

- O–H is a **POLAR BOND** because the oxygen atom is much more **ELECTRONEGATIVE** than the hydrogen atom. The oxygen atom has a partial negative charge, and the hydrogen atom has a partial positive charge.
- H_2O is a **POLAR MOLECULE** because it has a **DIPOLE**. Whether a molecule is polar also depends on its geometry (or overall shape). Because the H_2O molecule is bent at an angle of 104.5°, it has a dipole—the "side" of the molecule with the oxygen's unshared pairs of electrons is more negative than the side with the hydrogen atoms. [Compare H_2O with CO_2. Both have polar bonds, but CO_2 is nonpolar. The polarity of the two C=O bonds cancels out because CO_2 is linear: O=C=O.]

The polar nature of water allows it to interact with a variety of other molecules. For example, table salt (NaCl) dissolves completely in water because of ion-dipole interactions. Alcohol dissolves in water via hydrogen bonding and dipole-dipole interactions.

- **ELECTROSTATIC INTERACTIONS** occur between opposite charges (including the partial charges on atoms in a polar bond).

3.2 NONCOVALENT BONDING

IONIC INTERACTIONS

Example: In proteins, **SALT BRIDGES** are ionic interactions between oppositely charged amino acid side groups. Example: $R–CO_2^-$ ····· $^+H_3N–R$

HYDROGEN BONDS

- Occur between a hydrogen that is attached to an oxygen (or nitrogen) and a lone pair of electrons (on O, N, or S).
- Each H_2O molecule can form hydrogen bonds with up to four other H_2O molecules.
- Hydrogen bonding explains water's relatively high melting and boiling points, heat of vaporization, heat capacity, surface tension, and viscosity.

VAN DER WAALS FORCES

- The more easily an atom can be polarized, the stronger the van der Waals forces.
- Types of van der Waals forces:

- **DIPOLE-DIPOLE** interactions
- **DIPOLE-INDUCED DIPOLE** interactions
- **INDUCED DIPOLE-INDUCED DIPOLE** interactions (London dispersion forces); individually, these are the weakest, but they become significant when a great many of these interactions are present

3.3 THERMAL PROPERTIES OF WATER

- Higher-than-expected melting and boiling points due to hydrogen bonding
- High heat of vaporization—water does not boil easily
- High heat capacity—water can absorb and store heat and release it slowly.
- Living organisms use these properties to regulate temperature: high water content helps to retain heat because of water's high heat capacity; evaporation is used as a cooling mechanism.

3.4 SOLVENT PROPERTIES OF WATER

Water is polar and can form hydrogen bonds.

HYDROPHILIC MOLECULES, CELL WATER STRUCTURING, AND SOL-GEL TRANSITIONS

- Hydrophilic = "water loving"; soluble in water
- Polar molecules and ionic substances are hydrophilic
- Solvation spheres (shells of H_2O molecules formed by water around solutes):
  —depend on charge density: the smaller and more highly charged the ion, the larger the solvation sphere
  —an ion with a larger solvation sphere moves more slowly (example: Na^+ moves more slowly than K^+)

STRUCTURED WATER

- Layers of H_2O, constantly rearranging, between adjacent macromolecules
- H_2O molecules that are closer to polar surfaces move more slowly
- Helps to stabilize macromolecular structure, yet its dynamics allow the macromolecule to function

SOL-GEL TRANSITIONS

- Cytoplasm has gel-like properties because the polar surfaces of biopolymers have highly structured solvation layers of water.
- Sol-gel transitions may be caused by temperature changes, matrix architecture, or inclusion of solutes.
- Contributes to cell movement (and other cell functions); actin-binding proteins; amoeboid motion

HYDROPHOBIC MOLECULES AND THE HYDROPHOBIC EFFECT

- Hydrophobic molecules are **NONPOLAR.**
- Hydrophobic effect or hydrophobic interactions: Why do hydrophobic molecules group together in water?

  Water molecules maximize the formation of hydrogen bonds with other water molecules and minimize any association with the nonpolar molecules. H_2O does this by forming a large "cage" around a group of nonpolar molecules, as opposed to many small cages around individual nonpolar molecules.

Those weak van der Waals forces may contribute a little, but it's the *water* that *excludes* nonpolar molecules, *not* the attraction between nonpolar molecules that causes them to separate.

So—water hydrogen bonds with itself as much as possible, *excluding* the hydrophobic molecules.[1]

Amphipathic Molecules (contain both hydrophobic and hydrophilic ends)

Molecules with a hydrophilic "head" and a long hydrophobic "tail" form micelles or bilayers (example: phospholipids form bilayers to create biological membranes).

Osmotic Pressure (π)

- **Osmosis, osmometer**
- Osmotic pressure = $\pi = iMRT$ — i = degree of ionization (van't Hoff factor); M = molarity, R = 0.082 L·atm/K·mol; T = temperature in Kelvin (recall K = °C + 273)
- **Osmolarity** = iM
- **Isotonic** vs. **hypertonic** vs. **hypotonic**; **hemolysis** (cell swells and may burst) vs. **crenation** (cell shrinks)
- **Cell membrane potential**—cytoplasmic side of the cell membrane is negatively charged due to charged amino acid R groups on proteins.
- Cells regulate their osmolarity, usually by pumping ions across the membrane.
- Calculation of molecular weight using mass and osmotic pressure data: Use the equation (above) to solve for moles, then divide: No. of g/No. of moles = molecular weight

3.5 Ionization of Water

Acids, Bases, and pH

$$HA \rightleftharpoons H^+ + A^-$$

- Strong vs. weak acids and bases
- weak acid (HA) and conjugate base (A^-)
- K_a is the equilibrium constant for the loss of an H^+
- pK_a: the lower the pK_a, the stronger the acid (see below)

$$K_a = \frac{[H^+][A^-]}{[HA]}$$

- The stronger the acid, the more dissociated the acid. That means there is more H^+ in solution, so the K_a will be larger. But remember that the pK_a = –log K_a. If the K_a is 10^{-4}, the pK_a is 4. For a K_a that's 100 times greater at 10^{-2}, the pK_a is 2. So, the lower the pK_a, the stronger the acid.

Buffers

Buffer = solution of a weak acid (HA) and its conjugate base (A^-)

---

[1] Also, in Chapter 4 you will learn that the overall disorder of the water increases when a lipid micelle or membrane is formed (or when a protein folds) and that the water''s disorder outweighs the resulting increased order of the micelle.

Buffers resist pH changes when H^+ or OH^- is added (Le Chatelier's principle)

ACIDOSIS vs. **ALKALOSIS**

BUFFERING CAPACITY depends on total buffer concentration and ratio of $[A^-]/[HA]$

(total buffer concentration = $[HA] + [A^-]$)

HENDERSON-HASSELBALCH EQUATION: $\mathrm{pH} = \mathrm{p}K_\mathrm{a} + \log \dfrac{[\mathrm{A}^-]}{[\mathrm{HA}]}$

Buffers are most effective when $[A^-] = [HA]$ or in the pH range of $\mathrm{p}K_\mathrm{a} \pm 1$. Titration curves show relatively flat areas when pH = $\mathrm{p}K_\mathrm{a}$. These flat areas indicate good buffer ranges because a relatively large amount of OH^- may be added with very little change in pH.

WEAK ACIDS WITH MORE THAN ONE IONIZABLE GROUP

Examples: amino acids, H_3PO_4 (phosphoric acid)

PHYSIOLOGICAL BUFFERS

BICARBONATE BUFFER

The bicarbonate buffer system is slightly more complicated than one would expect because CO_2 reacts with H_2O to form H_2CO_3:

$$\mathrm{CO_2} + \mathrm{H_2O} \rightleftharpoons \mathrm{H_2CO_3} \rightleftharpoons \mathrm{H^+} + \mathrm{HCO_3^-}$$

This can be simplified to the following:

$$\mathrm{CO_2} + \mathrm{H_2O} \rightleftharpoons \mathrm{H^+} + \mathrm{HCO_3^-} \qquad \mathrm{p}K_\mathrm{a} = 6.37$$

Carbonic anhydrase is the enzyme in the blood that catalyzes this reaction (otherwise it would be much too slow). Note: Of course you know that CO_2 itself is not an acid, but this equation is a shortcut to take into account the chemistry of carbonic acid, the action of carbonic anhydrase, and the resulting capability of the blood to use CO_2 to control $[HCO_3^-]$ and $[H^+]$ (i.e., to control pH).

Given that its $\mathrm{p}K_\mathrm{a}$ differs from blood pH (7.4) by more than one pH unit, how can bicarbonate serve as an important buffer in the blood?

Both CO_2 and HCO_3^- can be regulated by the lungs and kidneys. CO_2 can be exhaled, and the kidneys can remove H^+ from the blood. Both of these actions are needed to keep the ratio $[HCO_3^-]/[CO_2]$ high.

Why does the ratio $[HCO_3^-]/[CO_2]$ need to be high in order to be an effective buffer at a pH of 7.4?

Compare the buffer pH with the $\mathrm{p}K_\mathrm{a}$. Because the buffer pH is more basic than the $\mathrm{p}K_\mathrm{a}$, the buffer needs more conjugate base than acid. Alternatively, we could use the Henderson-Hasselbalch equation, which gives a ratio of about 11:1.

PHOSPHATE BUFFER

$$\mathrm{H_2PO_4^-} \rightleftharpoons \mathrm{H^+} + \mathrm{HPO_4^{2-}} \qquad \mathrm{p}K_\mathrm{a} = 7.2$$

Given that the $\mathrm{p}K_\mathrm{a}$ of dihydrogen phosphate is so close to blood pH (7.4), why *is* $H_2PO_4^-/HPO_4^{2-}$ not an important buffer system in the blood?

Their concentrations are too low. Phosphate buffers *are* important in intracellular fluids, where [$H_2PO_4^-$] and [HPO_4^{2-}] are higher.

PROTEIN BUFFER

Some amino acid side groups are weak acids or bases. As a result many proteins—like serum albumins—can act as buffers to help regulate the pH of the blood.

SOLVING BUFFER PROBLEMS: USE THE HENDERSON-HASSELBALCH EQUATION

$$\text{pH} = \text{p}K_a + \log\frac{[\text{A}^-]}{[\text{HA}]}$$

$$\frac{[\text{A}^-]}{[\text{HA}]} = \frac{\text{conjugate base}}{\text{weak acid}}$$

Total buffer concentration = [A^-] + [HA]

One way to remember this equation is

H comes before K, and *AHA*!!

pH is equal to pK_a plus the log of AHA!!

Yell the "Aha!" with enthusiasm and you will always remember this equation!

A buffer is a mixture of a weak acid and its conjugate base. The toughest part of solving buffer problems is often identifying the weak acid (HA) and its conjugate base (A^-). Remember that an acid donates an H^+ and a base accepts the H^+. So, the weak acid will always have an extra H^+ when compared to its conjugate base, and the conjugate base will have *one* extra negative charge.

$$\text{HA} \rightleftharpoons \text{H}^+ + \text{A}^-$$

EXAMPLES:

HA (WEAK ACID)	A^- (CONJUGATE BASE)
H_2CO_3	HCO_3^-
$H_2PO_4^-$	HPO_4^{2-}

These examples were chosen for two reasons: (1) they are present in living systems as physiological buffers, and (2) they can be confusing because one of the forms listed in each example (above) can serve as either an acid or a conjugate base, depending on what else is present (and the pH). Pay close attention to the problem, and be sure that you identify the acid as the one with the extra H^+. For example, HCO_3^- is the conjugate base here, but if CO_3^{2-} had been present, then HCO_3^- would have been the weak acid.

HA (WEAK ACID)	A^- (CONJUGATE BASE)
HCO_3^-	CO_3^{2-}
H_3PO_4	$H_2PO_4^-$

Hint: It helps to write out the acid-base reaction in this form. Be sure that both the number of Hs and the charges balance.	$HA \rightleftharpoons H^+ + A^-$
Check out these examples. Writing reactions like this helps to make it more clear which is HA and which is A^-. Note that $H_2PO_4^-$ is the conjugate base in the first equation, but it is the weak acid in the second equation.	$H_3PO_4 \rightleftharpoons H^+ + H_2PO_4^-$ $H_2PO_4^- \rightleftharpoons H^+ + HPO_4^{2-}$
This carboxylic acid group is the weak acid, and its carboxylate is the conjugate base.	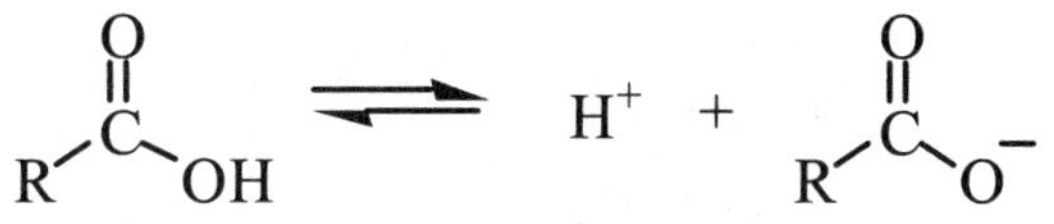
An amine group is a conjugate base, and its conjugate acid is shown here on the left.	$R{-}NH_3^+ \rightleftharpoons H^+ + R{-}NH_2$

How to Prepare a Buffer Given a Target pH and Total Concentration

Use the Henderson-Hasselbalch equation to solve for $[A^-]/[HA]$. Rearrange your answer in the form: $[HA] = x[A^-]$ With the equation for the total concentration of a buffer, you can solve for $[A^-]$ by substituting "$x[A^-]$" for the value of [HA]. Once you have $[A^-]$, you can use either equation to solve for [HA]. See Example Problem 3 below.

Also, note that "A^-" typically comes as a sodium or potassium salt (NaA or KA). For example, the buffer system $HPO_4^{2-}/H_2PO_4^-$ would be prepared by dissolving the appropriate amount of NaH_2PO_4 and Na_2HPO_4 in water and then diluting to the final volume. So, the final step often involves converting moles to grams of salt, using the molar mass.

What Happens When You Add a Strong Acid to a Conjugate Base,

either in a buffer solution or alone? Write out the acid-base equation and think about what is happening. [*Hint*: Be sure to work in moles (or millimoles).] For example, think about what would happen if you added 0.02 mol of strong acid (H^+) to 0.05 mol of conjugate base (A^-). The 0.02 mol of H^+ would react completely with 0.02 mol of A^-, forming 0.02 mol of HA and leaving $(0.05 - 0.02) = 0.03$ mol of A^- left over.

Examples of Buffer Problems *(Detailed Solutions Follow)*

1. What is the pH of a phosphate buffer if $[H_2PO_4^-] = 20$ mM and $[HPO_4^{2-}] = 15$ mM? The $pK_a = 7.20$ *(from Table 3.4, page 93 in your text).*

2. Calculate the ratio $\dfrac{[HPO_4^{2-}]}{[H_2PO_4^-]}$ at pH 6.2, 7.2, and 8.2. The pK_a is 7.2.

3. Describe how you would prepare 1 L of 0.20 M lactate buffer with a pH of 4.2. What ratio of lactate salt to lactic acid would you use? The pK_a of lactic acid is 3.86.

4. a. What is the pH of a solution prepared by mixing 100 mL of 1.00 M HCl with 300 mL of 0.500 M sodium succinate? The pK_{a1} of succinic acid is 4.21.

   b. Is this buffer system effective? If so, why? If not, how could you correct it?

c. 80 mL of 1.0 M NaOH have just been accidentally added into your carefully made solution! What is the final pH? Did the buffer work? How do you know? Is the solution still a good buffer?

EXAMPLES OF BUFFER PROBLEMS: *SOLUTIONS*

1. *What is the pH of a phosphate buffer when* $[H_2PO_4^-] = 20$ *mM and* $[HPO_4^{2-}] = 15$ *mM? The* $pK_a = 7.20$.

First, identify the acid and the conjugate base: $H_2PO_4^- \rightarrow H^+ + HPO_4^{2-}$
Since $H_2PO_4^-$ donates an H^+, it is the acid and HPO_4^{2-} is the conjugate base.
Use the Henderson-Hasselbalch equation to determine the pH of the buffer solution.

$$\text{pH} = \text{p}K_\text{a} + \log\frac{[\text{HPO}_4^{\,2-}]}{[\text{H}_2\text{PO}_4^{\,-}]} = 7.20 + \log 0.75 = 7.20 - 0.12 = 7.08$$

Always think about your answer to be sure that it makes sense. Compare the amount of acid vs. base present with the pH vs. pK_a. For example, in this problem we have more acid than base, and at 7.08, the final pH is more acidic than the pK_a.

2. *Calculate the ratio of* $\frac{[\text{HPO}_4^{\,2-}]}{[\text{H}_2\text{PO}_4^{\,-}]}$ *at pH 6.2, 7.2, and 8.2. The* pK_a *is 7.2.*

Identify the acid and the conjugate base: $H_2PO_4^- \rightleftharpoons HPO_4^{2-} + H^+$

As before, $H_2PO_4^-$ is the acid and HPO_4^{2-} is the conjugate base. Plug the pH and pK_a values into the Henderson-Hasselbalch equation and solve for the ratio $\frac{[\text{A}^-]}{[\text{HA}]}$.

$$\text{pH} = \text{p}K_\text{a} + \log\frac{[\text{A}^-]}{[\text{HA}]}$$

At pH 6.2: $$\log\frac{[\text{A}^-]}{[\text{HA}]} = \text{pH} - \text{p}K_\text{a} = 6.2 - 7.2$$

$$\frac{[\text{A}^-]}{[\text{HA}]} = 10^{(6.2-7.2)} = 10^{-1} = 0.1 \text{ or } \frac{1}{10}$$

At pH 7.2: $$\frac{[\text{A}^-]}{[\text{HA}]} = 10^{(7.2-7.2)} = 10^{0} = 1 \text{ or } \frac{1}{1}$$

At pH 8.2 $$\frac{[\text{A}^-]}{[\text{HA}]} = 10^{(8.2-7.2)} = 10^{1} = 10 \text{ or } \frac{10}{1}$$

3. *Describe how you would prepare a 0.20 M lactate buffer with a pH of 4.2. What ratio of lactate salt to lactic acid would you use? The* pK_a *of lactic acid is 3.86.*

Use the Henderson-Hasselbalch equation to calculate the ratios of the salt and acid, where the salt is "A^-" and the acid is "HA."

$$pH = pK_a + \log\frac{[A^-]}{[HA]}$$

Substituting these values into the equation gives:

$$4.2 = 3.86 + \log [A^-]/[HA]$$
$$0.34 = \log [A^-]/[HA]$$
$$100.34 = 2.19 = [A^-]/[HA]$$
$$[A^-] = (2.19)[HA] \qquad \text{(Equation 1)}$$

Since the total concentration of the lactate buffer is 0.20 M, we know that:

$$[A^-] + [HA] = 0.2 \text{ M} \qquad \text{(Equation 2)}$$

Use simultaneous equations (i.e., substituting the value for $[A^-]$ from Equation 1 into Equation 2) to determine the concentrations of the salt and acid for this particular buffer solution to give:

$$(2.19)[HA] + [HA] = 0.20 \text{ M}$$
$$(3.19)[HA] = 0.20 \text{ M}$$
$$[HA] = 0.063 \text{ M}$$

Take this value and insert it into Equation (1) to solve for $[A^-]$:

$$[A^-] = (2.19)[HA] = (2.19)(0.063) = 0.14 \text{ M}$$

To prepare this buffer, place 0.14 mol of lactate (or sodium lactate) and 0.063 mol of lactic acid in a 1-L volumetric flask and dilute with water to the 1-L mark.

And now to double-check: We are adding more base than acid to make this buffer. Since our final pH, 4.2, is more basic than our pK_a, our answer makes sense.

4. a. *What is the pH of a solution prepared by mixing 100 mL of 1.00 M HCl with 300 mL of 0.500 M sodium succinate? The pK_{a1} of succinic acid is 4.21.*

First, identify the weak acid and its conjugate base. Sodium succinate would give Na^+ and (succinate)$^-$ in solution. HCl is a strong acid that would react with the (succinate)$^-$ to form H-succinate, or succinic acid. So, the weak acid would be succinic acid and the conjugate base would be (succinate)$^-$.

The number of moles of (succinate)$^-$ initially is (300 mL)(0.500 M) = 150 mmol.

The number of moles of HCl added is (100 mL)(1.00 M) = 100 mmol.

It is a good assumption that HCl will react completely with the (succinate)$^-$, so that would give 150 mmol (succinate)$^-$ – 100 mmol reacted with HCl

= 50 mmol (succinate)$^-$ left over and 100 mmol of succinic acid formed.

Now, use the Henderson-Hasselbalch equation: $pH = pK_a + \log\frac{[\text{succinate}^-]}{[\text{succinic acid}]}$

$$pH = 4.21 + \log\frac{[50 \text{ mmol}]}{[100 \text{ mmol}]} = 4.21 - 0.30 = \mathbf{3.91}$$

A pH of 3.91 makes sense because more acid than base is present, and 3.91 is more acidic than 4.21, the pK_a.

Note that this method included a shortcut: millimoles rather than molarity was used. This is valid because the total volume is the same and would cancel out. To use molarity, divide both the numerator (0.0045 mol) and the denominator (0.0030 mol) by the total volume, 0.450 L (300 mL + 150 mL).

b. *Is this buffer system effective? If so, why? If not, how could you correct it?*
This is a good buffer because its pH is within the range of the pK_a ± 1 pH unit, or 3.21 to 5.21. For a pH lower than 3.21, more succinate should be added. For a pH higher than 5.21, more acid should be added.

c. *80 mL of 1.0 M NaOH have been accidentally added into your carefully made solution! What is the final pH? Did the buffer work? How do you know? Is the solution still a good buffer?*
pH = 4.61, yes (see below), yes
From (a), we know that before the addition of NaOH, we have 50 mmol of succinate and 100 mmol of succinic acid. OH^-, a strong base, will react completely with the succinic acid to form succinate. We have (80 mL)(1.0 M) = 80 mmol OH^-. So,

100 mmol succinic acid – 80 mmol reacted (with OH^-) = 20 mmol succinic acid

50 mmol succinate + 80 mmol formed = 130 mmol succinate

Plug these new values into the Henderson-Hasselbalch equation:

pH = 4.21 + log(130 mmol/20 mmol) = 4.21+0.81 = **5.02** (which makes sense!)

The buffer worked. If 80 mmol of OH^- were present in the same volume of pure water, the final pH would be 12.5. It is still a pretty good buffer since 5.02 is less than 1 pH unit from the pK_a. However, it will have a greater buffering capacity for acids than for bases.

Practice Problems to Promote Perfection

Solutions are at the end of this chapter, after the solutions to the Review, Fill-in-the-Blank, Short-Answer, and Thought Questions.

1. Describe how you would prepare 1.0 L of a 0.30 M acetate buffer with a pH of 5.4. What ratio of acetate salt to acetic acid would you use? The pK_a of acetic acid is 4.76.

2. An acetate buffer is prepared by adding 40.0 mL of 1.00 M acetic acid to 200 mL of an aqueous solution containing 2.00 g of sodium acetate. The final solution is diluted to 300 mL. (The molecular weight of sodium acetate is 82.0 g/mol, and the pK_a of acetic acid is 4.76.)
   a. What is the pH?
   b. What is the total buffer concentration?
   c. What would the final pH be if 20 mL of 1.0 M of hydrochloric acid were added to this buffer solution? Is this still a good buffer? What would the pH have been if the buffer had not been present? (Assume the same total volume of water.)

3. For laboratory experiments that are extremely sensitive to pH, why is it often recommended to use *freshly* distilled water?

4. An 8-oz serving (240 mL) of a popular cola contains 27 g of sugars, listed as "high fructose corn syrup and/or sugar." What is the osmotic pressure that would be exerted by 27 g of fructose in 240 mL of water at 37°C? If the sugars were 27 g of sucrose, how would that affect the osmotic pressure? The molecular weight of fructose is 180 g/mol, and the molecular weight of sucrose is 342 g/mol.[2]

---

[2] Obviously, this is not a very good approximation for the cola, since the sugars are a mixture and there are other ingredients, such as caffeine, phosphoric acid, citric acid, carbonation, and "natural flavors."

5. That same 240 mL serving of cola contains 25 mg sodium. If we assume that the sodium is present as sodium chloride, that corresponds to 4.53×10^{-3} M NaCl. What would be the osmolarity and osmotic pressure of an aqueous solution of NaCl at this concentration at 37°C? Assume 100% ionization of the NaCl.

6. A solution of 0.200 g of an unknown molecule in 100 mL of water exerts an osmotic pressure of 0.465 atm at 25°C. Calculate the molecular weight of this nonelectrolyte.

AFTER STUDYING THIS CHAPTER, YOU SHOULD BE ABLE TO:

NONCOVALENT INTERACTIONS

- Identify the types of noncovalent interactions that can occur between given molecules (or between a given molecule and water).
- State whether a given molecule has a dipole moment.
- State whether a micelle will be able to form in a solution of a given molecule.

ACIDS AND BASES

- Identify weak acid-conjugate base pairs.
- Calculate pH from a given $[H^+]$ or $[H^+]$ from a given pH.

BUFFERS: USE THE HENDERSON-HASSELBALCH EQUATION

- Identify mixtures that can form buffer systems and give the pH range where each would be most effective.
- Calculate the pH of a buffer given the pK_a and amounts of a weak acid and its conjugate base. (Calculate the amounts of a weak acid and its conjugate base given the amounts of conjugate base and HCl *or* given the amounts of weak acid and NaOH.)
- Solve for the ratio of conjugate base to acid ($[A^-]/[HA]$).
- Determine how to prepare a specific buffer given the target pH, total buffer concentration, and total volume.
- Calculate the pH of a buffer solution after HCl or NaOH has been added.
- Titration curves: Estimate the pK_a and the effective buffer range.

OSMOTIC PRESSURE

- Calculate osmotic pressure given the amount of solute and volume using $\pi = iMRT$
- Calculate osmolarity (iM).
- Given osmotic pressure data, calculate the molar mass.
- Predict the direction that water will flow during dialysis. Predict whether a cell will shrink or swell, given various changes in osmotic pressure.

Use this space to note additional objectives provided by your instructor.

CHAPTER 3: SOLUTIONS TO REVIEW QUESTIONS

3.1 a. polar—uneven distribution of electrons in a molecule

b. hydrogen bond—occurs when electron-deficient hydrogens of one water molecule are attracted to the unshared pairs of electrons of an oxygen atom in another water molecule

c. electrostatic interaction—occurs between two opposite partial charges of full charges

d. salt bridge—forms as a result of attraction between positively and negatively charged amino acid side chains

e. dipole—molecules in which charge is separated

3.2 a. heat of fusion—the energy required to melt a solid

b. solvation sphere—shells of water molecules that cluster around both positive and negative ions

c. amphipathic—a molecule that contains both polar and nonpolar groups

d. micelle—formed when amphipathic molecules are mixed with water with the polar surface exposed to water and the nonpolar surface internalized

e. hydrophobic effect—results because nonpolar molecules are attracted to each other by van der Waals forces and are unable to hydrogen bond with water, resulting in a water cage surrounding the nonpolar molecules

3.4 a. acid—a species that donate protons

b. base—a specied that accept protons

c. weak acid—acid that does not completely dissociate in water

d. weak base—base that does not completely protonate in water

e. conjugate base—the deprotonated product of the dissociation of an acid

3.5 a. buffer—often composed of weak acids and their conjugate bases to help maintain a relatively constant hydrogen ion concentration

b. acidosis—a condition that occurs when human blood pH falls below 7.35

c. alkalosis—a condition that occurs when human blood pH rises above 7.45

d. pH—the negative log of the hydrogen ion concentration

e. pK_a—the negative log of the dissociation constant, K_a

3.7 $\text{pH} = -\log [\text{H}^+]$

$8.3 = -\log [\text{H}^+]$

$[\text{H}^+] = 10^{-8.3} = 5.0 \times 10^{-9}$ M

3.8 To prepare a 0.1 M phosphate buffer with pH 7.2, use the Henderson-Hasselbalch equation to calculate the ratio of the conjugate base to acid, where the conjugate base is "A⁻" and the acid is "HA":

$$\text{pH} = \text{p}K_a + \log \frac{[\text{A}^-]}{[\text{HA}]}$$

From a table of ionization constants, choose the phosphate conjugate acid-base pair that has a pK_a closest to 7.2:

$$\underset{\text{acid}}{H_2PO_4^-} \rightleftharpoons \underset{\text{conjugate base}}{H^+ + HPO_4^{2-}}$$

$$pK_a = 7.2$$

Substituting these values into the equation gives

$$7.2 = 7.2 + \log [A^-]/[HA]$$

$$0 = \log [A^-]/[HA]$$

$$10^0 = 1 = [A^-]/[HA]$$

$$[A^-] = [HA] \qquad \text{(Equation 1)}$$

The concentrations of the conjugate base and acid must be equal. We also know that the total concentration of the phosphate buffer is 0.1 M. Therefore,

$$[A^-] + [HA] = 0.1 \text{ M}. \qquad \text{(Equation 2)}$$

Using simultaneous equations (i.e., substituting the value for [HA] from Equation 1 into Equation 2) to determine the concentrations of the conjugate base and acid for this particular buffer solution gives

$$[A^-] + [A^-] = 0.1 \text{ M}$$

$$2[A^-] = 0.1 \text{ M}$$

$$[A^-] = 0.05 \text{ M}.$$

Taking this value and inserting it into Equation (1) gives: [HA] = 0.05 M.

To prepare this buffer, place 0.05 mol of the acid and 0.05 mol of the conjugate base in a 1-L volumetric flask and dilute with water to the 1-L mark.

3.10 In a solution of 1 M sodium lactate, water flows into the dialysis bag. In solutions of 3 M or 4.5 M sodium lactate, water flows out of the dialysis bag.

3.11

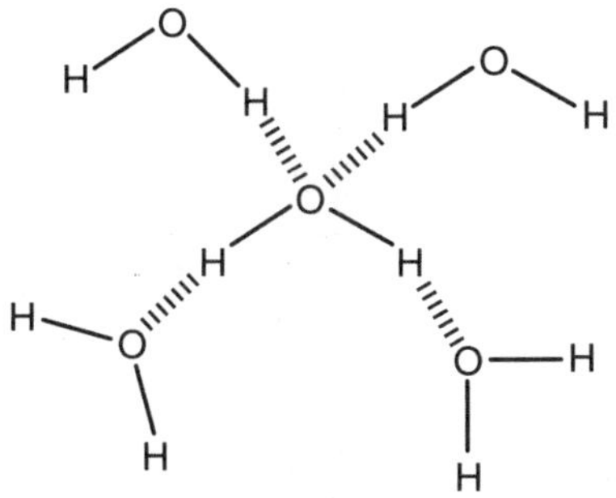

3.13 pH = pK_a + log [acetate]/[acetic acid]

4.76 = pK_a + log [0.1]/[0.1]

4.76 = pK_a + log 1

4.76 = pK_a + 0

4.76 = pK_a

$pK_a = \log K_a$

take the antilog of both sides

$4.76 = K_a$

$0.0000173 = K_a$

$1.73 \times 10\text{-}5 = K_a$

3.14 a. water and ammonia—hydrogen bonds

b. lactate and ammonium ion—ionic interactions

c. benzene and octane—van der Waals forces

d. carbon tetrachloride and chloroform—van der Waals forces

e. chloroform and diethyl ether—van der Waals forces

3.16 Molecules b, c, and d all would be expected to have a dipole moment.

3.17 When they are very close together, molecules of d are capable of forming micelles because one end of the molecule is polar and the other end is nonpolar.

3.19 The buffering capacity of a system is increased by raising the concentrations of both buffer components but not changing their ratio ($[A^-]/[HA]$). Increasing the concentration of only the weak acid, for example, would increase the buffer capacity for added base, but would lower the buffer capacity for added acid.

3.20 No. Not without knowing the pK_a of the buffer and the concentrations of the weak acid and the conjugate base.

3.22 Molecules b, c, and e are all weak acids because they are only partially ionized; a and d are strong acids (a is hydrochloric acid and d is nitric acid).

3.23 A buffer is composed of a weak acid and its conjugate base. Only c is a buffer.

3.25 No. The carbonic acid and carbonate react to produce bicarbonate. It is possible to have either a buffer system of carbonic acid and bicarbonate or a buffer system of bicarbonate and carbonate.

3.26 $K_a = 6.3 \times 10^{-8}$, therefore, $pK_a = 7.2$

$pH = pK_a + \log[A^-]/[HA]$

(where A^- = conjugate base of the weak acid HA)

$7.4 = 7.2 + \log[A^-]/[HA]$

$\log[A^-]/[HA] = 7.4 - 7.2 = 0.2$

$[A^-]/[HA] = 1.58{:}1$ or $1.6{:}1$

3.28 The contribution from the ionization of water must be considered. The hydrogen ion concentration is 1×10^{-8} M from acid and 10^{-7} M from water for a total acid concentration of 1.1×10^{-7} M acid. Therefore, the $pH = -\log(1.1 \times 10^{-7}) = 6.96$.

Note: Had the problem stated that $[H^+] = 1 \times 10^{-8}$ M, then the pH would be 8. However, without information to the contrary, only HCl and H_2O are present, which could be true only if the HCl had been added to pure water. (If NaOH were added to a more concentrated solution of HCl, then NaCl would also be in solution.) If acid is added to pure water, the pH will be acidic—perhaps only very slightly acidic—but definitely *not* basic.

3.29 For a substance to be used as a compatible solute, it must have solvent properties (polarity, ability to hydrogen bond) that are similar to those of water; it must be nontoxic even in high concentrations; it must provide freezing point depression and osmoprotection; and it must interact with structured water in a way that stabilizes the three-dimensional structures of the cell membrane and of proteins.

3.31 Detergents that readily form micelles in water have structures similar to those of the amphipathic lipid molecules that make up a cell membrane. These detergents, therefore, should be able to associate with—and form micelles with—the lipid components of the cell membrane, resulting in membrane disruption and cell death. (Bacteria that are resistant to the action of detergents have a more complex membrane structure, outer coating(s), and/or additional structures that protect their inner cell membrane.)

3.32 When the concentrations of the weak acid and the conjugate base are equal, the Henderson-Hasselbalch equation simplifies to pH = pK_a (because $[A^-]/[HA] = 1$, and $\log(1) = 0$). The pK_a of acetic acid is 4.75; therefore, the pH is 4.75.

$$\text{pH} = \text{p}K_a + \log\frac{[A^-]}{[HA]}$$

When $[A^-] = [HA]$,

$\text{pH} = \text{p}K_a + \log(1) = \text{p}K_a + 0$

$\text{pH} = \text{p}K_a = 4.75$ for acetic acid

3.34 When 1 mL of 1 M HCl is added to 1 L of water, the new $[H^+]$becomes:

$(0.001 \text{ L}) \times (1 \text{ M } H^+) = 0.001 \text{ mol } H^+$

$(0.001 \text{ mol } H^+)/(1.001 \text{ L total}) = 9.99 \times 10^{-4} \text{ M } H^+$

$\text{pH} = -\log[H^+] = -\log(9.99 \times 10^{-4} \text{ M } H^+)$

$\text{pH} = 3$

Compare this pH change (from 7 to 3) to the addition of the same amount of HCl to a buffer solution (Review Problem 3.33).

3.35 The extreme electronegativity of the oxygen polarizes the O–H bond of water and makes the hydrogen electron deficient. Because the unshared pairs of electrons on the oxygen are available for bonding, an electrostatic interaction occurs.

CHAPTER 3: SOLUTIONS TO FILL-IN-THE-BLANK QUESTIONS

3.37 Carbonic acid

3.38 Le Chatelier's principle

3.40 Alkalosis

3.41 Conjugate

3.43 Gel

3.44 Hydrogen

3.46 $4Fe_2O_3 + 3CH_4 \rightarrow 3CO_2 + 8Fe + 6H_2O$

CHAPTER 3: SOLUTIONS TO SHORT-ANSWER QUESTIONS

3.47 For hydrogen bonding to occur, the bond must be very polar (i.e., the atoms must have different electronegativity values). In the first three compounds the electronegativity difference between hydrogen and the other atom (fluorine, oxygen, or nitrogen, respectively) is very different. The electronegativity of carbon and hydrogen is about the same; hence, the bond is nonpolar and the hydrogens are incapable of hydrogen bonding.

3.49 The three chlorines are strongly electronegative and polarize the H—C bond toward the carbon. This makes the hydrogen electron deficient and a weak hydrogen bond occurs.

3.50 Lithium ions have a larger hydration sphere than sodium. Sodium does not easily diffuse into the cell because of its large hydration sphere. Lithium has an even larger hydration sphere and would have an even greater tendency to remain outside of the cell.

CHAPTER 3: SOLUTIONS TO THOUGHT QUESTIONS

3.52 Na^+, because the hydrated volume of K^+ is much smaller than that of Na^+. The smaller hydrated ion will diffuse through the gelatin more readily than Na^+, thereby lowering the amount of K^+ remaining in the well.

3.53 The size of an ion's solvation sphere is inversely related to its charge density (size of charge per unit volume). Sodium has a diameter of 1.96 Å and potassium has a diameter of 2.66 Å. Therefore, the hydrated volume of sodium is larger than that of potassium.

3.55 The highly concentrated sugar solution pulls water out of any bacterial cells present, which kills them, thereby preserving the fruit.

3.56 The regular crystal lattice of the ice crystal is more open than the tightly hydrogen-bound liquid water. If ice were more dense than water, ice formed in lakes and oceans would sink to the bottom. Eventually, only a narrow layer at the surface would be liquid. This environmental condition is incompatible with life (for most aquatic life, and they would not be able to survive).

3.58 The blood is so highly buffered by the bicarbonate buffer and the large amounts of blood proteins that under normal physiological conditions the transport of weak acids in the blood does not appreciably change its pH. For example, in the presence of bicarbonate, any acid that ionizes produces carbon dioxide (which is exhaled). The pH of the blood then remains virtually unchanged.

$$HCO_3^- + H^+ \rightleftharpoons H_2CO_3 \rightleftharpoons CO_2 + H_2O$$

3.59 The pH scale is derived using the ionization constant of water. To establish the pH scale for another solvent, the ionization constant of that solvent would have to be used, and the pH scale would be different from the pH scale for water.

3.61 No. The structure of cells is based on the phase separation of hydrophobic and hydrophilic substances. The function of the cell membrane is possible only because lipids are insoluble in water. If water dissolved every molecule, living organisms would not be able to create a barrier (membranes!) between themselves and their surroundings, and living organisms would not be possible.

3.62 The small water molecules can crowd closely around the ions and effectively disperse the charge, thereby facilitating solution. The bulky R group of the alcohol prevents this close interaction of solvent and solute. As a result, the ionic compound does not dissolve as easily.

3.64 Hydration tends to make ionization easier. The hydrated acid group on the protein surface would have a higher K_a than one in the anhydrous interior of the protein.

3.65 Water weakens ionic interactions by forming a solvation sphere around each ion. As the distance increases between the cation and the anion, the attractive force decreases between them. (See Figure 3.10.) In other words, the polar water molecules crowd around the ions, interacting with the ions and weakening the interactions between oppositely charged ions.

3.67 Magnesium, with a double positive charge, forms a strong hydration sphere. For Mg^{2+} to move into the structured water of macromolecules, its hydration sphere must be removed in a process that requires a large amount of energy. Chloride ion, on the other hand, has only a single negative charge and is a larger ion. Its hydration sphere is not as tightly held and it would take less energy to remove. As a result, chloride would be more easily incorporated.

3.68 The micelle that forms would be U shaped, with the carboxyl groups projecting into the water and the hydrophobic "bottoms" of the U toward the center. It could also form a membrane-like structure with the chains being straight, with the carboxyl groups positioned on either side of the membrane.

3.70 The titration curve for tyrosine is as follows:

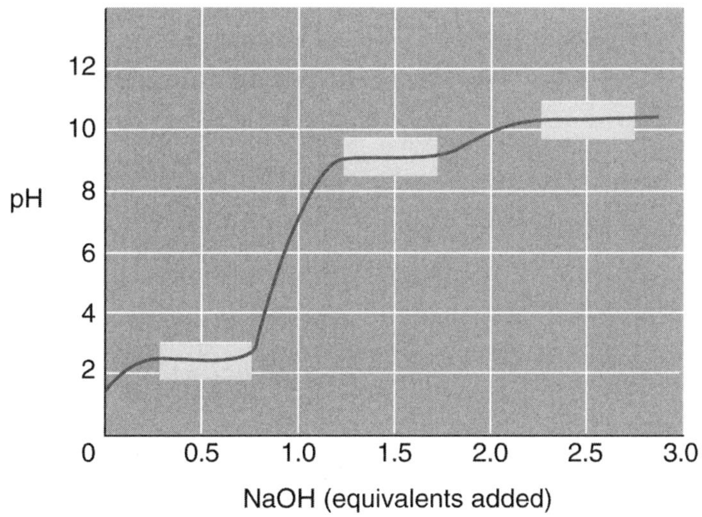

3.71 The calculation is as follows: 335 J/g(1 g) + 4.25 J/g(1 g)((100 − 0) + 2258 J/g(1 g) = 2597 J.

3.73 The solvation sphere of the potassium ion in methyl alcohol is as follows:

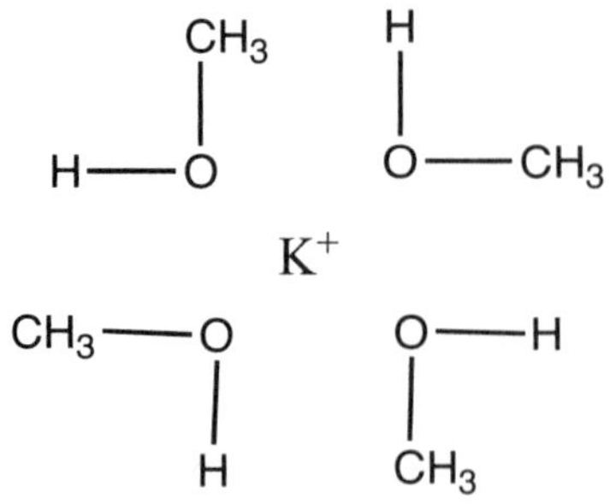

3.74 The molecule has polar and nonpolar components and would be expected to form a lipid bilayer. Alkyl chlorides are hydrophobic. As a result, one would expect a bilayer with the carboxyl groups projecting into the water and the chloride in the interior.

3.76 The energy of solvation stabilizes both the departing proton and the acetate ion. The extra energy released helps drive the process and makes ionization easier. As a result, the pK_a of acetic acid in water should be lower than in the absence of water.

Solutions to Practice Problems to Promote Perfection

1. *Describe how to prepare 1.0 L of a 0.30 M acetate buffer with pH 5.4. What ratio of acetate salt to acetic acid would you use? (The pK_a of acetic acid is 4.76.)*

$$pH = pK_a + \log[A^-]/[HA]$$

$$5.4 = 4.76 + \log[A^-]/[HA]$$

$$0.64 = \log[A^-]/[HA]$$

$$\mathbf{[A^-]/[HA] = 4.4} = \text{ratio of acetate salt to acetic acid}$$

$$[A^-] = (4.4)[HA]$$

Substitute this equation for $[A^-]$ into the total buffer concentration equation:

Total buffer concentration = $[A^-] + [HA] = 0.30$ M

$$(4.4)[HA] + [HA] = 0.30 \text{ M}$$

$$(5.4)[HA] = 0.30 \text{ M}$$

$$\mathbf{[HA] = 0.056\ M}$$

Substitute this value for [HA] into the total buffer concentration equation:

$$[A^-] + 0.056 \text{ M} = 0.30 \text{ M}$$

$$[A^-] = 0.24 \text{ M}$$

To prepare the buffer, mix 0.24 mol of acetate salt with 0.074 mol of acetic acid in a volumetric flask and dilute to the mark with water. (To be more specific, dissolve 20 g of sodium acetate in water in a 1-L volumetric flask. Add 74 mL of 1 M acetic acid. Dilute to the 1-L mark with freshly distilled water.)

2. a. millimoles of acetic acid = (40.0 mL)(1.00 M) = 40.0 mmol

millimoles of acetate = (2.00 g)/(82.0 g/mol) = 24.4 mmol

$pH = pK_a + \log[A^-]/[HA] = 4.76 + \log(24.4/40.0) = \mathbf{4.55}$

(Note that we do not need to use the total volumes to get the correct answer. If you calculated molarity, the equation would be:

pH = 4.76 + log([0.0813 M A^-]/[0.133 M HA]) = 4.55

b. Total buffer concentration = (40.0 mmol + 24.4 mmol)/300 mL = 0.215 M

c. HCl: (10 mL)(1 M) = 10 mmol HCl, which would react with the acetate to form acetic acid

acetate: 24.4 mmol – 10 mmol (reacted with HCl) = 14.4 mmol acetate

acetic acid: 40.0 mmol + 10 mmol (formed) = 50 mmol acetic acid

pH = 4.76 + log(14.4/50) = 4.76 – .54 = **4.22**

It is still a good buffer! And if 10 mmol HCl were in a total volume of 310 mL, the pH would be **1.5**. Buffers work!

3. Distilled water that has been exposed to air for any length of time will probably contain dissolved CO_2, which would lower the pH.

$$CO_2 + H_2O \rightleftharpoons H_2CO_3 \rightleftharpoons H^+ + HCO_3^-$$

4. First, calculate the molarity of the fructose: 27 g/180 g/mol = 0.15 mol

0.15 mol/0.240 L = 0.625 M

T = 37 + 273 = 310 K

$\pi = iMRT$ = (1)(0.625 M)(0.082)(310 K) = 16 atm

The molarity of sucrose would be 27 g/342 g/mol = 0.0790 mol

0.0790 mol/0.240 L = 0.329 M

$\pi = iMRT$ = (1)(0.329 M)(0.082)(310 K) = 8.4 atm

5. osmolarity = iM = (2)(4.53×10^{-3} M) = 9.06×10^{-3} M

osmotic pressure = $\pi = iMRT$ = (2)(4.53×10^{-3} M)(0.082)(310 K) = 0.23 atm

6. $\pi = iMRT$, R = 0.082 L·atm/K·mol T = 25°C + 273 = 298 K

i = 1 (for a nonelectrolyte) M = mol/L = No. mol/(0.100 L)

First, use $\pi = iMRT$ to solve for the number of moles:

0.465 atm = (1)(mol/0.100 L)(0.082 L·atm/K·mol)(298 K)

1.90×10^{-3} mol

molecular weight = g/mol = (0.200 g)/(1.90×10^{-3} mol) = 105 g/mol

4 Energy

Brief Outline of Key Terms and Concepts

OVERVIEW OF THERMODYNAMICS

All living organisms unrelentingly require energy. **BIOENERGETICS**, the study of energy transformations, can be used to determine the direction and extent to which biochemical reactions proceed. **ENTHALPY** (a measure of heat content) and **ENTROPY** (a measure of disorder) are related to the first and second laws of thermodynamics, respectively. **FREE ENERGY** is the portion of total energy that is available to do work and is related to enthalpy and entropy.

4.1 THERMODYNAMICS

FIRST LAW OF THERMODYNAMICS

INTERNAL ENERGY CHANGE = HEAT + WORK

At constant pressure, a system's **ENTHALPY CHANGE** ΔH is equal to the flow of **HEAT ENERGY**. **EXOTHERMIC** processes release heat energy and have a negative (–) ΔH. **ENDOTHERMIC** reactions have a positive (+) ΔH and require heat energy from the surroundings. In **ISOTHERMIC** processes no heat is exchanged with the surroundings. Total energy, free energy, enthalpy, and entropy are state functions.

$$\Delta H^\circ_{(\text{reaction})} = \Sigma\Delta H_f^\circ{}_{(\text{products})} - \Sigma\Delta H_f^\circ{}_{(\text{reactants})}$$

SECOND LAW OF THERMODYNAMICS

The second law of thermodynamics states that the universe tends to become more disorganized. **ENTROPY** increases may take place anywhere in the system's universe. For bioprocesses, entropy increases take place in the surroundings.

4.2 FREE ENERGY

FREE ENERGY is a thermodynamic function that can be used to predict the spontaneity of a process. **FREE ENERGY**, a state function that relates the first and second laws of thermodynamics, represents the maximum useful work obtainable from a process.

$$\Delta G = \Delta H - T\Delta S_{sys}$$

SPONTANEOUS reactions are **EXERGONIC** (–ΔG); they release energy. Nonspontaneous reactions are **ENDERGONIC** (+ΔG); they need energy input to proceed.

STANDARD FREE ENERGY CHANGES

ΔG° is the standard free energy change (ΔG) at standard conditions: 25°C, 1 atm, and 1 M solute.

In bioenergetics, $\Delta G^{\circ\prime}$ = the standard free energy change at pH 7. For: $a\text{A} + b\text{B} \rightleftharpoons c\text{C} + d\text{D}$,

$$\Delta G = \Delta G^\circ + RT\ ln\frac{[\text{C}]^c[\text{D}]^d}{[\text{A}]^a[\text{B}]^b} \qquad R = 8.315\ \frac{\text{J}}{\text{mol·K}}$$

(Remember to convert temperature into Kelvin and to convert J to kJ if necessary.)

$$K_{eq} = \frac{[\text{C}]^c[\text{D}]^d}{[\text{A}]^a[\text{B}]^b}$$

When the system is at equilibrium, $\Delta G = 0$, so:

$$\Delta G^\circ = -RT\ ln\ K_{eq}$$

COUPLED REACTIONS

The hydrolysis of ATP immediately and directly provides the free energy to drive an immense variety of endergonic biochemical reactions.

$$\Delta G^{\circ\prime}_{(\text{total})} = \Delta G^{\circ\prime}_{(\text{reaction 1})} + \Delta G^{\circ\prime}_{(\text{reaction 2})}$$

THE HYDROPHOBIC EFFECT REVISITED

To understand why biomolecules spontaneously self-assemble into complex ordered systems (e.g., lipid bilayers), consider the total entropy (disorder) of the *entire system*, including the huge increase in entropy of the surrounding water molecules.

4.3 THE ROLE OF ATP

ATP hydrolysis provides most of the free energy required for living processes. ATP is ideally suited to its role as universal energy currency, as ATP is easily hydrolyzed into products that have the following stabilizing advantages over ATP: (1) less electrostatic repulsion between adjacent (–) charges, (2) more **RESONANCE HYBRIDIZATION**, (3) more easily solvated, (4) greater entropy (more molecules). Since ATP has an ***INTERMEDIATE* PHOSPHORYL GROUP TRANSFER POTENTIAL**, it can carry phosphoryl groups from high-energy compounds to low-energy compounds.

BIOCHEMISTRY IN PERSPECTIVE: NONEQUILIBRIUM THERMODYNAMICS

Living organisms are far-from-equilibrium **dissipative** systems. They create internal organization via a continuous flow of energy.

4.1 THERMODYNAMICS

ENTHALPY (*H*), ENTROPY (*S*), FREE ENERGY (*G*)
Open vs. closed systems
State functions—values are independent of path
Enthalpy, entropy, and free energy are state functions, but work and heat are not.
Energy is the capacity to do work.
Energy is exchanged (between system and surroundings) as work and/or heat.

FIRST LAW OF THERMODYNAMICS

Energy can neither be created nor destroyed, but it can be transformed from one form into another.

ENTHALPY $\Delta H_{reaction} = H_{products} - H_{reactants}$

EXOTHERMIC *vs.* ENDOTHERMIC
$\Delta H = (-)$ $\Delta H = (+)$

Since enthalpy is a state function, the reaction mechanism (i.e., how you get from the reactants to the products) does not affect ΔH. That means that we can calculate the standard enthalpy of any reaction by summing up the ΔH_f° (the standard enthalpy of formation per mole) of the products and subtracting the sum of the ΔH_f° of the reactants:

$$\Delta H^\circ_{(reaction)} = \Sigma\Delta H_f^\circ{}_{(products)} - \Sigma\Delta H_f^\circ{}_{(reactants)}$$

Do not forget to take the stoichiometry of the reaction into consideration. You will need to multiply the standard enthalpy of formation of each molecule by its coefficient in the chemical reaction.

SECOND LAW OF THERMODYNAMICS

The disorder (entropy, ΔS) of the universe always increases.

Spontaneous reactions or processes are **EXERGONIC**, i.e., energy is released.

Systems *can* spontaneously become more ordered (decrease in entropy) if the surroundings become more disordered (increase in entropy) and the overall disorder (entropy) of the universe *increases*.

4.2 FREE ENERGY = ΔG = GIBBS FREE ENERGY CHANGE

$$\Delta G = \Delta H - T\Delta S \text{ (at constant temperature and pressure)}$$

EXERGONIC (**SPONTANEOUS**, releases energy)
vs. **ENDERGONIC** (nonspontaneous, requires energy)

The sign of ΔG allows us to predict whether a chemical reaction can occur.

IF	*THEN THE REACTION IS:*
$\Delta G = -$	spontaneous (exergonic, favorable)
$\Delta G = +$	nonspontaneous (endergonic, not favorable)
$\Delta G = 0$	at equilibrium (no change)

Consider a general reaction, A + B → C + D. If the ΔG for this reaction is negative (energy is released and it is spontaneous), then the reverse reaction, C + D → A + B, has to be nonspontaneous (ΔG is positive). Since ΔG is a state function, the magnitude of ΔG in the forward reaction is the same as the reverse reaction. So, when the direction of a

chemical reaction is reversed, the sign of ΔG is also reversed.
When ΔG = zero, there is no net change in the chemical reaction. The rate forward (A+B $\rightarrow$ C+D) equals the rate of the reverse reaction (C+D $\rightarrow$ A+B), and the reaction is at equilibrium.

STANDARD FREE ENERGY CHANGES: $\Delta G°$

Since ΔG depends on temperature, pressures, and concentrations, there needs to be some reference point. The *standard free energy*, $\Delta G°$, is defined as ΔG at standard state conditions: 25°C, 1 atm pressure, and 1 M reactant concentration. However, nearly all biochemical reactions occur in dilute, aqueous mixtures, so biochemists have developed their own reference point, $\Delta G°'$, which is $\Delta G°$ at pH = 7.[1]

ΔG	=	free energy change of a reaction under actual conditions
$\Delta G°$	=	free energy change at standard state conditions: 25°C, 1.0 atm, 1.0 M
$\Delta G°'$	=	free energy change at biochemical standard state conditions: $\Delta G°$ at pH 7.

So, the symbol (°) indicates standard state conditions, and a prime (′) indicates biochemical standard state conditions.

ΔG is related to the reference $\Delta G°$ (or $\Delta G°'$) by the following equation[2]:

$$\Delta G = \Delta G° + RT\ ln\frac{[C]^c[D]^d}{[A]^a[B]^b}$$

for the reaction: $aA + bB \rightarrow cC + dD$

$R = 8.315$ J/mol·K
T (in Kelvin) = °C + 273

At equilibrium, $\Delta G = 0$, so: $\boldsymbol{\Delta G° = -RT\ ln\ K_{eq}}$

$$K_{eq} \text{ is the equilibrium constant: } K_{eq} = \frac{[\text{products}]_{\text{at equilibrium}}}{[\text{reactants}]_{\text{at equilibrium}}} = \frac{[C]^c[D]^d}{[A]^a[B]^b}$$

The sign and value of $\Delta G°$ indicate the direction and magnitude of a particular reaction at equilibrium (and under standard state conditions). For a given $\Delta G°$, the K_{eq} can be calculated.

EXAMPLE:

Calculate K_{eq} for the hydrolysis of ATP, given the following:

$ATP + H_2O \rightarrow\ ADP + P_i$ $\qquad \Delta G° = -30.5$ kJ/mol $\qquad K_{eq} = \dfrac{[ADP][P_i]}{[ATP]}$

(Since water is the solvent, $[H_2O]$ is omitted from K_{eq}.)

SOLUTION: We can solve for K_{eq} using $\Delta G° = -RT\ ln\ K_{eq}$

-30.5 kJ/mol = $-(8.315$ J/mol·K$)(298$ K$)\ ln\ K_{eq}$

$12.31 = ln\ K_{eq}$

$K_{eq} = 2.22 \times 10^5$ or 222,000/1

In this example, you can see that a negative $\Delta G°$ produces a large K_{eq}. The large K_{eq} indicates

[1] Note that if $[H^+]$ does not affect the chemical reaction, then $\Delta G° = \Delta G°'$.

[2] Equilibrium constants are really defined for “activities” rather than “concentrations.” However, concentrations are simpler to use and adequate for our purposes.

that at equilibrium (at 25°C), there is 222,000 times more product than reactant, and the reaction will proceed in the forward direction. Another way to look at this is as follows: At equilibrium, for each molecule of ATP present, there will be 471 ADP molecules and 471 P_i molecules (since $[ADP][P_i] = 222{,}000$, and the square root of 222,000 is 471).

What if the $\Delta G°$ had been +30.5 kJ/mol? Then the K_{eq} would be 4.5 x 10^{-6} 1/222,000.

$ATP + H_2O$	$\rightarrow$	$ADP + P_i$	$\Delta G° = -30.5$ kJ/mol, spontaneous
$ADP + P_i$	$\rightarrow$	$ATP + H_2O$	$\Delta G° = +30.5$ kJ/mol, nonspontaneous

Although a reaction with a positive $\Delta G°$ is nonspontaneous, that same reaction may have a negative ΔG under different (nonstandard state) conditions (e.g., in a living cell). The actual direction of a reaction can be altered by changing the concentration of products or reactants. This is described by the general equation:

$$\Delta G = \Delta G° + RT\, ln\frac{[\text{products}]}{[\text{reactants}]} \qquad \text{or} \qquad \Delta G = \Delta G° + RT\, ln\frac{[C]^c[D]^d}{[A]^a[B]^b}$$

EXAMPLE:

glucose-1-phosphate → glucose-6-phosphate $\Delta G° = -7.1$ kJ/mol

From the negative $\Delta G°$, we already know that the reaction proceeds forward to form glucose-6-phosphate under biochemical standard state conditions. But, can this reaction be reversed? Intuitively, if we were to add a high concentration of glucose-6-phosphate, we would think that this reaction could be "pushed" backward. What if the concentration of glucose-6-phosphate was 100 mM and the concentration of glucose-1-phosphate was 1.0 mM? In which direction would the reaction proceed?

Use the equation $\Delta G = \Delta G° + RT\, ln\frac{[\text{glucose-6-phosphate}]}{[\text{glucose-1-phosphate}]}$

[glucose-6-phosphate] = 100 mM
[glucose-1-phosphate] = 1.0 mM

$$\Delta G = -7100 \text{ J/mol} + (8.315 \text{ J/mol K})(298 \text{ K})\, ln\frac{[100 \text{ mM}]}{[1 \text{ mM}]}$$

$\Delta G = +4.3$ kJ/mol

The positive ΔG tells us that this reaction will proceed in the opposite direction, as written. So, by increasing the concentration of the product, we can reverse the direction of the reaction.

As you can see, the direction of a reaction under actual cellular conditions depends not only on $\Delta G°$, but also on the concentrations of reactants and products.

COUPLED REACTIONS

A nonspontaneous reaction ($+\Delta G$) can be driven forward by coupling it with a spontaneous reaction ($-\Delta G$) to produce a negative ΔG overall.
Coupled reactions allow the cell to harness the energy produced by catabolism.

EXAMPLE:

Compare the reactions for PEP (phosphoenolpyruvate) hydrolysis vs. ATP synthesis:

$PEP + H_2O$	$\rightarrow$	pyruvate + P_i	$\Delta G°' = -61.9$ kJ/mol
$ADP + P_i$	$\rightarrow$	$ATP + H_2O$	$\Delta G°' = +30.5$ kJ/mol

Note that the hydrolysis of PEP is spontaneous ($\Delta G°$ is negative, and the reaction proceeds in the forward direction as written), but the reaction to form ATP is not ($\Delta G°$ is positive). To capture some of the energy of this PEP hydrolysis, consider coupling (adding) these two reactions:

PEP + H_2O	→	pyruvate + P_i	$\Delta G°' = -61.9$ kJ/mol
ADP + P_i	→	ATP + H_2O	$\Delta G°' = +30.5$ kJ/mol
PEP + ADP	→	pyruvate + ATP	$\Delta G°' = -31.4$ kJ/mol

When we add these reactions, we also add the individual $\Delta G°'$ values to obtain the $\Delta G°'$ of the overall reaction. (This is valid because ΔG is a state function.)

$$\Delta G°'_{\text{(total)}} = \Delta G°'_{\text{(reaction 1)}} + \Delta G°'_{\text{(reaction 2)}}$$

The overall reaction is still thermodynamically favorable ($-\Delta G°'$), but we used some of the free energy to "drive" an unfavorable chemical reaction forward. Some of the energy released from the hydrolysis of PEP (phosphoenolpyruvate) was captured by ADP to form ATP.

THIS CAPTURED ENERGY CAN BE RELEASED LATER (WHEN IT IS NEEDED) BY HYDROLYZING ATP TO REGENERATE ADP.

For example, consider the phosphorylation of glucose:

Glucose + P_i	→	Glucose-6-phosphate + H_2O	$\Delta G°' = +13.8$ kJ/mol

When this reaction is coupled to ATP hydrolysis, the overall $\Delta G°'$ is negative. The energy from ATP hydrolysis drives this endergonic reaction forward.

Glucose + P_i	→	Glucose-6-phosphate + H_2O	$\Delta G°' = +13.8$ kJ/mol
ATP + H_2O	→	ADP + P_i	$\Delta G°' = -30.5$ kJ/mol
Glucose + ATP	→	Glucose-6-phosphate + ADP	$\Delta G°' = -16.7$ kJ/mol

Since cells use ATP synthesis and hydrolysis to store and use chemical energy, ATP is often referred to as the "energy currency" of the cell.

In summary, a reaction that is not spontaneous can be driven forward if it is coupled with a spontaneous reaction and the overall ΔG is negative. In other words, the nonspontaneous reaction needs the energy from another reaction to drive it forward.

How can we get a nonspontaneous reaction to go forward (assuming constant temperature)?

- Couple it with a spontaneous reaction that will supply enough energy to give an overall negative ΔG, or
- Change the relative concentrations of reactant and product so that the actual ΔG will be negative.

THE HYDROPHOBIC EFFECT REVISITED

The hydrophobic effect explains why nonpolar molecules aggregate in water, why micelles and bilayers form, and why proteins fold. Although these processes seem to result in a decrease in entropy (i.e., an increase in order), the overall entropy (including that of the surrounding water molecules) is higher.[3]

[3] This is an oversimplification. Surface tension effects, van der Waals interactions, and the ordered arrangement of the water molecules immediately adjacent to a nonpolar molecule (i.e., the nature of structured water) should also be

4.3 THE ROLE OF ATP

ATP is produced using the energy released by breaking down nutrient molecules (catabolism) and by the light reactions of photosynthesis. Hydrolysis of ATP releases 30.5 kJ/mol of energy that is used to drive endergonic processes such as:

1. Biosynthesis (anabolic pathways)
2. Active transport of substances across cell membranes
3. Mechanical work (e.g., muscle contraction)

Why is ATP hydrolysis so exergonic? The products are more stable than the reactants, because the final products:

1. have less electrostatic repulsion (of negative charges)
2. have more resonance structures than the reactants[4]
3. are more easily solvated than the reactants

PHOSPHORYL GROUP TRANSFER POTENTIAL is the tendency of a phosphoryl-containing molecule to hydrolyze, resulting in its phosphoryl group being released as HPO_4^{2-} or transferred to another molecule.

The greater the phosphoryl group transfer potential, the more energy is released when a phosphoryl group is hydrolyzed (and the more stable a molecule would be without its phosphoryl group).

Since ATP has an *intermediate* phosphoryl group transfer potential, it can take a phosphoryl group from a higher-energy compound and transfer it to a lower-energy compound. For instance, recall the two examples given earlier for coupled reactions:

PEP + ADP	$\rightarrow$	Pyruvate + ATP	$\Delta G^\circ = -31.4$ kJ/mol
Glucose + ATP	$\rightarrow$	Glucose-6-phosphate + ADP	$\Delta G^\circ = -16.7$ kJ/mol

Essentially, a phosphoryl group could be transferred from PEP to ATP and then from ATP to glucose, because both of these coupled reactions are spontaneous. PEP has a higher phosphoryl group transfer potential than ATP, and ATP has a higher phosphoryl group transfer potential than glucose-6-phosphate.

THERMODYNAMICS VS. KINETICS

THERMODYNAMICS tell us whether a reaction will be **SPONTANEOUS**.

If a reaction is spontaneous, there will be more products than reactants when the reaction is in equilibrium. That is, the reaction proceeds forward until it reaches equilibrium. Spontaneous reactions are **EXERGONIC**, i.e., they release energy.

KINETICS tell us **HOW *FAST*** the reaction will go.

Consider the combustion of glucose: $C_6H_{12}O_6 + 6O_2 \rightarrow 6CO_2 + 6H_2O$

Glucose reacts with oxygen to form carbon dioxide and water. Thermodynamics tell us that this reaction is spontaneous and highly exothermic, but kinetics tell us that the rate at room temperature is incredibly slow. That is why glucose in contact with air doesn't simply burst into flames.

So, the chemical term *spontaneous* has a different meaning from the common dictionary definition. Glucose will *spontaneously* combust, but it will not *suddenly* combust (unless enough energy is applied to overcome its activation energy and/or the activation energy is

---

taken into consideration. In general, at biological temperatures, these processes are driven by an increase in entropy (disorder) of the surroundings.

[4] Of course, this assumes that the H^+ of ADP dissociates.

lowered with the help of a catalyst or enzyme).[5]

THERMODYNAMICS	**KINETICS**
SPONTANEITY: *Can* a reaction happen? Is a reaction *spontaneous*? Gibbs free energy change, ΔG exergonic vs. endergonic enthalpy, ΔH, and entropy, ΔS $\Delta G = \Delta H - T\Delta S$ equilibrium K_{eq} (equilibrium constant) Chapter 4: **ENERGY**	**RATE**: *How fast* will a reaction happen? E_a, activation energy k (rate constant) order of a reaction depends on mechanism (path) *Enzymes* are biochemical catalysts that reduce the activation energy and cause a reaction to go *faster*. Enzymes affect the rate; they do not affect the thermodynamics of a reaction. Chapter 6: **ENZYMES**

FOOD FOR THOUGHT: What is wrong with the following statement?
ATP releases energy when its high-energy phosphoryl bond is broken.

ANSWER:

1. The "high-energy phosphoryl bond" is not simply "broken." It is hydrolyzed (cleaved with the addition of water). Some bonds are broken, and others are formed.
2. The term "high-energy bond" suggests that the bond is unstable. Stability really refers to its ability to participate in reactions as opposed to the magnitude of the bond energy. The phosphoanhydride bond of ATP is actually more stable relative to compounds that have a higher phosphoryl group transfer potential.

EQUATIONS

$$\Delta H^\circ_{(reaction)} = \Sigma\Delta H_f^\circ{}_{(products)} - \Sigma\Delta H_f^\circ{}_{(reactants)}$$

$$\Delta G = \Delta H - T\Delta S$$

$$K_{eq} = \frac{[\text{products}]_{\text{at equilibrium}}}{[\text{reactants}]_{\text{at equilibrium}}} = \frac{[C]^c[D]^d}{[A]^a[B]^b}$$

(for the reaction $aA + bB \rightleftharpoons cC + dD$)

(For the K_{eq} term, remember to use exponents if the coefficients in the chemical equation are greater than 1.)

$$\Delta G = \Delta G^\circ + RT\,ln\frac{[\text{products}]}{[\text{reactants}]}$$

$R = 8.315$ J/mol K; T (in Kelvin) = °C + 273

$$\Delta G = \Delta G^\circ + RT\,ln\frac{[C]^c[D]^d}{[A]^a[B]^b}$$

Hint: Be careful to use ***either* J *or* kJ** in your final equation, since R contains joules (J) and ΔG is typically in kilojoules (kJ). (1000 J = 1 kJ)

$$\Delta G^\circ = -RT\,ln\,K_{eq} \text{ (at equilibrium)}$$

$$\Delta G^{\circ\prime}_{(total)} = \Delta G^{\circ\prime}_{(reaction\ 1)} + \Delta G^{\circ\prime}_{(reaction\ 2)}$$

(for coupled reactions)

[5] "Spontaneous combustion" as defined by popular usage is beyond the scope of this discussion.

AFTER STUDYING THIS CHAPTER, YOU SHOULD BE ABLE TO:

- Calculate $\Delta H_{\text{reaction}}$ given $\Delta H_f°$ data.
- Determine whether a reaction is endothermic or exothermic, exergonic or endergonic, spontaneous or nonspontaneous.
- Use $\Delta G = \Delta G° + RT\ ln\frac{[\text{products}]}{[\text{reactants}]}$ to determine whether a reaction will proceed forward at particular concentrations. That is, will it be spontaneous under different cellular conditions?
- Coupled reactions: Determine whether the overall reaction will be spontaneous using the equation: $\Delta G°'_{(\text{total})} = \Delta G°'_{(\text{reaction 1})} + \Delta G°'_{(\text{reaction 2})}$.
- Coupled reactions: Choose a reaction that will provide enough energy to drive a nonspontaneous reaction.
- Use $\Delta G° = -RT\ ln\ K_{eq}$ to calculate either $\Delta G°$ or K_{eq}.

Use this space to note any additional objectives provided by your instructor.

CHAPTER 4: SOLUTIONS TO REVIEW QUESTIONS

4.1 a. thermodynamics—the study of the heat and energy transformations in a chemical reaction

b. bioenergetics—study of energy transformations in living organisms
c. enthalpy—a measure of the heat evolved during a reaction
d. entropy—a measure of the disorder during a reaction
e. free energy—a measure of the tendency of a reaction to occur

4.2 a. work—a physical change caused by a change in energy

b. exothermic reaction—a reaction that releases heat

c. endothermic reaction—a reaction that requires an energy input

d. isothermic process—a reaction that has no heat exchanged with the surroundings

e. spontaneous process—reactions that occur with the release of energy

4.4 a. redox reaction—oxidation-reduction reaction where electrons are transferred from an electron donor to an electron acceptor

b. resonance hybrid—occurs when a molecule has two or more alternative structures that differ only in the position of electrons

c. chemolithotroph—hetertrophs that generate ATP by oxidizing inorganic compounds

d. electron donor—a molecule that provides electrons to an oxidation-reduction reaction

e. biogeochemical cycle—a pathway driven by solar and geothermal energy in which a chemical element moves through Earth's biotic and abiotic compartments

4.5 State functions are those that are independent of path. Entropy, enthalpy, and free energy are all path independent.

4.7 Given that the ionization constant for formic acid is 1.8×10^{-4}, the $\Delta G°$ for the reaction would be calculated as follows:

$$\Delta G° = -RT \ln K_{eq}$$

$$\Delta G° = -(8.315 \text{ J/mol} \cdot \text{K})(298 \text{ K}) \ln(1.8 \times 10^{-4})$$

$$\Delta G° = -(8.315)(298\text{K})(-8.62) \text{ J/mol}$$

$$\Delta G° = +21{,}366 \text{ J/mol or } 21.4 \text{ kJ/mol}$$

4.8 When temperature = 0 K.

4.10 The first law of thermodynamics concerns the conservation of energy where energy cannot be created or destroyed. Energy can be transformed from one form to another.

$$\Delta E = q + w$$

where ΔE = the change in energy of the system
q = the heat absorbed or released by the system
w = the work done by or to the system

The second law of thermodynamics concerns the spontaneity of reactions where spontaneous reactions occur in the direction that increases the total disorder of the universe.

$\Delta S_{univ} = \Delta S_{surr} - \Delta S_{sys}$
where S is entropy in the universe (univ), surroundings (surr), and system (sys).

4.11 Given that the ionization constant for acetic acid is 1.8×10^{-5}, the $\Delta G°$ for the reaction would be calculated as follows:
$\Delta G = - RT \ln K\text{eq}$
$= - (8.315 \text{ J/mol} \cdot \text{K})(298 \text{ K}) \ln(1.8 \times 10^{-5})$
$= - 27{,}071 = 27.1 \text{ kJ/mol}$

4.13 Statements (a) and (b) are undetermined, since no rate information was provided.

Statement (c) is false. Both reactions have a negative $\Delta G°'$, so both reactions are spontaneous.

Statement (d) is true.

4.14 Work is defined as a change in energy that produces a physical change. Physiological examples include the maintenance of concentration gradients across membranes, biomolecule synthesis, active transport across membranes, and muscle contraction.

4.16 AMP hydrolysis involves cleavage of an ester bond and therefore releases the least energy. Hydrolysis of the other phosphate linkages involves the hydrolysis of either an anhydride or an enol bond.

4.17 The following statements are true: a, b, c, and f.

d. The sign and magnitude of ΔG give important information about the direction of a reaction, but no information about the rate.

e. At equilibrium, $\Delta G = 0$.

4.19 $\Delta G°' = -RT \ln K_{eq}$

$-7{,}100 \text{ J/mol} = -(8.315 \text{ J/mol K})(298 \text{ K})(\ln K_{eq})$

$\ln K_{eq} = 2.865$

$K_{eq} = 17.56$

4.20 With an intermediate phosphoryl group transfer potential, ATP can accept a phosphate group from compounds that have a higher phosphoryl group transfer potential and transfer it to lower energy compounds. In other words, the number of compounds that can transfer a phosphate to or from ATP is maximized.

4.22 glucose-1-phosphate → glucose + P_i $\quad \Delta G°' = -20.9 \text{ kJ/mol}$

$[\text{glucose}] = [P_i] = 4.8 \text{ mM} = 4.8 \times 10^{-3} \text{ M}$

$\Delta G°' = -RT \ln K_{eq} \qquad K_{eq} = \dfrac{[\text{glucose}][P_i]}{[\text{glucose-1-phosphate}]}$

$-20.9 \text{ kJ/mol} = -(8.315 \times 10^{-3} \text{ kJ/mol K})(298 \text{ K})(\ln K_{eq})$

$\ln K_{eq} = 8.43$

$$K_{eq} = 4.60 \times 10^3 = \frac{[\text{glucose}][P_i]}{[\text{glucose-1-phosphate}]}$$

$$4.60 \times 10^3 = \frac{[4.8\times10^{-3} \text{ M}][4.8\times10^{-3} \text{ M}]}{[\text{glucose-1-phosphate}]}$$

$$[\text{glucose-1-phosphate}] = \frac{[2.3\times10^{-5}]}{[4.6\times10^{3}]} = 5.0 \times 10^{-9}\text{ M}$$

4.23 At equilibrium, $\Delta G^{\circ\prime} = -RT \ln K_{eq}$

$$-9700 \text{ J/mol} = -(8.315 \text{ J/mol} \cdot \text{K})(298 \text{ K}) \ln K_{eq}$$

$$3.915 = \ln K_{eq}$$

$$K_{eq} = 50.1 = [\text{glycerol}][P_i]/[\text{glycerol-3-phosphate}]$$

$$50.1 = (1 \times 10^{-3}\text{ M})^2/[\text{glycerol-3-phosphate}]$$

$$[\text{Glycerol-3-phosphate}] = (1 \times 10^{-3}\text{ M})^2/50.1 = 2 \times 10^{-8}\text{ M}$$

4.25 Yes, the value of $\Delta G^{\circ\prime}$ for the hydrolysis of pyrophosphate (PP_i) would change as the pH changes from 7 to 9. Although H^+ is neither a reactant nor a product in the hydrolysis reaction, the pK_a values for PP_i must be considered to determine its degree of ionization. With pK_a values of 6.70 ($H_2P_2O_7^{2-}$) and 9.32 ($HP_2O_7^{3-}$), the predominant charge at both pH 7 and pH 9 would be –3. At pH 7, PP_i exists as a 2:1 mixture of $HP_2O_7^{3-}$ and $H_2P_2O_7^{2-}$, whereas at pH 9, PP_i exists as a 2:1 mixture of $HP_2O_7^{3-}$ and $P_2O_7^{4-}$. Since these reactant ions differ in the number of negative charges, their stabilities would differ due to the varying amounts of repulsion between adjacent negative charges.

Their hydrolysis products differ in stability as well:

$$H_2P_2O_7^{2-} + H_2O \rightarrow 2H_2PO_4^{-}$$

$$HP_2O_7^{3-} + H_2O \rightarrow H_2PO_4^{-} + HPO_4^{2-}$$

$$P_2O_7^{4-} + H_2O \rightarrow 2HPO_4^{2-}$$

These ratios were calculated using the Henderson-Hasselbalch equation. However, the same conclusion may be obtained using a more qualitative approach. At pH 8.01, the midpoint between the two pK_a values, 100% $HP_2O_7^{3-}$ exists. Above and below pH 8.01, different reactant mixtures exist that differ in charge, stability, and hydrolysis products.

With different degrees of stability in both the reactants and the hydrolysis products, the $\Delta G^{\circ\prime}$ values would be expected to differ as well.

4.26 At equilibrium: $\Delta G^{\circ\prime} = -RT \ln K_{eq}$

$$-13{,}800 \text{ J/mol} = -(8.315 \text{ J/mol} \cdot \text{K})(298 \text{ K}) \ln K_{eq}$$

$$5.57 = \ln K_{eq}$$

$$K_{eq} = 262$$

$$K_{eq} = [\text{glucose}][P_i]/[\text{glucose-6-phosphate}]$$

$$[\text{glucose-6-phosphate}] = 4 \text{ mM} = 4 \times 10^{-3}\text{ M}$$

$$262 = [\text{glucose}][P_i]/(4 \times 10^{-3}\text{ M})$$

$$1.05 = [\text{glucose}][P_i]$$

Assuming that [glucose] = $[P_i]$, $[P_i]$ = 1.02 M.

4.28 $$\Delta G = \Delta G^{\circ\prime} + RT \ln \frac{[\text{glucose}][P_i]}{[\text{glucose-6-phosphate}]}$$

$$\Delta G^{\circ\prime} = -13.8 \text{ kJ/mol}$$

$$\frac{[\text{glucose}][\text{P}_i]}{[\text{glucose-6-phosphate}]} = 9.2 \times 10^{-2} \text{ (from Problem 4.22)}$$

$\Delta G = -13.8$ kJ/mol + $(8.315 \times 10^{-3}$ kJ/mol K)(298K)(ln 9.2×10^{-2})

$\Delta G = -13.8$ kJ/mol + (2.48)(–2.39) kJ/mol

$\Delta G = -19.7$ kJ/mol

4.29 ATP → AMP + PP_i $\Delta G^{\circ\prime} = -32.2$ kJ/mol

PP_i → 2 P_i $\Delta G^{\circ\prime} = -33.5$ kJ/mol

ATP → AMP + $2P_i$ $\Delta G^{\circ\prime} = -65.7$ kJ/mol

$\Delta G^{\circ\prime} = -RT \ln K_{eq}$

–65.7 kJ/mol = $-(8.315 \times 10^{-3})$(298 K)($\ln K_{eq}$)

$\ln K_{eq} = 26.52$

$K_{eq} = 3.29 \times 10^{11}$

Chapter 4: Solutions to Fill-in-the-Blank Questions

4.31 Work

4.32 Disorder

4.34 Heat/work

4.35 Entropy

4.37 Energy

4.38 Zero

Chapter 4: Solutions to Short-Answer Questions

4.40 Although it is thermodynamically favorable for methane autoignition, this will not occur unless the temperature is very high (in this case about 600°C). Below this temperature the process is not kinetically favorable.

4.41 The third law states that the entropy of a pure substance is zero at absolute zero. Because most biochemical reactions occur about 280 Kelvin above this temperature, the third law is significantly less useful than the first two laws.

4.43 In an exothermic reaction ΔH is positive. In an exergonic reaction ΔG is positive.

4.44 The two reactions are functioning independently. However, if the product of the first reaction is a reactant in a second, strongly exothermic reaction, the second reaction will consume the common intermediate. Le Chatelier's principle states that when the product of an equilibrium reaction is consumed, the reaction will shift to produce more of that product.

Chapter 4: Solutions to Thought Questions

4.46 The change in free energy is negative.

4.47 The energy liberated by the hydrolysis of 12.5 mol of ATP is

(12.5 mol)(–30.5 kJ/mol) = – 381.3 kJ

The energy required to produce 12.5 mol of ATP is 1142.2 kJ. The apparent efficiency of the process is

$(381.3/1142.2) \times 100 = 33.4\%$

4.49 Although only a few molecules are involved, the laws of thermodynamics are still obeyed.

4.50 In the case of endothermic solutions, the enthalpy may be negative but the entropy is sufficiently positive to make the overall $\Delta G°$ favorable.

4.52 ATP has an intermediate phosphoryl group transfer potential. This makes it possible for ATP to serve as a carrier of phosphoryl groups from-high energy compounds to those of lower energy.

4.53 Glucose-1-phosphate has a higher phosphoryl group transfer potential than glucose-6-phosphate because it contains an anhydride bond, which is more reactive than the ester bond in glucose-6-phosphate.

4.55 $\Delta H° = -88$ kJ/mol $\quad \Delta S = 0.3$ kJ/mol · K $\quad T = 298$ K

$\Delta G° = \Delta H° - T\Delta S°$

$\Delta G° = (-88 \text{ kJ/mol}) - (298)(0.3 \text{ kJ/mol} \cdot \text{K})$
$= (-88 - 89.4)\text{kJ/mol}$

$= -177.4$ kJ/mol

$\Delta G° = -RT \ln K_{eq}$

$(-177.4 \text{ kJ/mol})(1000 \text{ J/kJ}) = -(8.315 \text{ J/mol} \cdot \text{K})(298 \text{ K})\ln K_{eq}$

$71.6 = \ln K_{eq}$ and $K_{eq} = 1.2 \times 10^{31}$

The K_{eq} value indicates that the denaturation of the protein is virtually irreversible.

4.56 When magnesium ions coordinate with the phosphate groups of ATP, the repulsion between adjacent oxygen anions is decreased. Consequently, ATP is stabilized and the free energy of hydrolysis is reduced. [Conditions that stabilize the reactant (ATP) would make the ΔG less negative.]

4.58 Both acetate and phosphate are resonance stabilized. However, phosphate has more resonance forms than acetate. Hydrolysis reduces the repulsions between nearby oxygen anions in phosphate. There is no such repulsion in acetate.

4.59 The data provided are as follows:

$\Delta H° = -88$ kJ/mol $\qquad \Delta S° = 0.3$ kJ/mol·K $\qquad \Delta G° = 0$

$\Delta G° = \Delta H° - T\Delta S°$

$0 = -88 \text{ kJ/mol} - T(0.3 \text{ kJ/mol})$

$88 \text{ kJ/mol} = -T(0.3 \text{ kJ/mol})$

$T = -88/0.3 = -293.3$ K

This is *below* absolute zero, an absolute impossibility. There is no temperature at which the denaturation of this protein can be in equilibrium.

4.61 Glucose-1-phosphate contains a high-energy anhydride bond. Glucose-6-phosphate, on the other hand, has a lower-energy ester bond.

4.62 CO_2 → Plants → Photosynthesis

↑ ↓

Respiration ← Animals ← O_2 + carbohydrate

5 Amino Acids, Peptides, and Proteins

Brief Outline of Key Terms and Concepts

OVERVIEW

POLYPEPTIDES are polymers composed of **AMINO ACIDS** linked by **PEPTIDE BONDS**. The **AMINO ACID SEQUENCE** is the order of the amino acids in a polypeptide. An **AMINO ACID RESIDUE** is an amino acid within a polypeptide chain. **PEPTIDES** have less than 50 amino acid residues; **PROTEINS** have more than 50. Proteins may contain more than one chain.

5.1 AMINO ACIDS

AMPHOTERIC, ZWITTERION

AMINO ACID CLASSES: NONPOLAR, POLAR, ACIDIC, and **BASIC**. Amino acids are classified according to their capacity to interact with water.

BIOLOGICALLY ACTIVE AMINO ACIDS

HORMONES, NEUROTRANSMITTERS, nucleotide (and other) precursors, metabolic intermediates

MODIFIED AMINO ACIDS IN PROTEINS

AMINO ACID STEREOISOMERS

All amino acids—except glycine—have a chiral carbon (asymmetric carbon) atom and, as a result, can exist as **STEREOISOMERS. ENANTIOMERS** are mirror-image forms of a molecule. Most asymmetric molecules in living organisms occur in only one stereoisomeric form. With few exceptions, only L-amino acids (not D-) are found in proteins.

TITRATION OF AMINO ACIDS

Titration is useful in determining the relative ionization potential of acidic and basic groups in an amino acid or peptide. The **ISOELECTRIC POINT** is the pH at which an amino acid has no net charge.

AMINO ACID REACTIONS

PEPTIDE BOND FORMATION

Peptide bonds have partial double-bond character, making them rigid and flat. Amino acid sequences are written from N-terminal to C-terminal.

CYSTEINE OXIDATION FORMS DISULFIDE BRIDGES that help stabilize polypeptide and protein structure.

SCHIFF BASE FORMATION

When amine groups react reversibly with carbonyl groups, they form Schiff bases (**ALDIMINES**).

5.2 PEPTIDES

Peptides have significant biological activity that includes a variety of signal transduction processes.

5.3 PROTEINS

Functions include catalysis, structure, movement, defense, regulation, transport, storage, and stress response (**HEAT SHOCK PROTEINS**). **MULTIFUNCTION PROTEINS, PROTEIN FAMILIES AND SUPERFAMILIES**

FIBROUS PROTEINS; GLOBULAR PROTEINS

CONJUGATED PROTEINS contain **PROSTHETIC GROUPS** (nonprotein components). **HOLOPROTEINS** have their prosthetic groups; in **APOPROTEINS**, prosthetic groups are **ABSENT.**

PROTEIN STRUCTURE

INVARIANT amino acid residues are essential to its function. **HOMOLOGOUS** polypeptides have similar amino acid sequences and a common origin.

PRIMARY STRUCTURE is the sequence of amino acid residues connected by peptide bonds. Molecular diseases arise from mutations that result in changes in key protein's primary structure.

SECONDARY STRUCTURE describes hydrogen-bond stabilized polypeptide segments with 3D order, such as α-helices and β-pleated sheets. **MOTIF, SUPERSECONDARY STRUCTURE**

TERTIARY STRUCTURE is the unique three-dimensional conformation that a protein assumes (**PROTEIN FOLDING**) because of the interactions between amino acid side chains. (**FOLD**—the core 3D structure)

Interactions that stabilize tertiary structure: the hydrophobic effect, electrostatic interactions (**SALT BRIDGES**), hydrogen bonds, covalent bonds, and hydration. **MODULAR** (or **MOSAIC**) **PROTEINS, LIGANDS**

QUATERNARY STRUCTURE describes proteins with several separate polypeptide **SUBUNITS** held together by noncovalent and covalent bonds. **OLIGOMER, PROTOMER; ALLOSTERY, ALLOSTERIC TRANSITION, EFFECTOR, MODULATOR**

UNSTRUCTURED PROTEINS: INTRINSICALLY UNSTRUCTURED PROTEINS (IUPS), NATIVELY UNFOLDED PROTEINS

LOSS OF PROTEIN STRUCTURE – DENATURATION

THE FOLDING PROBLEM: The primary structure contains all the information required for each newly synthesized polypeptide to fold into its biologically active conformation Some relatively simple polypeptides fold spontaneously; other larger molecules require the assistance of **MOLECULAR CHAPERONES** such as the **hsp70S, hsp90S AND CHAPERONINS. MOLTEN GLOBULE**

FIBROUS PROTEIN function usually involves structural support.

GLOBULAR PROTEIN function usually involves binding to small **LIGANDS** or to other macromolecules. Examples: myoglobin, hemoglobin. **COOPERATIVE BINDING**

5.4 MOLECULAR MACHINES

Proteins and proteins complexes function as molecular machines. Examples include the muscle, sarcomere and DNA polymerase.

STUDY STRATEGY:

I *highly recommend* learning the amino acids now. *Knowing* the structures without having to think twice will be a tremendous advantage as you build understanding and work the protein problems later in the chapter. You will continue to see the positive effects of knowing these structures as you move forward. Specific amino acid structures are critical to enzyme function (Chapter 6). It follows, then, that every pathway that includes discussion of the mode of action or the regulation of an enzyme mentions specific amino acid structures (Chapters 8, 9, 10, 12, etc.). Check out the HINTS FOR LEARNING THE 20 STANDARD AMINO ACIDS on the next page.

5.1 AMINO ACIDS

There are 20 standard amino acid structures and abbreviations

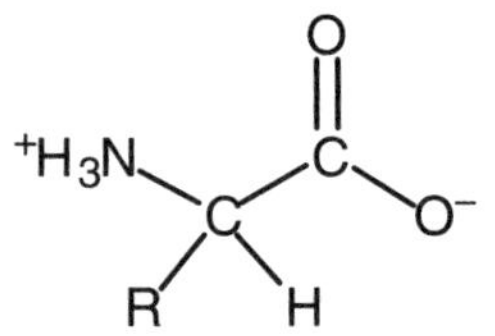

STANDARD vs. **NONSTANDARD** amino acids (standard: commonly found in proteins; nonstandard: chemically modified after a peptide or protein was synthesized)[1]

AMPHOTERIC (can act as an acid or a base)

at pH 7, amino acids exist as **ZWITTERIONS** (two ions on two different atoms in the same molecule)

AMINO ACID CLASSES: DETERMINED BY HOW THE R GROUP INTERACTS WITH H_2O

All amino acids have an acidic carboxyl group $-CO_2H$ and a basic amino group $-NH_2$, so amino acids are classified by the *one thing* that makes *each* amino acid *unique*: its R group.

NONPOLAR	R contains aromatic, aliphatic, and/or sulfur groups
POLAR, neutral	R contains –OH or an amide
ACIDIC	R contains $-CO_2H$ that can donate an H^+
BASIC	R contains an N that can accept an H^+

Be aware that amino acid R groups that are acidic or basic are almost always written as they exist at pH 7. So, the R group of an acidic amino acid such as aspartate, with $R = -CH_2CO_2H$, will have lost its H^+ at pH 7, so it's written as $-CH_2CO_2^-$. Even though it is written in this form (i.e., its conjugate base), its official amino acid classification is still, and will always be, ACIDIC.

[1] Note that the definitions of *standard* vs. *nonstandard* amino acids differ from the dietary *essential* vs. *nonessential* amino acids.

THREE-LETTER ABBREVIATIONS = FIRST THREE LETTERS EXCEPT FOR THESE:

Ile	=	Isoleucine	Asn	=	Asparagine
Trp	=	Tryptophan	Gln	=	Glutamine

HINTS FOR LEARNING THE 20 STANDARD AMINO ACIDS

Set out to LEARN the amino acids, not just to memorize them. Flash cards help. Write the name on one side *and the structure on the other side* (to make the learning active rather than passive. It does make a difference).

Write out the structures for yourself. As you learn the structures, make a shorter list of the ones that continue to be troublesome for you.

Group similar amino acids together and compare them.

Group confusing amino acids together and compare them. For example, leucine and isoleucine are troublemakers, since leucine has the isobutyl group, not isoleucine. We're all stuck with the chore of keeping those two straight. Also, there's aspartate and glutamate, but it's helpful that aspartate has the shorter chain and also comes before glutamate in the dictionary.

Try drawing the structures differently. For example, comparing tryptophan and histidine drawn as shown below may help you to learn their structures.

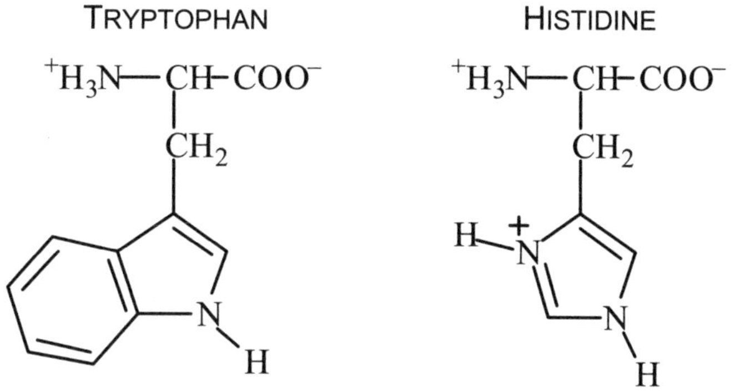

BIOLOGICALLY ACTIVE AMINO ACIDS

Chemical messengers; examples:

NEUROTRANSMITTERS: glycine, glutamate, GABA, serotonin, melatonin;
HORMONES: thyroxine, indole acetic acid (derived from amino acids)

Precursors to complex N-containing molecules (nucleotides, heme, chlorophyll)

Metabolic intermediates (ornithine and citrulline in the biosynthesis of urea)

MODIFIED AMINO ACIDS IN PROTEINS

Examples of how amino acids can be modified to form **AMINO ACID DERIVATIVES** include carboxylation, hydroxylation, and phosphorylation. Phosphorylation of Ser, Thr, and/or Tyr residues in proteins is used to regulate entire metabolic pathways.

What do the R groups of serine, threonine, and tyrosine all have in common?

– an –OH group that can be phosphorylated

Amino Acid Stereoisomers

Amino acid stereochemistry affects protein structure, which affects function

All of the α-amino acids—except for glycine—has at least one chiral carbon. (Glycine's R group is just an H, so the α-carbon is $-CH_2-$ instead of –CHR–.)

In most organisms proteins contain L-amino acids, not D-.

Connection to Organic Chemistry:

Remember these terms?

Chiral carbon (or chiral center), **stereoisomers**, **enantiomers** (R vs. S), **optical isomers** [dextrorotary(+) vs. levorotary(–) (how a sample rotated a plane of polarized light)]

How Biochemistry Is the Same: Biochemistry uses D- and L- too, BUT it is not *quite* the same . . .

How Biochemistry Is Different: "L-" refers to a similarity with L-glyceraldehyde at the chiral carbon furthest from the C=O. So, L-alanine shares a structural similarity with L-glyceraldehyde.

Titration of Amino Acids

Amino acid titrations provide insight into the reactivity of side chains in processes such as protein folding and enzyme function. To solve titration problems, follow these steps:

1. Draw the amino acid in its most acidic form.
2. Draw the amino acid structure that occurs when it reacts with one OH–. Then, draw structures that react with further OH– until the amino acid is in its most basic form. (Groups with the lowest pK_a values donate their H^+ ions first.)
3. Determine the **isoelectric point** (**pI**), the pH where the overall charge is zero.

To determine the isoelectric point:

1. Identify the acid(s) and base(s) on the amino acid or peptide. Which functional groups can easily lose or gain an H^+? (*"Easily"* means within the pH range of 1–14.)
2. Rank the functional groups in the order in which they will lose an H^+. Remember that the lower the pK_a, the stronger the acid, so the acid with the lowest pK_a will lose its H^+ first.
3. Average the pK_a values on either side of the isoelectric (neutral) molecule. Note: Be aware that if an acidic or basic side group is present, you *never* average *all* of the pK_as. Only average these two pK_as: the pK_a to go from a net charge of +1 to 0 and the pK_a to go from a net charge of 0 to –1.

   If it is not clear which two pK_a values to average, first draw the amino acid or peptide at the lowest (acidic) pH and label each ionizable group with its pK_a. Then remove one H^+ at a time (in order of pK_a) and draw each structure. It helps to label the reaction arrows with their pK_a values and each structure with its net charge.

   Remember that the lower the pK_a, the stronger the acid. Also: when $pH = pK_a$, $[A^-] = [HA]$. Below the pK_a, there will be more HA. Above the pK_a, there will be more A^-.

A detailed example of lysine is located after the table of amino acids with ionizable R groups on the next page. Also, the examples of alanine and glutamic acid are in your text along with several worked problems (pp. 138-141.)

In addition to the Review and Thought Questions at the end of Chapter 5, **more pI practice problems are located at the end of this study guide chapter.**

AMINO ACIDS WITH IONIZABLE R GROUPS

Amino Acid	Structure at pH 7 and R group pK_a	Amino Acid	Structure at pH 7 and R group pK_a
Aspartate Asp	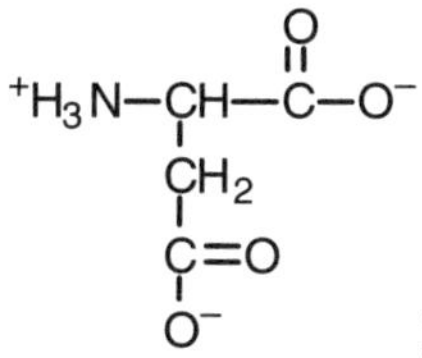 **3.86**	**Glutamate** Glu	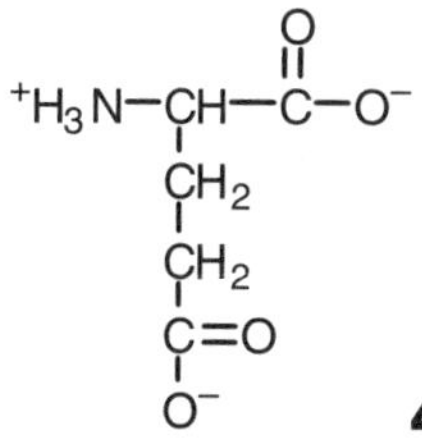 **4.25**
Tyrosine Tyr	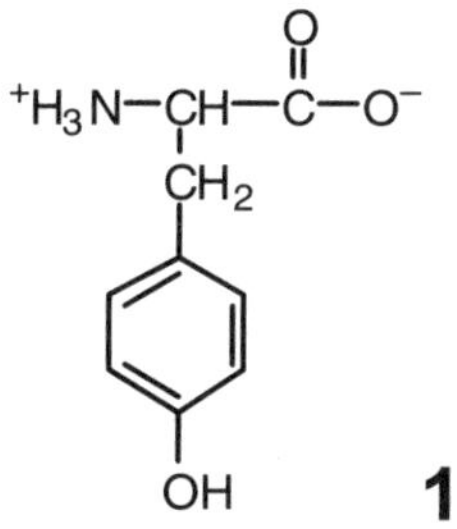 **10.07**	**Lysine** Lys	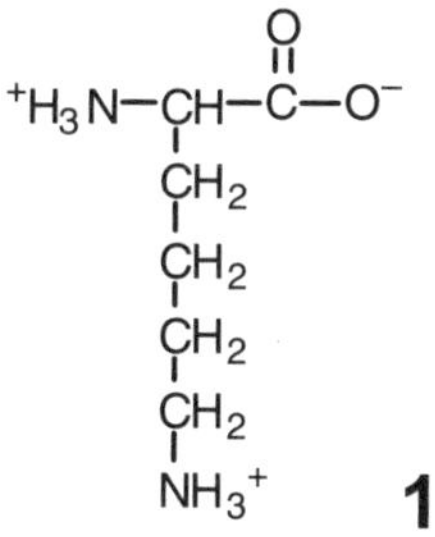 **10.79**
Histidine His	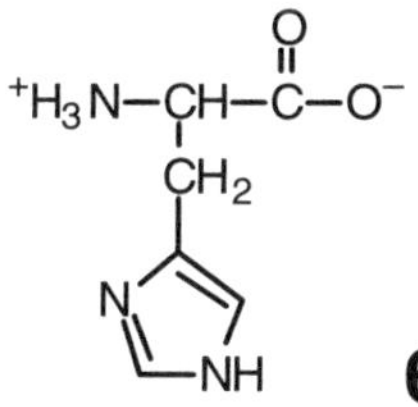 **6.0**	**Cysteine** Cys	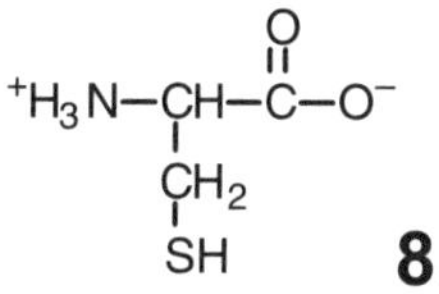 **8.33**
Arginine Arg	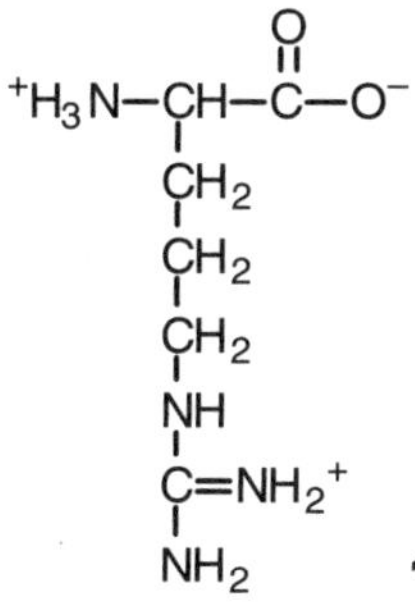 **12.48**		

Unusual Functional Groups

- Guanidino: in arginine"s R group
- Imidazole: in histidine"s R group
- Phenol: in tyrosine"s R group
- Sulfhydryl: in cysteine"s R group (thiol)

CALCULATING PI: THE EXAMPLE OF LYSINE IN MORE DETAIL.

1. Identify acidic and basic functional groups.

–COOH is a weak acid and can donate its H^+ to form –COO^-. (pK_a = 2.18)

–NH_2 is a weak base and can accept an H^+ to form –NH_3^+. (pK_a = 8.95)

The R group also has an –NH_2 that can accept an H^+ to form –NH_3^+. (pK_a = 10.07)

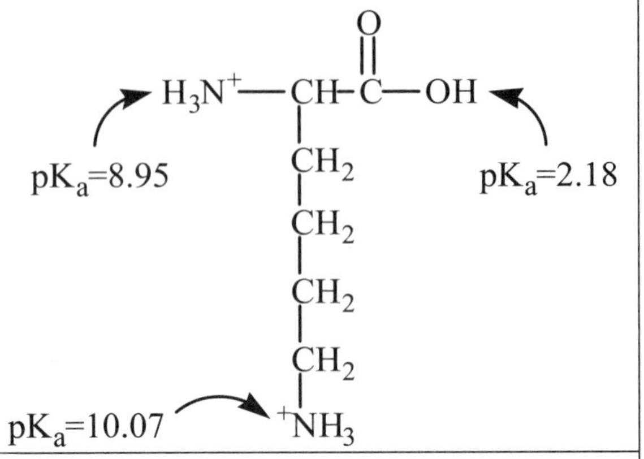

2. Rank functional groups according to pK_a.

What happens to each acid/base group at very low pHs (acidic conditions)? All acid and base groups will have their H^+s: –COOH, –NH_3^+.	Which group will donate its H^+ first (that is, at the lower pH values)? The one with the lowest pK_a, in this case, the –COOH with a pK_a of 2.18.	The next group to give up its H^+ is the one with the next higher pK_a value, in this case, the amino group with a pK_a of 8.95.	The third and last group to lose its H^+ is the R group, with a pK_a of 10.07.

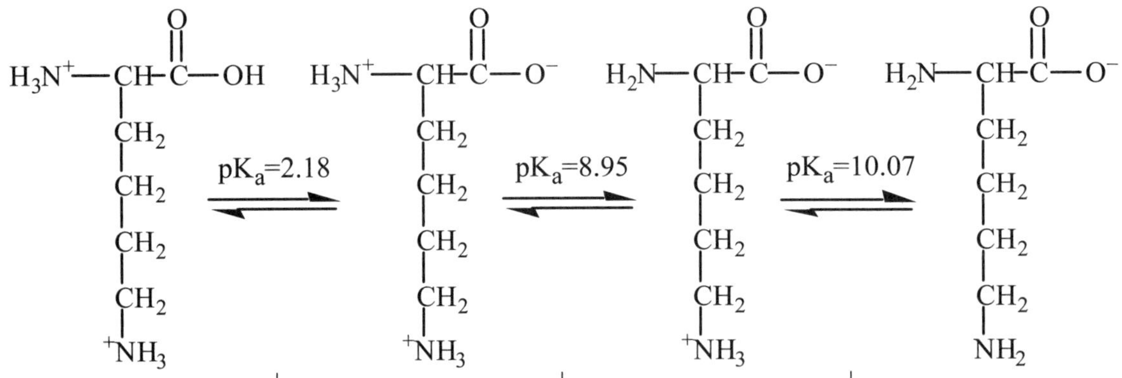

+2	**+1**	**0**	**–1**
Net charge	Net charge	Net charge	Net charge
pH < 2.18	2.18 < pH < 8.95	8.95 < pH < 10.07	pH > 10.07

3. To calculate the pI, average the pK_as on either side of the neutral molecule. For lysine, this would be the average of 8.95 and 10.07 = (8.95 + 10.07)/2 = **9.52.**

Amino Acid Reactions

Peptide Bond Formation

—peptide bond has partial double-bond character due to resonance, so the peptide bond is rigid and planar
  —α-carbon (Cα) is next to the peptide-bonded C=O
  —ψ = rotation around the Cα–N bond; ϕ = rotation around the Cα–C bond
—amino acid residues are amino acids that are part of a peptide or protein chain. Once an amino acid reacts to form a peptide or protein, it is more accurate to refer to a glycine residue rather than a glycine, since it is that part of the amino acid that has become part of a peptide or protein
—draw peptides from N-terminal to C-terminal
—name peptides (amino acid sequence from N-terminal to C-terminal)

Peptide Bond = An Amide Bond Linking Two Amino Acids

Amino acids are linked by a covalent bond between the α-amino (–NH_2) of one amino acid and the α-carboxyl (–COOH) of another amino acid.

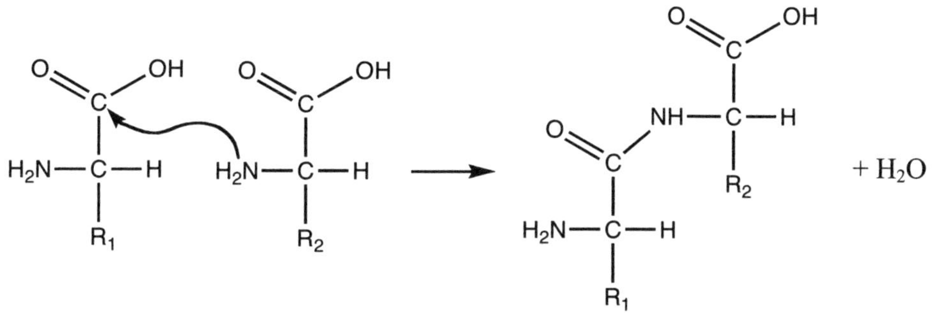

The curved arrow above shows where the new peptide bond will form. The amino acids are drawn in their neutral forms to illustrate the loss of an –OH from the carboxyl group and an –H from the amino group to produce a water molecule. [Note: Of course, "real life" is much more complex. The amino acids would be ionized at pH 7, the general mechanism is multistep, and protein synthesis in living organisms is quite complicated (see Chapter 19).]

The peptide C–N bond is shorter and much more rigid than a typical carbon-nitrogen single bond. This can be explained by drawing the resonance structure for an amide bond:

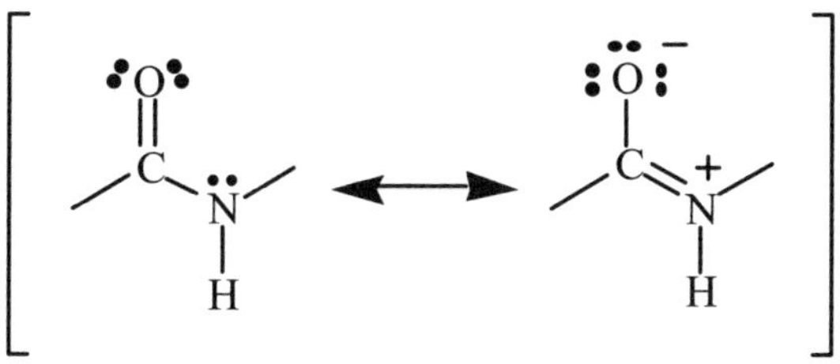

Although this resonance structure may seem unlikely because of the charge separation, it is indeed significant. Resonance explains the observed similarities of the peptide bond to a carbon-carbon double bond, namely, its rigidity and its shorter-than-expected bond length.

A carbon-nitrogen double bond is planar, or flat, and this makes the peptide bond rigid. The planar, rigid peptide bond has important consequences for protein structure.

The effect that this rigidity has on possible protein shapes is to limit somewhat the number of possibilities. This introduces an element of control. Picture a heavy chain with rather long links. Between the links there is free rotation, but the links themselves are rigid. Compared to a rope, there are fewer possible ways that the chain could be contorted. However, for the biological functions that proteins perform, having specific ways (rather than infinite ways) that proteins can fold and form shapes results in greater control and specificity in protein function.

CYSTEINE OXIDATION AND THE FORMATION OF DISULFIDE BRIDGES

—covalent S–S bond = **DISULFIDE BOND**, called a **DISULFIDE BRIDGE** in polypeptides
—formed by the oxidation of two cysteine R groups:

$$R\text{–}SH + HS\text{–}R \rightleftharpoons R\text{–}S\text{–}S\text{–}R \text{ (cystine)}$$

—can occur within a chain or between two separate chains

SCHIFF BASE FORMATION

Schiff base: **IMINE** (C=N) produced when R–NH_2 reacts reversibly with a carbonyl

Schiff bases are also called aldimines when the carbonyl is an aldehyde

Schiff bases are intermediates in transamination reactions (Chapter 14).

IMINE functional group: C=N

5.2 PEPTIDES

PEPTIDE FUNCTIONS	SPECIFIC PEPTIDES THAT PERFORM THESE FUNCTIONS:
Reducing agent	glutathione (GSH, where "SH" indicates a cysteine R group) $2\ GSH + H_2O_2 \rightarrow GSSG + 2\ H_2O$ Note that GSH forms a disulfide bond with another GSH.
Appetite control	α-melanocyte stimulating hormone, cholecystokinin, galanin, neuropeptide Y
Blood pressure control	vasopressin, atrial natriuretic factor
Pain perception	opioid peptides (Met-enkephalin, Leu-enkephalin) vs. substance P, bradykinin

WRITING PEPTIDE AND PROTEIN SEQUENCES: N-TERMINUS TO C-TERMINUS

Peptide and protein sequences are written from left to right as amino, or N-terminus, to carboxyl, or C-terminus. The amino/carboxyl refers to free amino or free carboxyl groups, meaning that they are not part of a peptide bond.

Example:

Draw the dipeptides Phe–Asp and Asp–Phe. They are different! Not only do they have different structures, but their pI values will also be different. Calculate their pI values.

Solution:

Phe–Asp

$^{+}H_3N$–Phe–Asp–COO^{-}

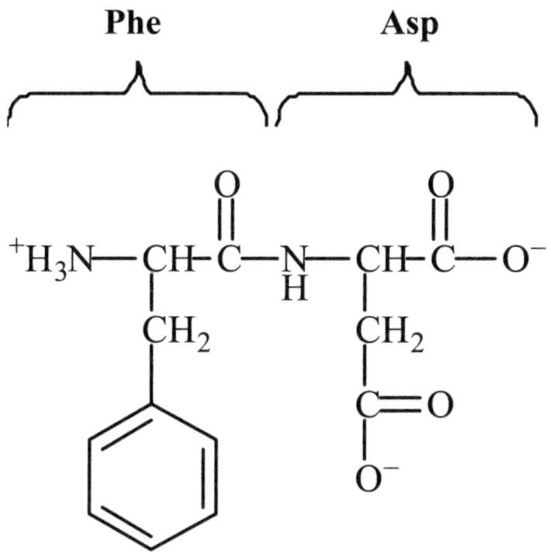

pI = (2.09+3.86)/2

pI = 2.98

Asp–Phe

$^{+}H_3N$–Asp–Phe–COO^{-}

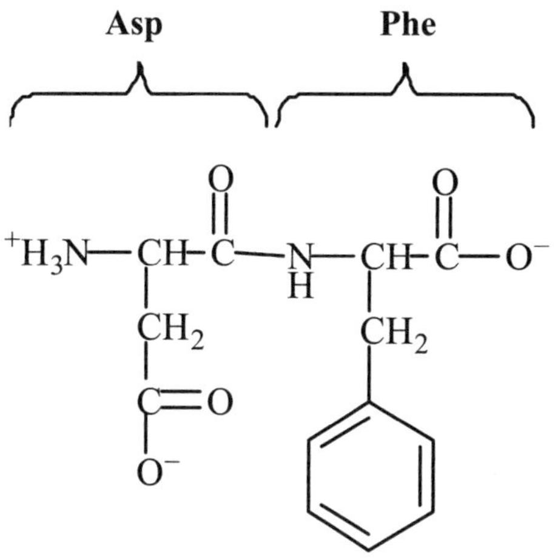

pI = (1.83+3.86)/2

pI = 2.85

Aspartame

Asp–Phe is an important commercial dipeptide. Its methyl ester (on the C-terminus) is Aspartame, an artificial sweetener that is 200 times sweeter than sugar. Both amino acids have the L-configuration around the α-carbon. If either is in the D-configuration, then the peptide is bitter rather than sweet.

5.3 PROTEINS: CLASSIFIED BY FUNCTION, SHAPE, OR COMPOSITION

PROTEIN FUNCTIONS: catalysis, structure, movement, defense, regulation, transport, storage, stress response, toxins

PROTEIN SHAPES: **FIBROUS** proteins vs. **GLOBULAR** proteins

PROTEIN COMPOSITIONS: **SIMPLE** vs. **CONJUGATED**

Conjugated protein = simple protein + prosthetic group.

Holoproteins ***H***AVE THEIR PROSTHETIC GROUPS, BUT

PROSTHETIC GROUPS ARE ***A***BSENT IN ***A***POPROTEINS.

Examples of conjugated proteins: **GLYCOPROTEINS, LIPOPROTEINS, METALLOPROTEINS, PHOSPHOPROTEINS, HEMOPROTEINS**

PROTEIN STRUCTURE

PRIMARY STRUCTURE = AMINO ACID SEQUENCE

HOMOLOGOUS POLYPEPTIDES have similar amino acid sequences and have arisen from the same ancestral gene

INVARIANT vs. **VARIABLE RESIDUES**; mutations that change the amino acid sequence and their connection to evolution and molecular diseases

HOMOZYGOUS vs. **HETEROZYGOUS**

MOLECULAR DISEASES result from amino acid substitutions (caused by DNA mutations) at invariant residues of key proteins; example: sickle-cell anemia caused by mutant hemoglobin

SECONDARY STRUCTURE = REPEATING PATTERNS OF LOCALIZED STRUCTURE

α-HELIX: right-handed helix, 3.6 residues per turn, pitch = 54 nm
Hydrogen bonds between N–H and C=O are four residues apart, and R groups extend *outward* from the helix.
Amino acids that are incompatible with the α-helix: Gly, Pro, and sequences with large numbers of charged and/or bulky R groups.

β-PLEATED SHEET; β-strand; parallel vs. antiparallel
Stabilized by H-bonds between N–H and C=O of adjacent chains
Each β-strand is fully extended. Antiparallel is more stable than parallel because the hydrogen bonds between chains are more direct (*colinear* and shorter).

SUPERSECONDARY STRUCTURES or **MOTIFS**

βαβ unit, β-turn, β-meander, αα-units, β-barrel, Greek key

TERTIARY STRUCTURE AND PROTEIN FOLDING OF GLOBULAR PROTEINS

Features:
1. Amino acids that are far apart in the primary structure may be close together once folded.
2. Globular proteins are compact.
3. Large globular proteins often contain **DOMAINS,** compact units with specific functions. (Examples: *EF hand* binds Ca^{2+}; *leucine zipper* and *zinc finger* domains found in DNA-binding proteins); **FOLD** = core 3--D structure of a domain
4. **MODULAR** or **MOSAIC PROTEINS** in eukaryotes contain numerous duplicate or imperfect copies of one or more domains that are linked in series

Stabilizing factors:

HYDROPHOBIC INTERACTIONS

ELECTROSTATIC INTERACTIONS (SALT BRIDGES)

HYDROGEN BONDING

COVALENT BONDS (DISULFIDE BRIDGES)

HYDRATION (structured water forms a dynamic hydration shell that stabilizes tertiary protein structure)

QUATERNARY STRUCTURE—SEVERAL SUBUNITS (POLYPEPTIDE CHAINS)

Oligomers and protomers; why multisubunit proteins are common

Noncovalent and covalent interactions hold the subunits in place. The most important interaction is the hydrophobic effect.

Covalent crosslinks: disulfide bridges, desmosine and lysinonorleucine

Allostery, ligand binding, allosteric transitions, effectors or modulators

UNSTRUCTURED PROTEINS

IUPS are **INTRISICALLY UNSTRUCTURED PROTEINS.**

NATIVELY UNFOLDED PROTEINS have no ordered structure.

IUP functions include the regulation of signal transduction, transcription, translation, and cell proliferation. The disorder of the IUPs allows them to be more flexible and "search" for binding partners; IUPs tend to become more ordered upon binding with a target molecule.

LOSS OF PROTEIN STRUCTURE: PROTEIN DENATURATION—loss of 3-D structure. Interactions between amino acid residues are disrupted, but peptide bonds are NOT broken.

STUDY HINT: Review the bonding, interactions, and forces that stabilize 3-D structure. Anything that can disrupt these forces can cause denaturation.

Denaturing agents include strong acids or bases, organic solvents, detergents, reducing agents (reduce disulfide bonds), salt concentration (salting out), heavy metal ions, temperature changes, and mechanical stress.

PROTEIN DYNAMICS AND FLEXIBILITY IN A PROTEIN'S 3-D STRUCTURE; why flexibility in a protein''s 3-D structure is essential to most protein functions

BIOCHEMISTRY IN PERSPECTIVE: MOLECULAR MACHINES

Molecular machine function is made possible by conformational changes triggered by the hydrolysis of nucleotides bound to protein subunits called motor proteins.

MOTOR PROTEINS

1. Classical motors: myosins, kinesins, dyneins
2. Timing devices
3. Microprocessing switching devices
4. Assembly and disassembly factors

THE FOLDING PROBLEM

THE TRADITIONAL FOLDING MODEL: Interactions between amino acids side chains alone force the molecule to fold into its final shape.

LIMITATIONS OF THE TRADITIONAL FOLDING MODEL

1. Time constraints: Folding happens on the order of seconds (or a few minutes), not in years, as calculated by the traditional folding model.
2. Complexity: Think about the number of bonds that can rotate, not only in the backbone, but in the side groups.

RECENT ADVANCES IN PROTEIN FOLDING RESEARCH HAVE REVEALED:

- secondary structure (α-helix, β-sheet) forms early in the process
- hydrophobic interactions are very important
- larger polypeptides have partially folded intermediate structures; molten globules are partially organized globular structures that resemble the final protein structure
- **MOLECULAR CHAPERONES** assist in protein folding:
  —bind to denatured or unfolded proteins
  —protect unfolded proteins from inappropriate interactions that would lead to incorrect structures
  —assist proteins in folding rapidly, precisely, and correctly
  —promote protein degradation when refolding isn't possible
- most molecular chaperones are hsps (heat shock proteins); there are four groups of molecular chaperones involved in *de novo* protein folding.

  —**Ribosome-associated chaperones—**bind to and promote folding of polypeptides as they emerge from ribosomes.

  —**Hsp70s**—bind to short hydrophobic segments in unfolded polypeptides to prevent their aggregation. ATP hydrolysis releases the polypeptide, which is then passed on to an hsp60.

  —**Hsps90s**—finalize the folding of a limited, but diverse, set of "client proteins."

  —**Chaparonins**—family of chaperones that mediate protein folding and release the polypeptide upon ATP hydrolysis

FIBROUS PROTEINS TYPICALLY SERVE STRUCTURAL FUNCTIONS

– have high proportions of regular secondary structures; rodlike or sheetlike shapes

Examples: α-keratin, collagen, silk fibroin
Take a close look at the descriptions and figures of these examples in your text. The structures are rigid and well suited to their functions. Where are each of these located, and what are their specific functions?

Note that the helical structure in collagen is not an α-helix, but a triple left-handed helix. Earlier in the chapter, Gly and Pro were described as being likely to disrupt an α-helix. Review that section, and compare the α-helix with the description of collagen on p. 169 Note how the odd features of Gly and Pro work together well in the structure of collagen.

GLOBULAR PROTEINS TYPICALLY SERVE DYNAMIC FUNCTIONS

- Function: usually involves binding ligands or large biomolecules that induce a conformational change "linked to a biochemical event."
- Myoglobin (in skeletal and cardiac muscle), hemoglobin (in red blood cells)
- Heme protein decreases heme"s affinity for O_2 and protects Fe^{2+} from irreversibly oxidizing to Fe^{3+} (hematin), so O_2 can bind reversibly (like a sticky note that can be placed or removed when needed).
- Fetal hemoglobin (HbF) has greater affinity for O_2 than HbA, maternal hemoglobin. (How else could a growing baby receive O_2 from the mother?)

- Myoglobin has a greater affinity for oxygen than hemoglobin. So, oxygen moves from blood to muscle, and myoglobin only gives up its oxygen when the muscle cell"s O_2 concentration is very low.
- **TAUT STATE** vs. **RELAXED STATE** (T, R states)
- **COOPERATIVE BINDING**: The binding of the first ligand (example – O_2 to hemoglobin) causes a conformational change that facilitates binding of more ligands (3 more O_2 molecules to hemoglobin).
- **BOHR EFFECT**: Dissociation of O_2 from hemoglobin is enhanced at lower pH. Why? Higher levels of CO_2 cause higher $[H^+]$, since $CO_2 + H_2O \rightarrow HCO_3^-$ and H^+. H^+ stabilizes the deoxy form of hemoglobin, so it is formed faster.

THE IMPORTANCE OF ALLOSTERY

Allostery pulls together the complexity and importance of protein 3-D structure and ties them directly to function. It all boils down to those intra- and intermolecular forces (as well as covalent bonds) that determine a protein"s overall shape. The shape affects (some might say determines) the protein"s function.

The main idea is that a small molecule (or ion) called a **LIGAND** binds to a protein and induces a conformational change (an **ALLOSTERIC TRANSITION)**, which changes its shape, which changes the protein's affinity for other ligands. So, binding one ligand results in the protein having an increased or decreased affinity for binding more ligands. Ligands that trigger allosteric transitions are called **EFFECTORS** or **MODULATORS**. Allosteric enzymes can be regulated by ligand binding. (We will see more of this in Chapter 6.)

PRACTICE PROBLEMS TO PROMOTE PERFECTION

Solutions are at the end of this chapter, after the solutions to the Review, Fill-in-the-Blank, Short-Answer, and Thought Questions.

1. Given the peptide: Val-Arg-Ala-Tyr-Gly
    a. Draw the structure of the peptide in its most acidic form.
    b. Name it.
    c. Would you expect this peptide to have a high or a low pI? Calculate the pI. (Refer to Table 5.2 in your text for pK_a values.)
    d. Draw the titration curve for this peptide.
    e. What is the net charge of this peptide at pH 7? In gel electrophoresis, would it move toward the positive anode or the negative cathode?

2. Given the peptide: Pro-His-Met-Ser-Phe
    a. Draw the structure of the peptide in its most acidic form.
    b. Calculate the pI. (Refer to Table 5.2 in your text for pK_a values.)
    c. What is the net charge of this peptide at pH 12? In gel electrophoresis, would it move toward the positive anode or the negative cathode?

3. Given the following peptide, write out the proper sequence of amino acids using their three-letter abbreviations.

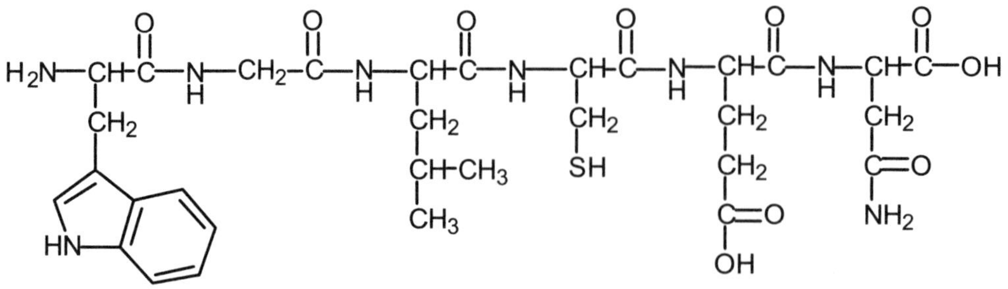

(Here's a bonus question: There is an error in this structure that doesn't interfere with the problem. Can you find the error? What is wrong with this structure?)

AFTER STUDYING THIS CHAPTER, YOU SHOULD BE ABLE TO:

- Classify amino acids as polar (neutral), nonpolar, acidic, or basic
- Draw the structure and give the net charge of an amino acid or peptide at a specific pH. Will it move toward the anode or the cathode in gel electrophoresis?
- Draw (or use) titration curves; determine (or use) pK_a and pI values.
- Name and draw peptides, given the three-letter abbreviations. Given a structure of a peptide, write it out using the three-letter abbreviations.
- Calculate the pI values of peptides.
- Determine the amino acid sequence of a peptide.
- Predict the secondary structure of a given polypeptide: which amino acids tend to stabilize or disrupt each type of secondary structure?
- Describe interactions, bonding, and other forces that contribute to stabilizing protein tertiary structure. Identify amino acids that would participate in a given interaction.

CHAPTER 5: SOLUTIONS TO REVIEW QUESTIONS

5.1 a. supersecondary structure—combinations of α-helix and β-pleated sheet secondary structures

b. protomer—individual subunits of multisubunit protein

c. phophoprotein—proteins that contain a phosphate group

d. denaturation—process of structure disruption

e. ion exchange chromatography—a chromatographic method that separates proteins based on their charge

5.2 a. protein motif—patterns of α-helix and β-pleated sheet secondary structures

b. conjugated protein—proteins that contain a prosthetic group

c. dynein—motor protein that moves vesicles and organelles along microtubules

d. zwitterions—molecules that possess both positive and negative charges

e. electrophoresis—process where proteins are separated in an electric field

5.4 a. aliphatic hydrocarbon—nonaromatic hydrocarbons

b. neurotransmitter—substances released from one nerve cell that influences the function of a second nerve or a muscle cell

c. asymmetric carbon—a carbon that is bonded to four different functional groups

d. chiral carbon—a carbon that is bonded to four different functional groups

e. stereoisomer—molecules that differ only in the spatial arrangement of their atoms

5.5 a. optical isomer—molecule that rotates polarized light in opposite directions from its enantiomer

b. isoelectric point—pH where an amino acid is electrically neutral

c. peptide bond—the bond formed when the α-carboxyl group of one amino acid reacts with the amino group of another

d. disulfide bridge—occurs when two cystene amino acids form a disulfide bond

e. Schiff base—imine products of amine groups that reaction with carbonyl groups

5.7 a. moonlighting protein—a protein with multiple functions in the cell

b. fibrous protein—long, tough rod-shaped molecules often insoluble in water

c. globular protein—compact spherical molecules that are usually water soluble

d. prosthetic group—a nonprotein component of a protein

e. apoprotein—a protein that lacks its prosthetic group

5.8 a. homologous polypeptide—peptides that contain high amino acid sequence homology

b. molecular disease—a disease that results from a change on the molecular level (i.e., amino acid change)

c. protein fold—a core three-dimensional structure of a protein domain

d. mosaic protein—contain numerous duplicate or imperfect copies of one or more domains that are linked in series

e. ligand—molecules that are bound to specific sites

5.10 a. molecular chaperone—heat shock proteins that function to promote proper protein folding

b. chaperonin—a family of molecular chaperones of the Hsp60 (60 kDa) family of heat shock proteins

c. cooperative binding—the binding of the first ligand enhances the binding of the next ligand

d. motor protein—proteins that bind and hydrolyze nucleotides to induce conformational changes

e. primary structure—the amino acid sequence of a polypeptide

5.11 a. aldol condensation—aldol cross-link formed when two aldehydes form an α,β-unsaturated aldehyde linkage

b. α-carbon—the carbon atom in an amino acid that is bonded to an amino group and a carboxyl group

c. hydrophobic amino acid—an amino acid with a nonpolar side chain

d. secondary structure—localized arrangement of α-helices and β-pleated sheets of adjacent amino acids

e. β-pleated sheet—form when two or more polypeptide chains of β strands line up side by side

5.13

a. nonpolar	b. polar	c. acidic	d. basic	e. nonpolar
f. basic	g. nonpolar	h. polar	i. nonpolar	j. nonpolar.

5.14 Structure and net charge of arginine at various pH levels:

pH	Structure	Net Charge
1	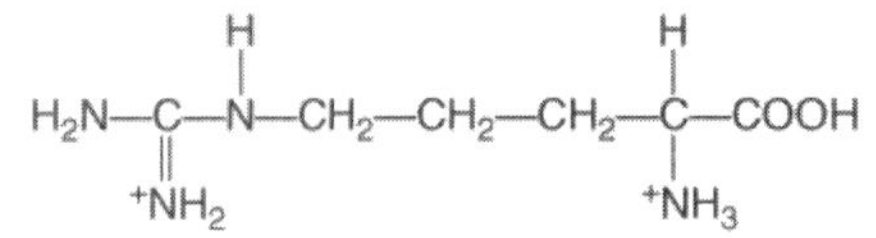	+ 2
4	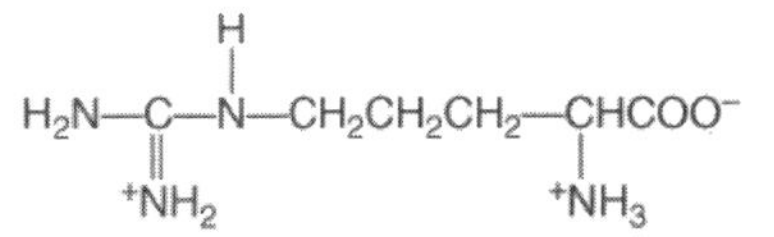	+ 1
7	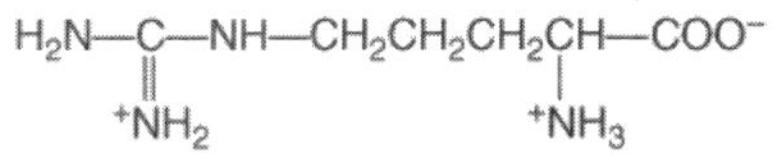	+ 1
10	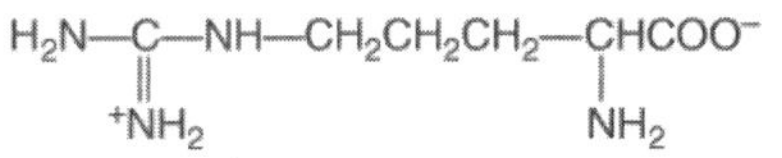	0
12	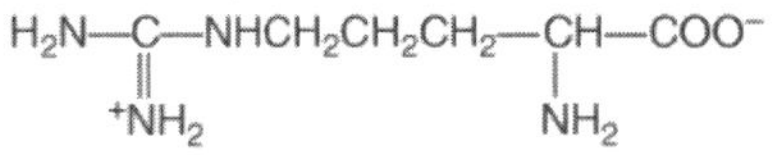	0

5.16 The resonance forms of the peptide bond in glycylglycine are as follows:

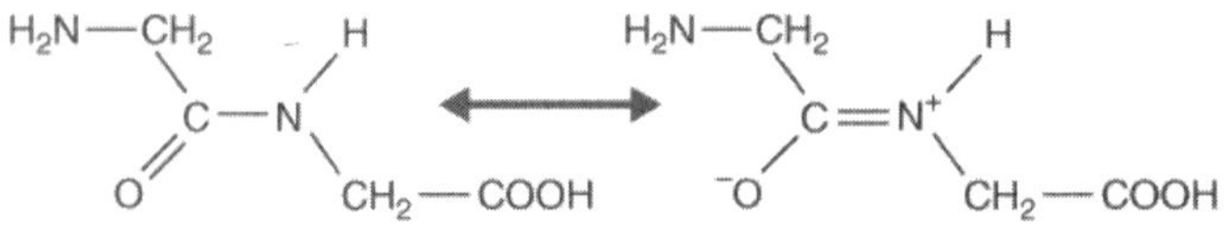

The partial double-bond character of the peptide bond makes it rigid and planar. Rotation around this bond is therefore hindered.

5.17 Six examples of the major functions of protein in the body are catalysis, structure, movement, defense, regulation, and transport.

5.19 a. Polyproline—left-handed helix

b. Polyglycine—β-pleated sheet

c. Ala-Val-Ala-Val-Ala-Val—α-helix

d. Gly-Ser-Gly-Ala-Gly-Ala—β-pleated sheet

5.20 The structural features of several amino acids do not foster *α*-helix formation. Because the R group of glycine is too small, the polypeptide chain becomes too flexible. Proline's rigid ring prevents the required rotation of the N—C bond. Sequences with larger numbers of amino acids with charged side chains (e.g., glutamate) and bulky side chains (e.g., tryptophan) are also incompatible with α-helix formation.

5.22 Amino acids with basic side groups such as lysine, arginine, or tyrosine would contribute to a high p*I* value.

5.23 Refer to Biochemistry in the Lab (pp. 172–177) for protein purification techniques.

5.25 Refer to p. 173 for a discussion of chromatographic separation techniques.

5.26 The primary structure (the amino acid sequence) of a polypeptide provides opportunities for an exceedingly large number of interactions between side groups, in addition to hydrogen bonding that can take place between amide groups along the backbone. As a result, the number of three-dimensional shapes that are theoretically possible is too high to be able to predict the one specific conformation the polypeptide will adopt after synthesis.

5.28 As components of molecular machines, motor proteins undergo conformational changes that, in turn, trigger additional conformational changes in adjacent subunits. Motor proteins bind nucleotides that are hydrolyzed to drive the initial conformational changes. Organisms use motor proteins as classical motors to move a load along a protein filament, as timing devices to provide a delay period for complex processes, as on-off molecular switches in signal transduction pathways, and as assembly/disassembly factors to reversibly form larger molecular complexes from protein subunits.

5.29 In addition to mediating protein folding, molecular chaperones help protect newly synthesized proteins from inappropriate, premature protein-protein interactions by binding to and stabilizing proteins during early stages of folding. Molecular chaperones also direct the refolding of proteins that have partly unfolded; if refolding is not possible, they promote protein degradation. Ribosome-associated chaperones bind to a nascent polypeptide as it emerges from a ribosome. They prevent folding until an entire domain or polypeptide has exited. Hsp70s bind to, stabilize and promote the folding of nascent polypeptides and rescue misfolded and aggregated polypeptides. When a hydrophobic peptide segment binds within a hsp70 binding pocket, it is refolded as a result of ATP hydrolysis. Hsp90s finalize the folding of a limited, but diverse set of client proteins, in collaboration with hsp70. Hsp90s also facilitate the assembly of protein complexes such as RNA polymerase II. The chaperonins are large double-ring complexes that promote fast and efficient polypeptide refolding.

5.31 The exceptionally high concentrations of the oxygen storage molecule myoglobin in the muscles of diving mammals allows them to hold their breath for long periods of time, thus enabling prolonged dives.

CHAPTER 5: SOLUTIONS TO FILL-IN-THE-BLANK QUESTIONS

5.32 Proteins

5.34 Zwitterions

5.35 Enantiomers

5.37 Disulfide

5.38 Imines

5.40 Structural

5.41 Oligomers

CHAPTER 5: SOLUTIONS TO SHORT-ANSWER QUESTIONS

5.43 Because there is no water to stabilize either the anion or the cation, more energy will be required to remove the proton and the pK_a would increase.

5.44 The primary structure of a polypeptide is a unique amino-acid sequence. A change in this sequence changes the identity of the polypeptide. Secondary and tertiary structures are the results of protein folding. As a result, it is possible for a polypeptide to have more than one possible conformation, even though its primary structure has not changed.

5.46 Taurine is derived from the sulfur-containing amino acid cysteine. Note that the taurine residue is linked to cholic acid through an amide linkage. The synthesis of taurine from cysteine is relatively simple: a decarboxylation and the oxidation of the sulfhydryl group of cysteine to a sulfonic acid group.

CHAPTER 5: SOLUTIONS TO THOUGHT QUESTIONS

5.47

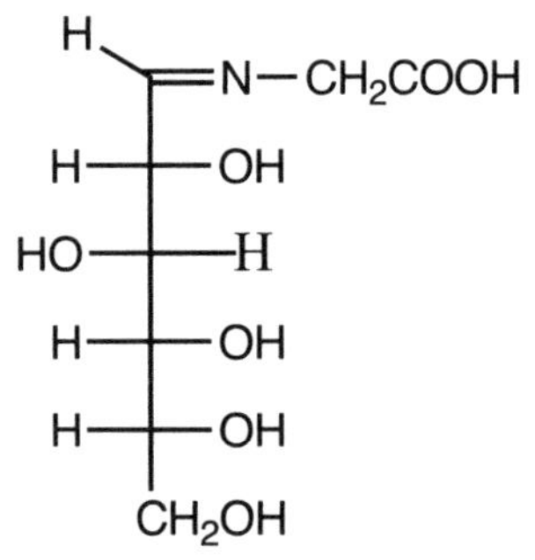

Adduct of Glucose and Glycine

5.49

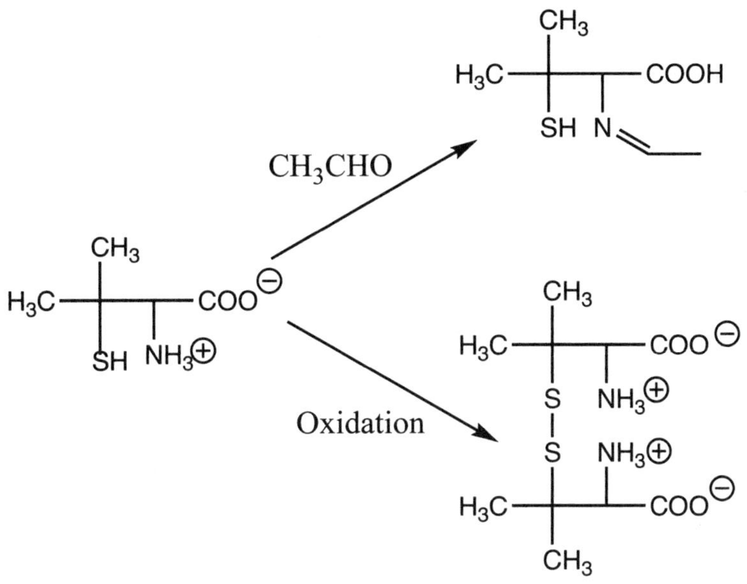

5.50 Protein folding is dependent upon a hydrophobic core. Ser, Gly, Lys, and Glu are less hydrophobic than other amino acids.

5.52 Hydrophobic amino acids such as valine, leucine, isoleucine, methionine, and phenylalanine usually occur within the anhydrous core of proteins because of the hydrophobic effect. Hydrophilic amino acids such as arginine, lysine, aspartic acid, and glutamic acid occur most often on or near the surface of proteins, where they interact with water molecules. Glycine and alanine are hydrophobic amino acids and so tend to occur in the interior of proteins. Glutamine has a polar side chain that can form hydrogen bonds and, therefore, is often on the surface of proteins.

5.53 Living cells possess complex mechanisms for assisting the proper folding of nascent polypeptides. These mechanisms are poorly understood and cannot yet be duplicated in the laboratory.

5.55 The immobilized water of protein molecules is locked into position by hydrogen bonding between polar and ionic groups and water molecules. This gives rise to a three-dimensional structure in which the water molecules have very little freedom of motion (i.e., they are frozen in place).

5.56 The peptide bond is stronger than the ester bond for two reasons. The N is closer to the size of C than the O is, which makes for greater covalency in the bond. Also, because the O and N differ in electronegativity and both have lone pair(s) of electrons, the amide has resonance hybridization. The amide bond, therefore, has partial double bond character.

5.58 a. The isoelectric point is calculated by taking the average pK_a values for the amino group of glycine (9.6) and the carboxyl group of valine (2.32). The pI is 5.96.

b. At pH 1 the tripeptide is positively charged and will move toward the negative electrode. At pH 5 the peptide has zero net charge and will not migrate. At pH 10 and 12 it will have a –1 charge and will move toward the positive electrode.

5.59 The properties that may be responsible for the different functions of a multifunctional protein include different ligand binding sites and the capacity to form homo- or heterocomplexes. In addition, the structural properties of such a protein may be affected by its cellular location.

5.61 The synthesis of protein requires large amounts of energy. Proteins capable of several functions reduce this energy cost.

5.62 The hydrophilic amino acids tyrosine, asparagine, serine, and histidine should all be on the surface.

5.64 The products are glycine, phenylalanine, methionine, valine, leucine, serine, histidine, and aspartic acid. The asparagine would be changed to aspartic acid by the hydrolysis conditions.

5.65 A glycine residue with its small R group (an H atom) within a peptide molecule increases the molecule's rotational freedom. A polypeptide with a series of three chain-stiffening prolines in the positions ordinarily occupied by three glycine residues would become less flexible, a circumstance that would cause a change in conformation and probably loss of function.

5.67 p53 is a major tumor suppressor protein. p53 performs this role because it integrates vast amounts of information related to cell cycle control and cell health. The unstructured domain of p53 binds to a diverse group of transcription factors activated by cell stress conditions such as DNA damage, ER stress, UV light and hypoxia.

5.68 In addition to causing reduced hemoglobin synthesis with a resulting decrease in O_2 transport, an iron deficient diet causes decreased synthesis of 1,3-BPG. Since the binding of 1,3-BPG to hemoglobin results in increased delivery of O_2 to the tissues, a deficit of this molecule would be expected to depress O_2 delivery even further.

5.70 The first step in the process is to develop an assay that allows the investigator to detect the peptides of interest during the purification protocol. Next, the peptides, as well as many other substances, are released from the pituitary gland by cell disruption. Low-molecular-weight substances can then be removed by dialysis. The resulting material is then passed through an exclusion chromatography column in which the high-molecular-weight materials elute first. The fractions containing the oxytocin and vasopressin are then passed though an affinity column, which will retain oxytocin or vasopressin, respectively. Electrophoresis can be used to assess the purity of each peptide.

SOLUTIONS TO PRACTICE PROBLEMS TO PROMOTE PERFECTION

1. a. The pK_a value for each ionizable group is shown in bold.

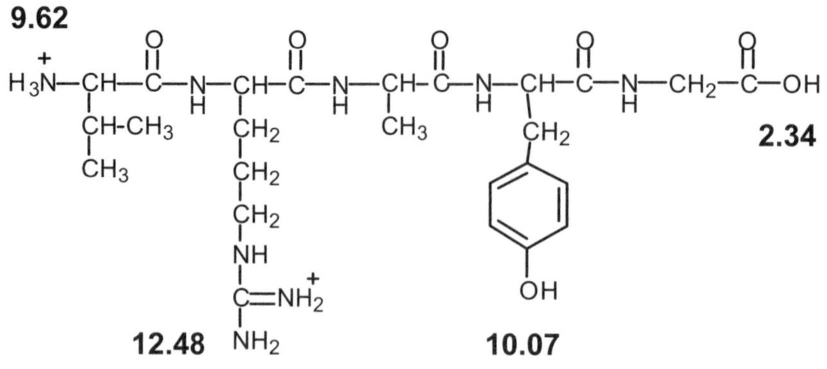

b. Valylarginylalanyltyrosylglycine (Remove the "ine" and replace it with "yl.")

c. We would expect this peptide to have a high pI because the side groups are either basic or neutral. To calculate the pI, determine the structure that has a net charge of zero and average the pK_a values on either side of that structure. The best way to do this is to draw each form in order of pK_a, that is, as the peptide is titrated.

The pI is the average of 9.62 and 10.07, or (9.62 + 10.07)/2 = 9.85. (Note that if this were a multiple-choice problem, one wrong choice would probably be 8.63, the average of *all* of the pK_a values.)

d. To draw the titration curve, first label the y-axis "pH" and the x-axis "Equivalents OH^-." Draw short plateaus at each pK_a value. Place a point at a pH value that is midway between each two pK_a values. Draw a smooth curve from the plateaus through each inflection point.

e. At pH 7, the peptide has a +1 charge and would move toward the cathode.

2. a. The pK_a value for each ionizable group is shown in bold. Pro-His-Met-Ser-Phe:

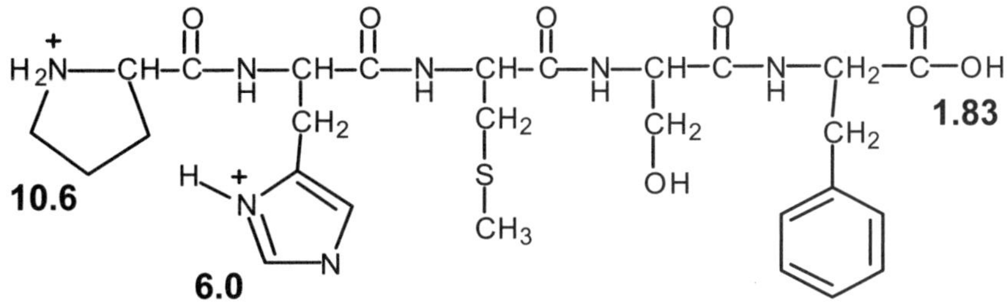

$$+2 \underset{}{\overset{1.83}{\rightleftharpoons}} +1 \overset{6.0}{\rightleftharpoons} 0 \overset{10.6}{\rightleftharpoons} \text{Š}1$$

b. The pI is 8.3 (the average of 6.0 and 10.6).

c. At pH 12, the peptide will have a –1 charge and will move toward the anode.

6 Enzymes

Brief Outline of Key Terms and Concepts

6.1 PROPERTIES OF ENZYMES

ENZYMES CATALYSTS: The Basis

ENZYMES: Activation Energy and Reaction Equilibrium

ENZYMES are catalysts; most are proteins.

CATALYSTS modify the rate of a reaction because they provide an alternative reaction pathway that requires less ACTIVATION ENERGY than the uncatalyzed reaction.

ACTIVE SITE	TRANSITION STATE
SUBSTRATE	ACTIVITY COEFFICENT
COFACTOR	APOENZYME
COENZYME	HOLOENZYME

6.2 CLASSIFICATION OF ENZYMES

OXIDOREDUCTASE	LYASE
TRANSFERASE	ISOMERASE
HYDROLASE	LIGASE

6.3 ENZYME KINETICS is the quantitative study of enzyme catalysis. Kinetic studies measure reaction rates (VELOCITY) and enzyme affinity for substrates and inhibitors.

Kinetics also provides insight into REACTION MECHANISMS: FIRST-ORDER, SECOND-ORDER, PSEUDO-FIRST-ORDER, ZERO-ORDER KINETICS; MOLECULARITY: UNIMOLECULAR, BIMOLECULAR

MICHAELIS-MENTEN KINETICS

MICHAELIS-MENTEN EQUATION

K_m (MICHAELIS CONSTANT)

TURNOVER NUMBER

SPECIFICITY CONSTANT, k_{cat}/K_m

DIFFUSION CONTROL LIMIT; CATALYTIC PERFECTION

INTERNATIONAL UNITS (IU), SPECIFIC ACTIVITY, KATAL

LINEWEAVER-BURK PLOTS

Lineweaver-Burk double-reciprocal plot: $1/v_0$ vs. $1/[S]$ to determine K_m and V_{max}

MULTISUBSTRATE REACTIONS

SEQUENTIAL REACTIONS

DOUBLE-DISPLACEMENT REACTIONS

ENZYME INHIBITION

REVERSIBLE VS.IRREVERSIBLE INHIBITION

REVERSIBLE INHIBITORS

COMPETITIVE INHIBITORS compete with substrate for the active site.

NONCOMPETITIVE INHIBITORS can bind to either the enzyme or the enzyme-substrate complex.

UNCOMPETITIVE INHIBITORS bind only to the enzyme-substrate complex.

KINETIC ANALYSIS OF ENZYME INHIBITION

ALLOSTERIC ENZYMES

ENZYME KINETICS, METABOLISM, AND MACROMOLECULAR CROWDING

METABOLONS; METABOLIC FLUX

6.4 CATALYSIS

ORGANIC REACTIONS AND THE TRANSITION STATE

REACTION MECHANISMS; INTERMEDIATE;

REACTIVE INTERMEDIATES:

CARBOCATION, CARBANION, FREE RADICAL

CATALYTIC MECHANISMS

PROXIMITY AND STRAIN EFFECTS

ELECTROSTATIC EFFECTS

ACID-BASE CATALYSIS

COVALENT CATALYSIS

THE ROLES OF AMINO ACIDS IN ENZYME CATALYSIS

The amino acid side chains in the active site of enzymes catalyze proton transfers and nucleophilic substitutions.

DYADS, TRIADS

THE ROLE OF COFACTORS IN ENZYME CATALYSIS

METAL IONS; COENZYMES; VITAMINS

EFFECTS OF TEMPERATURE AND pH ON ENZYME-CATALYZED REACTIONS

OPTIMUM TEMPERATURE; pH OPTIMUM

DETAILED MECHANISMS OF ENZYME CATALYSIS

SUBSTRATE SPECIFICITY, REACTION MECHANISM

CHYMOTRYPSIN; ALCOHOL DEHYDROGENASE

6.5 ENZYME REGULATION

GENETIC CONTROL; ENZYME INDUCTION

COVALENT MODIFICATION; PROENZYME, ZYMOGEN

ALLOSTERIC REGULATION

CONCERTED VS. SEQUENTIAL MODELS;

POSITIVE VS. NEGATIVE COOPERATIVITY

COMPARTMENTATION

ENZYMES: OVERVIEW

Enzymes are:
- proteins with specific, globular shapes (except for RNA molecules that are also catalytic; Chapter 18).
- molecular machines
- biological catalysts that increase the rates of reactions that are essential to life. Without enzymes, these essential reactions would take place too slowly. Enzymes make life possible!

Enzyme function depends upon:
- having a shape that is complementary to reactant molecules
- having catalytic interactions between reactant molecules and the more or less flexible binding sites on the enzyme
- internal protein motions that extend throughout the molecule (for certain enzymes)

Enzymes function in crowded, gel-like conditions.

6.1 PROPERTIES OF ENZYMES

Enzyme Catalysis: The Basics
Enzymes are biological catalysts. Like all catalysts, enzymes:

—increase reaction rates

—stabilize the transition state, which lowers its energy, thus lowering the activation energy (E_a, the energy a substrate must have before it can be transformed into product, also called the free energy of activation, $\Delta G^{\ddagger}$).

—do not change the equilibrium constant or the ΔG (the thermodynamics) of a reaction. Enzymes like all catalysts only affect the *kinetics*, or the rate, of a reaction. If a chemical reaction isn't *thermodynamically* favorable (i.e., if it is nonspontaneous with a positive (+) ΔG), an enzyme cannot change this.

ENZYMES— HELP SPONTANEOUS REACTIONS ACHIEVE EQUILIBRIUM *FASTER*.

—regain their original form by the end of a reaction, and are ready to catalyze another reaction. Enzymes function at very low concentrations (nanomolar or picomolar) under physiological conditions.

Enzymes: Activation Energy and Reaction Equilibrium
How enzymes are unique and amazing:

—Enzymes catalyze reactions to phenomenally high rates (10^7-10^{19} times greater)

—Side products are rarely formed.

—Enzymes are *highly specific* to the reactions they catalyze and the substrates that it can bind, and this specificity is due to its unique **ACTIVE SITE.** Enzyme-substrate binding at the active site is via noncovalent interactions such as hydrogen bonding, electrostatic, and hydrophobic forces between its amino acid side chains and the substrate. Contact points at the active site have a particular orientation and a distinct shape, like a right-handed glove. For example, a particular enzyme might catalyze a reaction with L-alanine but not D-alanine or any other amino acid. The charge distribution within the active site is due to the location of amino acid side chains, which also participate in the catalytic mechanism.

—Enzymes stabilize the transition state via conformational changes made possible by the active site''s unique shape and charge distribution that fit a specific substrate in a specific orientation. This fit forces the substrate to undergo conformation changes in a way that favors, or resembles, the transition state, so that the enzyme-substrate complex can proceed to product with a much lower activation energy.

—Enzymes can be regulated (because of their relatively large and complex structures) to conserve energy and raw materials.
—Enzymes work at moderate temperatures.

ENZYMES WORK UNDER NONIDEAL CONDITIONS. Equilibrium constants for such reactions use **ACTIVITY**(a), or **EFFECTIVE CONCENTRATION**, instead of concentration. Activity includes the effect of intermolecular interactions, and the activity coefficient (γ) is a correction factor for concentration: $a = \gamma c$ where c = concentration (mol/L) and γ depends on the size and charge of the species and the ionic strength of the reaction solution

Enzyme function can be very different in a dilute buffer solution versus within a crowded cell. The equilibrium constant for a reaction under nonideal conditions includes a **NONIDEALITY FACTOR** (the ratio of the activity coefficients of the products and reactants)

6.2 Classification of Enzymes According to Type of Reaction Catalyzed

(For further examples, see Table 6.2, page 197 of your text.)

OXIDOREDUCTASES	**HYDROLASES**	**ISOMERASES**
TRANSFERASES	**LYASES**	**LIGASES**

OXIDOREDUCTASE

Reaction type: Oxidation-reduction

Names include: dehydrogenase, oxidase, oxygenase, reductase, peroxidase, hydroxylase

Example: Succinate dehydrogenase catalyzes the transfer of electrons from succinate to FAD, producing fumarate and $FADH_2$ in the sixth step of the citric acid cycle. Electrons are removed from the carbons to form the alkene.

$$^{-}O_2C\text{-}CH_2\text{-}CH_2\text{-}CO_2^{-} + FAD \rightarrow {}^{-}O_2C\text{-}CH{=}CH\text{-}CO_2^{-} + FADH_2$$

Succinate Fumarate

TRANSFERASE

Reaction type: Transfer of functional groups

Names include: transferase, kinase (phosphoryl group transfer), transaminase [*trans* + (group transferred) + *ase*], transmethylase, transcarboxylase

Example: Catechol *N*-methyltransferase catalyzes the transfer of a methyl group from *S*-adenosylmethionine to norepinephrine. Used in the synthesis of epinephrine, a neurotransmitter.

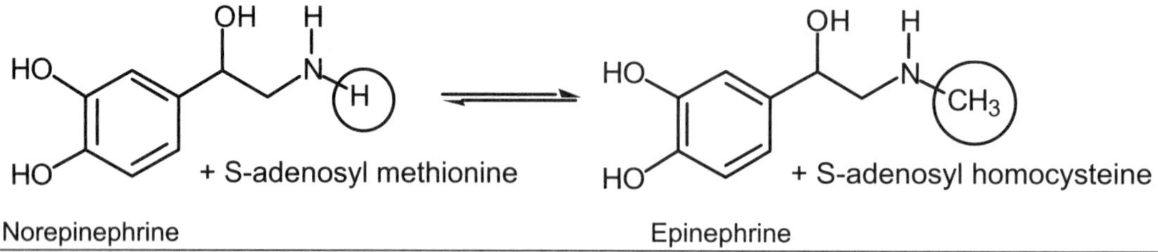

Norepinephrine Epinephrine

HYDROLASE

Reaction type: Hydrolysis (*hydro*—add water; *lysis*—cleave a bond)

Names include: esterase, phosphatase, peptidase, protease

Serine proteases use the $-CH_2OH$ side chain of serine to hydrolyze a peptide. Examples of serine proteases include trypsin, chymotrypsin, and thrombin

Example: Thrombin hydrolyzes the peptide bond after an arginine. Converts fibrinogen into fibrin; fibrin then polymerizes to form a blood clot.

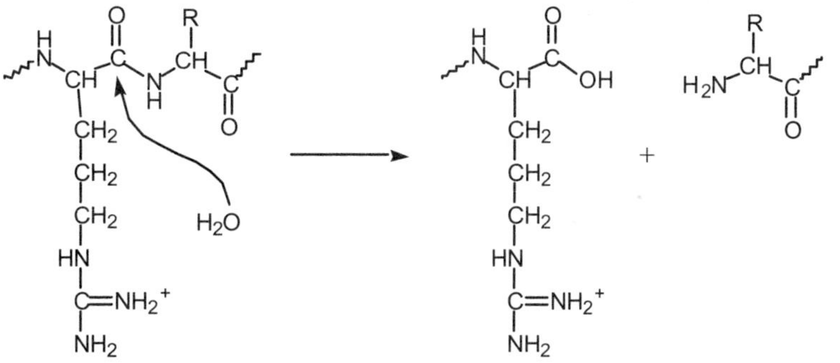

Fibrinogen Fibrin

LYASE

Reaction type: Removes a group and forms a double bond or adds a group to a double bond (the reverse reaction)

Names include: decarboxylase, hydratase, dehydratase, deaminase, synthase, (name of molecule + lyase)

Example: Argininosuccinate lyase converts argininosuccinate to arginine and fumarate in urea synthesis. The enzyme removes a four-carbon dicarboxylic acid group, forming a double bond between the two central carbons.

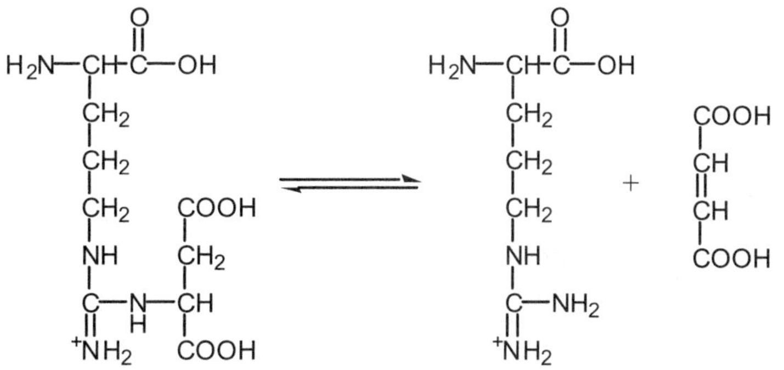

Arginosuccinate Arginine Fumarate

ISOMERASE

Reaction type: Isomerization

Names include: mutase, racemase, epimerase, (name of molecule + isomerase)

Example: Triose phosphate isomerase interconverts glyceraldehyde-3-phosphate and dihydroxyacetone phosphate, moving the carbonyl group from C-1 in glyceraldehyde to C-2 in dihydroxyacetone.

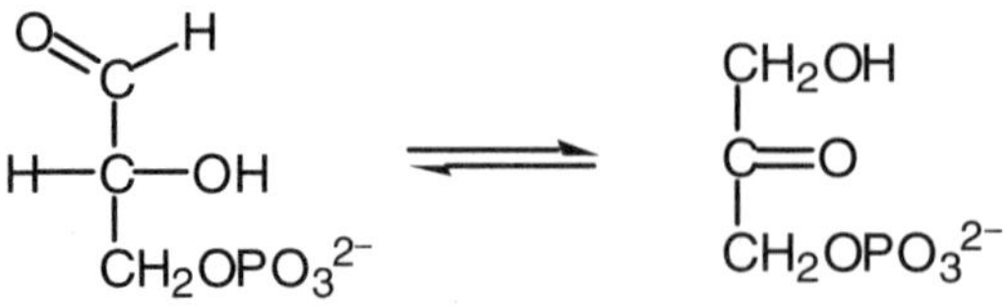

Glyceraldehyde-3-phosphate Dihydroxyacetone phosphate

LIGASE

Reaction type: Bond formation (often includes removal of water; the reverse of hydrolysis); requires an energy source (ATP hydrolysis)

Names include: synthetase, carboxylase (name of molecule formed + ligase)

Example: Glutamine synthetase catalyzes the formation of glutamine from glutamate and ammonia using ATP hydrolysis to make the reaction thermodynamically favorable.

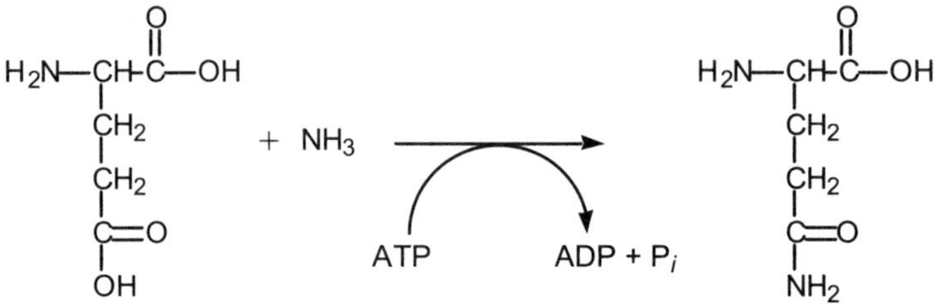

Glutamate Ammonia Glutamine

6.3 Enzyme Kinetics

The velocity (v), or rate, of any chemical reaction is the change in concentration of substrate (or product) as a function of time. The rate may also be expressed as follows:

$v_0 = k[S]^n$ where v_0 = initial velocity

n = the order of the reaction
k = the rate constant

The order of a reaction depends on the reaction mechanism, specifically, how many reactants (or substrates) are involved in the slowest step of the mechanism (since the rate cannot go faster than the slowest step).

FIRST-ORDER KINETICS: For S → P, a simple first-order reaction, $n = 1$, and $v_0 = k[S]$. If [S] is doubled, for example, v_0 should also double.

Consider the reaction: A + B → P. The general rate equation is:

$v_0 = k[A]^m[B]^n$ where m = the order with respect to A, n is the order with respect to B, and m + n is the order overall.

If the rate depends only upon the concentration of A, then $v_0 = k[A]^1[B]^0 = k[A]$, and the reaction is first-order in A, zero-order in B, and first-order overall.

SECOND-ORDER KINETICS: If the rate depends on the concentrations of both A and B, then $v_0 = k[A][B]$, and the reaction is first-order in A, first-order in B, and second-order overall.

Remember that **THE ORDER OF A REACTION MUST BE DETERMINED EXPERIMENTALLY.** There is no way to know the order of a reaction by just looking at the chemical equation.

PSEUDO-FIRST-ORDER KINETICS is a second-order reaction that behaves as if it is first order, typically because one reactant (such as water) is present in excess. (Doubling the excess reactant results in no change in the rate. Double excess is still excess.)

ZERO-ORDER KINETICS: the reaction rate doesn't depend on [S]. Adding more substrate will not increase the rate. Why not? If all of the active sites are filled (saturated) with substrate (and there is plenty of substrate around to hit any active site that becomes available), then adding more substrate will not increase the rate. The enzyme cannot work any faster as it is at its maximum velocity (V_{max}).

MOLECULARITY is the number of colliding molecules in a single-step reaction:
UNIMOLECULAR: 1 molecule (A → B) vs. **BIMOLECULAR:** 2 molecules (A + B → C + D)

MICHAELIS-MENTEN KINETICS

MICHAELIS-MENTEN EQUATION:	MICHAELIS-MENTEN PLOT:
$v_0 = \frac{V_{max}[S]}{[S] + K_m}$	v_0 vs. [S]

V_{max} = maximum velocity

K_m = Michaelis constant (experimentally determined) = [S] at half of V_{max}

When [S] = K_m, then the Michaelis-Menten equation becomes $v_0 = \frac{V_{max}[S]}{2[S]}$, or $v_0 = \frac{V_{max}}{2}$.

In other words,

K_m is equal to the substrate concentration when v_0 equals half of V_{max}.

So, K_m can be determined from the Michaelis-Menten plot of v_0 vs. [S].

Keep in mind that the Michaelis-Menten plot is actually a *series* of experiments that measure the *initial rate* of a reaction at a number of different substrate concentrations, with the same enzyme concentration in each experiment. Take a second look at Figure 6.3 on page 199 Figure (a) shows one experiment that measures the change in [P] over time. Figure (b) is a compilation of data from many experiments like (a), each data point representing one experiment that began with a different [S].

ASSUMPTIONS NEEDED TO DERIVE THE MICHAELIS-MENTEN EQUATION:

1. Rate of the reverse reaction, E + P → ES is ignored. This is valid as initial velocities are measured, when [P] is very low. (Recall that rate equations include the concentrations of reactants, not products. This assumption simplifies the derivation by removing [P].)
2. k_1 (the rate constant for E + S → ES) >>> k_{-1}(rate constant for ES → E + S)
3. Steady-state assumption: (the rate to form ES) = (the rate to degrade ES); the enzyme-substrate complex (ES) is in steady state, so [ES] remains constant as a function of time.

k_{cat} = **TURNOVER NUMBER** = V_{max}/(total enzyme concentration) = No. of substrate molecules that an enzyme molecule converts to product per unit time when enzyme is saturated

SPECIFICITY CONSTANT = k_{cat}/K_m = an indicator of substrate binding efficiency at low [S] ([S] << K_m), when the enzyme is saturated, and the reaction is second-order.

LOWER K_m = GREATER AFFINITY OF THE ENZYME FOR THE SUBSTRATE. Why is this true? Consider two enzyme-catalyzed reactions with the same V_{max} but different K_m values. For the reaction in which the enzyme has a greater affinity for substrate, it takes less substrate to get to the same rate (half of V_{max}). Conversely, an enzyme with a lower affinity for substrate needs a higher substrate concentration to work at the same velocity.

CATALYTIC PERFECTION For enzymes working at their **DIFFUSION CONTROL LIMIT**, their reaction rate is limited only by how fast the substrate can get to the active site.

INTERNATIONAL UNIT (IU) of enzyme activity = amount of enzyme to produce 1 μmol of product per minute; **SPECIFIC ACTIVITY** = No. of IU per mg of protein; a measure of enzyme purification

KATAL (kat) = 1 mol of substrate converted to product per second

LINEWEAVER-BURK PLOTS: DETERMINATION OF K_m AND V_{max}

SIGNIFICANCE:

- Accurate values of V_{max} and K_m are much easier to determine using a straight-line Lineweaver-Burk plot as opposed to a Michaelis-Menten plot, in which the velocity approaches (but never reaches) a maximum value.
- Comparing Lineweaver-Burk plots of inhibited vs. uninhibited reactions can indicate the mechanism of enzyme inhibition.

Lineweaver-Burk turned the Michaelis-Menten plot into a straight line by rearranging the *reciprocal* of the Michaelis-Menten equation to fit the equation for a straight line, $y = mx + b$.

Michaelis-Menten equation: $$v_0 = \frac{V_{max}[S]}{[S] + K_m}$$

Its Lineweaver-Burk reciprocal: $$\frac{1}{v_0} = \frac{[S] + K_m}{V_{max}[S]}$$

$$\frac{1}{v_0} = \frac{[S]}{V_{max}[S]} + \frac{K_m}{V_{max}[S]} = \frac{1}{V_{max}} + \frac{K_m}{V_{max}[S]}$$

After rearranging slightly: $$\frac{1}{v_0} = \frac{K_m}{V_{max}} \bullet \frac{1}{[S]} + \frac{1}{V_{max}}$$

Equation for a straight line: $$\boldsymbol{y = m \quad x + b}$$

A Lineweaver-Burk plot of: $\frac{1}{v_0}$ vs. $\frac{1}{[S]}$ is a straight line, with

$m = \text{slope} = \frac{K_m}{V_{max}}$ $\quad b = y\text{-intercept} = \frac{1}{V_{max}}$

Further algebra reveals: $x\text{-intercept} = -\frac{1}{K_m}$

So, to calculate K_m and V_{max}, all that is needed are the x and y intercepts from the Lineweaver-Burk plot.[1] A Lineweaver-Burk plot is also called a *double-reciprocal* plot, since it is $\frac{1}{v_0}$ vs. $\frac{1}{[S]}$ instead of the Michaelis-Menten plot of v_0 vs. [S]. [Hint: To keep the intercepts straight, remember that the x-axis is a measure of substrate concentration, from which we get K_m, and that the y-axis is a measure of velocity.]

MULTISUBSTRATE REACTIONS

SEQUENTIAL:

All substrates must bind (**ORDERED** or **RANDOM**) before the reaction can occur.

DOUBLE-DISPLACEMENT REACTIONS (PING-PONG MECHANISM)

The first substrate binds and the first product is released *before* the second substrate binds.

ENZYME INHIBITION AND INHIBITORS

- Inhibitors bind to an enzyme and interfere with its activity. Inhibitors help to regulate enzyme activity. Inhibiting a key enzyme can result in inhibition of an entire pathway.
- Reversible inhibition: competitive, noncompetitive (pure and mixed), and uncompetitive
- To determine the type of inhibitor, use a Lineweaver-Burk plot that includes both inhibited and uninhibited reaction data.
- Irreversible inhibitors bind to an enzyme covalently and permanently stop enzyme activity. Removing the inhibitor or increasing [S] will not restore enzyme activity.

MECHANISMS FOR REVERSIBLE ENZYME INHIBITION

COMPETITIVE: $E + I \rightleftharpoons EI \rightarrow$ *(no reaction)* $E + S \rightleftharpoons ES \rightarrow P + E$

Competitive inhibitors bind to the active site of the free enzyme; they "compete" or prevent the substrate from binding to the enzyme, resulting in a change in K_m (since it affects substrate binding) but not V_{max} (since the effect of the inhibitor can be overcome by increasing [S]). In other words, it takes much more substrate to reach V_{max}, so K_m will increase.

NONCOMPETITIVE: $E + I \rightleftharpoons EI$ $E + S \rightleftharpoons ES \rightarrow P + E$

$ES + I \rightleftharpoons EIS$ $EI + S \rightleftharpoons EIS \rightarrow$ *(no reaction)*

Noncompetitive inhibitors bind to a site other than the substrate binding site; they can remove both free enzyme (E) and ES. Increasing [S] can *sometimes partially*—but not completely—reverse noncompetitive inhibition.

PURE noncompetitive inhibitors change the V_{max} but not the K_m (the simplest case, but rare). **MIXED** noncompetitive inhibitors change both K_m and V_{max}.

UNCOMPETITIVE: $ES + I \rightleftharpoons EIS$ $E + S \rightleftharpoons ES \rightarrow P + E$

Uncompetitive inhibitors bind only to ES and not to E. Since some of the enzyme is always tied up as ESI, V_{max} cannot be attained and is lower. K_m also *decreases*.

[1] Note that the x-intercept is just a mathematical trick to calculate K_m, since a negative substrate concentration has no real physical meaning. The calculation of V_{max} is also a mathematical trick, since a reaction rate at [S] = 0 also has no physical meaning.

How can an inhibitor *increase* the affinity of an enzyme for its substrate and still inhibit the reaction? If the inhibitor binds to ES and removes ES, then according to LeChatelier's principle equilibrium will shift to compensate for the change in concentration. The reaction (E + S $\rightleftharpoons$ ES) is shifted to the right because the inhibitor binds to and removes ES. So, the enzyme binds more substrate than expected (decreasing K_m), but the ES complex is waylaid by inhibitor before it can be turned into product (decreasing V_{max}). Adding more [S] cannot overcome this—even though more ES would form, more ESI would also form. Uncompetitive inhibition is considered a type of noncompetitive inhibition.

SUMMARY OF REVERSIBLE ENZYME INHIBITION

TYPE:	BINDS TO:	AFFECTS:	L-B PLOT:[2]	NOTES:
COMPETITIVE	active site of E, not ES	higher K_m same V_{max}	lines intersect on *y*-axis	I is often similar to substrate's structure
NONCOMPETITIVE PURE	both E and ES (at a site other than the active site; need not resemble the substrate)	same K_m lower V_{max}	lines intersect on *x*-axis	rare; the rate to form EI equals the rate to form ESI
NONCOMPETITIVE MIXED		changes K_m lower V_{max}	*x*- and *y*-intercepts and slopes differ	the rate to form EI does not equal the rate to form ESI
UNCOMPETITIVE	ES, not E	lower K_m lower V_{max}	slopes are the same, *x*- and *y*-intercepts differ	typically for reactions with more than one S

ALLOSTERIC ENZYMES (Michaelis-Menten plot is sigmoidal rather than hyperbolic; Michaelis-Menten model does not explain the kinetic properties of allosteric enzymes.

ENZYME KINETICS, METABOLISM, AND MACROMOLECULAR CROWDING

Cell interiors are very crowded and highly heterogeneous; such numerous and varied intermolecular interactions result in lower rates of diffusion, higher effective concentrations, and other effects that are nonlinear and difficult to predict.

MACROMOLECULAR CROWDING: many macromolecules such as proteins, membranes, and cytoskeletal components impede molecular movement within a cell

Example of the complexity this introduces: Consider a slow substrate diffusion vs. greater efficiency via the formation of **MICROCOMPARTMENTS** (with high substrate concentrations) or **METABOLONS** (enzyme complexes that channel pathway intermediates from one enzyme to the next). Will the result be a greater or a lower efficiency? *In vivo* conditions are difficult to model *in vitro* and in silico.

METABOLIC FLUX—the rate of flow of metabolites (substrates, products, intermediates) along biochemical pathways

---

[2] Lineweaver-Burk plot of inhibited and uninhibited reactions: Increasing K_m makes the *x*-intercept ($-1/K_m$) less negative (-1/20 is less negative than –1/10). Lowering V_{max} makes the *y*-intercept ($1/V_{max}$) larger (1/2 is larger than 1/4). The slope is K_m/V_{max}.

6.4 Catalysis

Organic Reactions and the Transition State

Reaction mechanisms and related terminology are the same as in organic chemistry:

TRANSITION STATE VS. INTERMEDIATE;

REACTIVE INTERMEDIATES: CARBOCATION, CARBANION, FREE RADICAL

The difference? The active site stabilizes the transition state and has a greater affinity for the transition state than for substrates or products.

Catalytic Mechanisms

Proximity and Orientation Effects; Electrostatic Effects

The more tightly and efficiently that an active site can bind substrate while it is in its transition state, the faster the reaction rate will be.

Acid-Base Catalysis

Amino acids with side chains that can act as either proton donors or acceptors depending upon their pK_a and local environment: His, Asp, Glu, Tyr, Cys, Lys

Covalent Catalysis

Nucleophilic amino acid side chains (like Ser) form weak covalent bonds with substrate molecules to facilitate the reaction (example: serine proteases like chymotrypsin)

The Roles of Amino Acids in Enzyme Catalysis

Catalytic amino acids have polar or charged side chains:

Ser, Thr, Tyr, Cys, Gln, Asn, Glu, Asp, Lys, Arg, His

DYADS or **TRIADS** are groups of 2 or 3 amino acids that are positioned precisely for catalytic effects, such as polarizing a specific atom or group, or changing the pK_a of a nearby functional group.

Noncatalytic amino acids function in substrate orientation and transition state stabilization. Example: creation of a hydrophobic pocket of a specific size and shape.

The Role of Cofactors in Enzyme Catalysis

Cofactors are nonprotein components that add chemical functionality to enzymes and thus increase the kinds of reactions that enzymes can catalyze. (Otherwise, the enzymes would be limited to the amino acid residues' R groups.)

Apoenzyme vs. Holoenzyme

In ***A***poenzymes, the necessary cofactor or coenzyme is ***A***bsent. (An apoenzyme is just the protein component.) ***H***oloenzymes ***H***ave the cofactor or coenzyme.

METALS: transition metals (Fe^{2+}, Cu^{2+})
alkali and alkaline earth metals (Na^+, K^+, Mg^{2+}, Ca^{2+})

Why are the transition metals good cofactors? Their high concentration of positive charge binds small molecules well. They can act as Lewis acids (i.e., accept electron pairs). Because transition metals can interact with two or more ligands, they can form a substrate-metal ion complex in the active site, thus helping to orient the substrate properly, polarizing the substrate and promoting catalysis. Metals with two or more valence states can mediate redox reactions.

COENZYMES VS. VITAMINS

VITAMINS: water soluble vs. lipid soluble; many are precursors to coenzymes

The vitamin niacin is a precursor to the coenzyme NAD^+ (nicotinamide adenine dinucleotide), which functions as an intracellular electron carrier and as a hydride ($H:^-$) transfer agent.

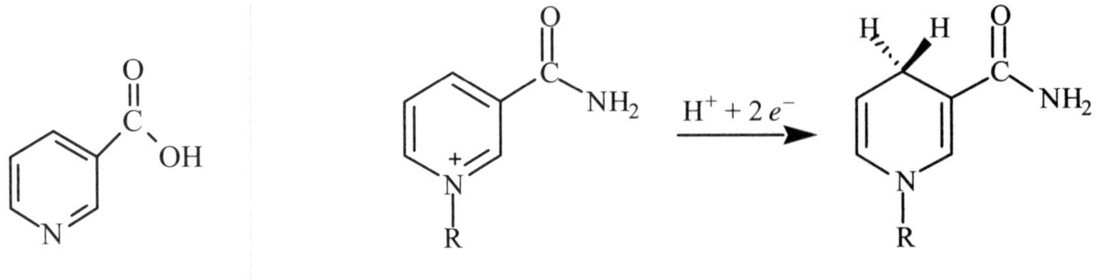

Niacin Oxidized NAD^+ Reduced NADH

The vitamin riboflavin is a precursor to the coenzymes FMN (flavin mononucleotide) and FAD (flavin adenine dinucleotide), which are components of flavoproteins (a subclass of the oxidoreductases).

EFFECTS OF TEMPERATURE AND pH ON ENZYME-CATALYZED REACTIONS

Optimum temperature; pH optimum

Temperature and pH changes affect the tertiary structure of proteins, which must be very specific for enzymes to function at their maximum velocity. Optima vary according to enzyme function (consider enzymes that function at the different pH environments of the stomach at pH ~ 3, or small intestine, at pH 5-8).

Higher temperatures generally increase rates, but denaturation of the enzyme can occur if the temperatures are too high.

DETAILED MECHANISMS OF ENZYME CATALYSIS: CHYMOTRYPSIN AND ALCOHOL DEHYDROGENASE

Note that these mechanisms resemble those in organic chemistry, but the proximity of certain functional groups to each other and the geometry of the active site encourages reactions to happen that would not normally happen. Enzymes cannot make the impossible happen. They just make spontaneous (i.e.,thermodynamically favorable) reactions happen *faster*.

6.5 ENZYME REGULATION

Enzyme regulation: maintains an ordered state, conserves energy, and helps the cell to respond to environmental changes (i.e., modulates specific pathways in response to a cell"s needs). Regulatory enzymes in a pathway are usually controlled by covalent modification or allosteric regulation.

GENETIC CONTROL

ENZYME INDUCTION: Enzymes are synthesized when needed.

REPRESSION: Enzyme synthesis is inhibited.

COVALENT MODIFICATION

- **PHOSPHORYLATION** (or **DEPHOSPHORYLATION**) to convert an enzyme between its active and inactive forms (whether the phosphorylated enzyme is active or inactive depends on the specific enzyme).
- **ZYMOGENS** (**PROENZYMES**) are converted into active enzymes by the irreversible cleavage of one or more peptide bonds.

ALLOSTERIC REGULATION

Effector molecules (ligands) bind to allosteric sites on the enzyme and trigger conformational changes.

Importance of the flexibility of proteins (binding of ligand to one subunit prompts a conformational change that is transmitted to other subunits)

HOMOTROPIC (ligand = substrate) vs. **HETEROTROPIC** allosteric effects

CONCERTED (symmetry) model vs. **SEQUENTIAL** model

T vs. **R** forms: **TAUT** vs. **RELAXED**
Inhibitors bind to and stabilize the Taut form;
Activators bind to and stabilize the Relaxed form

CONCERTED (SYMMETRY) MODEL: all T subunits change to R (or R → T) upon binding
SEQUENTIAL MODEL: upon binding of an effector, subunits change sequentially

POSITIVE COOPERATIVITY— – first ligand increases subsequent ligand binding
NEGATIVE COOPERATIVITY— – first ligand reduces the enzyme's affinity for similar ligands

Many allosteric proteins are more complex than either of these models.
There are no simple rules that explain metabolic regulation.

COMPARTMENTATION

Enzymes, substrates, and regulatory molecules are located in separate compartments (or areas) so that opposing pathways are physically separated, or located close together, so that a pathway can be more efficient.

Enzymes may also be attached to the cytoskeleton or to a membrane.

Compartmentation solves problems:

1. Divide and control: separation of enzymes into different regions or compartments to use resources efficiently (e.g., to prevent a just-synthesized molecule from being degraded). This also gives the cell an additional level of regulation by controlling the transport of substrates, products, effectors, etc., across organelle membranes (example: mitochondria membranes).
2. Diffusion barriers: microenvironments that concentrate enzymes and/or substrates and so reduce diffusion barriers. Enzymes located close to each other can increase the efficiency of coupled reactions and/or of entire pathways (such as the electron transport chain); **METABOLIC CHANNELING**
3. Specialized reaction conditions (e.g., low pH within lysosomes)
4. Damage control: segregation of molecules that may be toxic to the cell

THE ORGANIC CONNECTION: NICOTINAMIDE ADENINE DINUCLEOTIDE (NAD^+)

Take another look at the nicotinamide rings of NAD^+ and NADH. Which one would you expect to be more stable, that is, at a lower energy? Why?

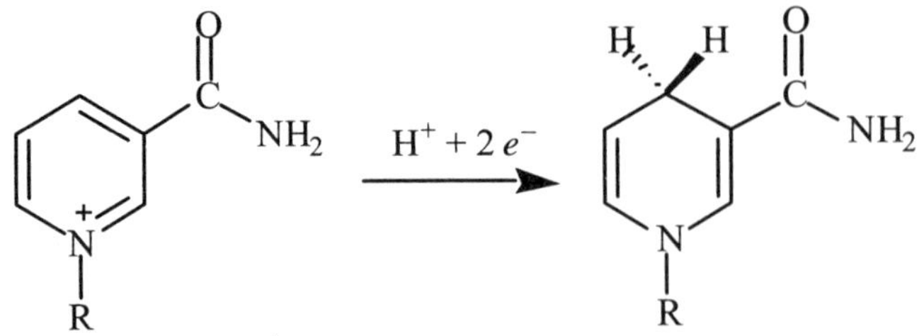

Oxidized NAD^+ Reduced NADH

When the hydride (or, the H+ with two electrons) is added, it creates a tetrahedral carbon that interrupts the aromaticity of the ring. The nicotinamide ring of NAD^+ is aromatic and thus more stable and lower energy than the nonaromatic ring of NADH. Since NADH is at a higher energy than NAD^+, this half-reaction would require an input of energy:

$$NAD^+ + H^+ + 2e^- \rightarrow NADH$$

Couple this with an oxidation half-reaction that releases energy, and energy can be captured in the form of electrons. (Refer to Chapter 4.)

AFTER STUDYING THIS CHAPTER, YOU SHOULD BE ABLE TO:

- Describe how enzymes work, general properties of enzymes, and factors that contribute to enzyme catalysis.
- Determine the class of an enzyme, given the reaction that the enzyme catalyzes.
- Given a biochemical reaction and reaction rate data at various substrate concentrations, determine the order for each substrate and the overall order of a reaction.
- Michaelis-Menten plots: Determine K_m.
- Lineweaver-Burk plots: Determine the type of enzyme inhibition, given data for both uninhibited and inhibited enzyme-catalyzed reactions. Determine V_{max} and K_m from the x- and y-intercepts.
- Compare enzyme efficiencies given k_{cat} and K_m data for various substrates.
- Describe methods of enzyme regulation.

Use this space to note any additional objectives provided by your instructor.

CHAPTER 6: SOLUTIONS TO REVIEW QUESTIONS

6.1 a. catalyst – a substance that enhances the rate of a chemical reaction but is not permanently altered by the reaction

b. transition state – in catalysis, the unstable intermediate formed by the enzyme that has altered the substrate so that it now shares properties of both the substrate and the product

c. substrate – the reactant in a chemical reaction that binds to an enzyme active site and is converted to a product

d. activation energy – the threshold energy required to produce a chemical reaction

e. active site – the cleft in the surface of an enzyme where a substrate binds

6.2 a. cofactor – the nonprotein component of an enzyme (either an inorganic ion or a coenzyme) required for catalysis

b. coenzyme – a small organic molecule required in the catalytic mechanism of a certain enzyme

c. apoenzyme – the protein portion of an enzyme that requires a cofactor to function in catalysis

d. holoenzyme – a complete enzyme consisting of an apoenzyme plus a cofactor

e. velocity – the rate of a biochemical reaction; the change in concentration of a reactant or product per unit time

6.4 a. competitive inhibitor – an inhibitor molecule that competes with the substrate for occupation of the active site

b. uncompetitive inhibitor – an inhibitor that binds only to the enzyme-substrate complex

c. noncompetitive inhibitor – an inhibitor that binds to both the free enzyme and the enzyme-substrate complex but not in the active site

d. reversible inhibitor – a form of enzyme inhibition in which the inhibitory effect of a compound can be counteracted by increasing substrate or removing the inhibitor while the enzyme remains intact

e. irreversible inhibition – a form of enzyme inhibition in which an inhibitor permanently impairs an enzyme, usually through binding via a covalent bond

6.5 a. Lineweaver-Burk plot – a graph in which K_m and V_{max} values for an enzyme are determined using the reciprocals of initial velocities and substrate concentrations

b. allosteric enzyme – an enzyme whose activity is affected by the binding of effector molecules

c. macromolecular crowding – the dense packing of an enormous variety of macromolecules and other molecules within the interior of cells

d. reaction intermediate – in a reaction a species that exists for a finite length of time and is then transformed into product

e. carbocation – a carbon with a positive charge

6.7 The important properties of enzymes are: high catalytic rates, a high degree of substrate specificity, negligible formation of side products, and capacity for regulation.

6.8 Cells regulate enzymatic reactions by using genetic control (certain key enzymes are synthesized in response to changing metabolic needs), covalent modification (certain enzymes are regulated by the reversible interconversion between their active and inactive forms, a process involving covalent changes in structure), allosteric regulation (binding effector molecules to pacemaker enzymes alters catalytic activity), and compartmentation (preventing wasteful "futile cycles" by physical separation of opposing biochemical processes within cells).

6.10 Three reasons why the regulation of biochemical processes are important are maintenance of an ordered state, conservation of energy, and responsiveness to environmental cues.

6.11 Negative feedback inhibition is a process in which the product of a pathway inhibits the activity of the pacemaker enzyme

6.13 In the concerted model, the substrate and activators bind to the relaxed conformation. This binding shifts the equilibrium to the R conformation. In the sequential model, binding the activator molecule changes the conformation of the enzyme to a shape more favorable to binding substrate. When oxygen binds to hemoglobin, the first oxygen molecule binds slowly. It, however, introduces a conformational change that makes the sequential binding of the second, third, and fourth oxygen molecules much easier.

6.14 a. vitamin – an organic molecule required by organisms in minute quantities; some vitamins are coenzymes required for the function of cellular enzymes

b. cofactor – the nonprotein component of an enzyme (either an inorganic ion or a coenzyme) required for catalysis

c. transition state – the unstable intermediate in catalysis in which the enzyme has altered the form of the substrate so that it now shares properties of both the substrate and the product

d. inhibitor – a molecule that reduces an enzyme's activity

e. allosteric enzyme – an enzyme whose activity is affected by the binding of effector molecules

6.16 The activation energy for the reaction of glucose with molecular oxygen is quite high and, consequently, the reaction is relatively slow to occur.

6.17 Enzymes decrease the activation energy required for a chemical reaction because they provide an alternate reaction pathway that requires less energy than the uncatalyzed reaction. They do so principally because of the unique, intricately shaped active sites, which possess strategically placed amino acid side chains, cofactors, and coenzymes that actively participate in the catalytic process.

6.18 At the start of a reaction, the concentrations of the reactants and products can be known precisely. Because equilibrium has not yet been established presumably only the forward reaction is taking place.

6.19 The pK_a of the imidazole group of histidine is approximately 6. Therefore, the histidine side chain ionizes within the physiological pH range. The protonated form of histidine is a general acid, and the unprotonated form is a general base.

6.20 The amino acid residues that compose the active site are stereoisomers. Consequently, the active site is chiral and can bind only one form of an optically active compound

6.22 The law of mass action assumes: linear reaction rates with respect to solute concentrations; a homogeneous reaction system; and random, independent interactions between molecules. However, macromolecular crowding increases the effective concentration, which impacts binding affinities, reaction rates, equilibrium constants, and diffusion rates. As a result, catalytic activities of enzymes *in vivo* are typically not predictable based upon *in vitro* studies of enzyme activity in dilute solutions.

6.23
a. oxygenase – an enzyme that catalyzes the addition of oxygen atoms into a substrate

b. epimerase – an enzyme that catalyzes the inversion of an asymmetric carbon atom of a sugar, converting it into one of its epimers (epimerization)

c. protease – an enzyme that catalyzes the cleavage of peptide bonds with the addition of water

d. hydroxylase – an enzyme that catalyzes the insertion of an oxygen atom into a C–H bond in order to form a new hydroxyl group

e. oxidase – an enzyme that catalyzes a reaction in which O_2 is reduced, but oxygen atoms do not appear in the product

6.25 The amino acid residues that most commonly participate in enzyme mechanisms are serine, threonine, tyrosine, cysteine, histidine, aspartate, glutamate, asparagine, glutamine, lysine, and arginine. Note that all of these amino acids have polar or charged side chains. See Figure 5.2 (p. 126) for the structures of these amino acids.

6.26 The roles played by amino acid side chains in enzymatic catalysis include: donating or accepting proton(s) in acid/base catalysis (His, Tyr, Cys, Glu, Asp, and Lys), acting as a nucleophile in covalent catalysis (Ser, Thr, Cys, Gln, Asn, Asp, and Glu), and having a polarizing effect on the substrate (Arg, Glu, Asp, Ser, His). Examples of enzymes that illustrate these roles are pepsin, the serine proteases, and adenylate kinase. Pepsin's active site contains two Asp carboxyl groups, one of which has a polarizing effect on the other, increasing its basicity. This initiates the acid-base hydrolysis of a peptide bond. The serine proteases also hydrolyze peptide bonds. This mechanism includes the deprotonation of serine's hydroxyl group and the use of the resulting oxyanion as a nucleophile. Adenylate kinase, which catalyzes the transfer of phosphoryl groups from ATP to other nucleotides, contains an Arg-Arg dyad whose polarizing effect converts phosphate into a good leaving group.

6.28 By physically separating enzymes that catalyze competing reactions, compartmentation prevents the inefficiency of futile cycles by allowing these enzymes to be regulated independently. Within microcompartments, metabolite channeling reduces the degree to which diffusion will limit reaction rates. Compartmentation can provide unusual environments for enzymes that require them, and can protect cellular components from damage due to toxic reaction products. Examples of compartmentation include: the separation of kinases and phosphatases, the attachment of multienzyme complexes to membranes or microfilaments, and the low pH within lysosomes.

6.29 The two major types of enzyme inhibitors are reversible and irreversible. Reversible inhibition does not destroy the enzyme. Adding substrate or removing the inhibitor can remove the inhibitory effect. Malonate, whose structure resembles that of succinate, is a reversible competitive inhibitor of succinate dehydrogenase. An irreversible inhibitor binds permanently (usually covalently) to the enzyme and destroys its catalytic ability. Removing irreversible inhibitors will not restore enzyme activity. Iodoacetate is an irreversible inhibitor that alkylates glyceraldehyde-3-phosphate dehydrogenase.

6.31 The three-dimensional shape of the active site is vital to enzyme function and its ability to stabilize the transition state. A very specific structure of an enzyme is required to ensure that the active site has the optimum shape, electrostatic environment, flexibility (to optimize proximity and strain effects), and precise locations of amino acid side chains that actively participate in the enzyme's catalytic mechanism. This complex precision of the active site structure is created and maintained by the intricate web of interactions between amino acid residues in the enzyme as a whole. The inefficiency of creating such a large molecule is tolerated because of the enormous advantage provided by the increase in reaction rates and enzyme efficiency that result.

6.32 The amount of energy required for the transition state to be reached during an enzyme-catalyzed reaction is lowered in part because the enzyme's active site constrains the motions and allowed conformations of the substrate. The active site, which optimally orients the substrate, forces it to adopt a conformation similar to that of the transition state. The energy gained upon substrate binding is also used to lower the activation energy.

6.34 An enzyme achieves catalytic perfection if it converts every substrate molecule that enters the active site into a product molecule.

6.35 Among the observed effects of macromolecular crowding on enzymes are increased values for substrate binding affinities, changes in reaction ratio and equilibrium constants and decreased substrate diffusion rates.

6.37 Catalysts do alter the reaction mechanism. A metal catalyst does so because it binds to reactant on its surface thereby orienting them to maximize their reactivity. Enzymes do so because amino acid side chains and other catalytic species actively participate in the reaction. Catalysts provide a different pathway for the reaction to occur.

6.38 Since the enzyme is unchanged during the course of the reaction it does not appear in the rate equation and hence does not affect the order of the reaction.

6.40 Mercuric ion inactivates the enzyme by reacting with sulfhydryl groups. Methanol, whose structure is very similar to that of ethanol, is acting as a competitive inhibitor.

6.41 The energy of a reaction is the difference between the energy of the reactants and product. The energy of activation is the energy difference between the reactants and the transition state.

Chapter 6: Solutions to Fill-in-the-Blank Questions

6.43 Capacity for regulation

6.44 Active site

6.46 Inhibitors

6.47 Competitive

6.49 Free radicals

6.50 Enzyme induction

6.52 Induced fit

CHAPTER 6: SOLUTIONS TO SHORT-ANSWER QUESTIONS

6.53 RNA molecules are long, single-stranded polynucleotides containing the sugar ribose with its catalytically active 2′-hydroxyl group. RNA molecules can fold into complex structures that contain groupings of atoms that act as catalytic centers.

6.55 Enzymes are composed of chiral amino-acid residues, a circumstance that creates a unique spatial arrangement within their active sites. One stereoisomer has a complementary spatial arrangement within the enzyme's active site and therefore can enter the site, where it undergoes the reaction. Its enantiomer will not fit properly into the active site. As a result, there is no reaction.

6.56 The basic concepts that have emerged from investigations of enzyme-catalyzed reactions in dilute solution (e.g., Michaelis-Menten kinetics and enzyme inhibition kinetics) are only applicable to the interior of a living cell to an unknown extent, because the principal assumption of enzyme kinetics, substrate and product diffusion, is apparently not applicable to the diffusion-resistant high-density gel that exists within cells. Cells use other means (e.g., metabolons) to transfer biochemical pathway intermediates between enzyme active sites.

CHAPTER 6: SOLUTIONS TO THOUGHT QUESTIONS

6.58 The data indicate that the reaction is first-order in pyruvate and ADP and second-order in P_i. The overall reaction is fourth-order.

6.59 Note the difficulty in estimating V_{max} for the uninhibited reaction. If 95 μmol/min is chosen for the maximum velocity, then K_m is 20 μM (the substrate concentration, [S], at half V_{max}, or 45 μmol/min). However, if 100 μmol/min is the maximum velocity, then K_m = 25 μM. The inhibited reaction appears to have a V_{max} of 16 μmol/min; K_m=15 μM.

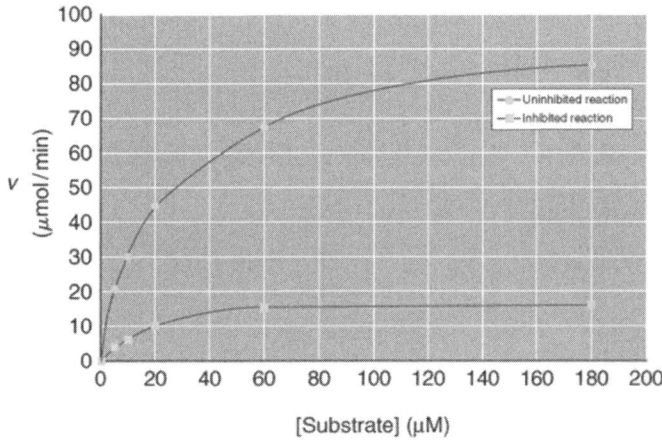

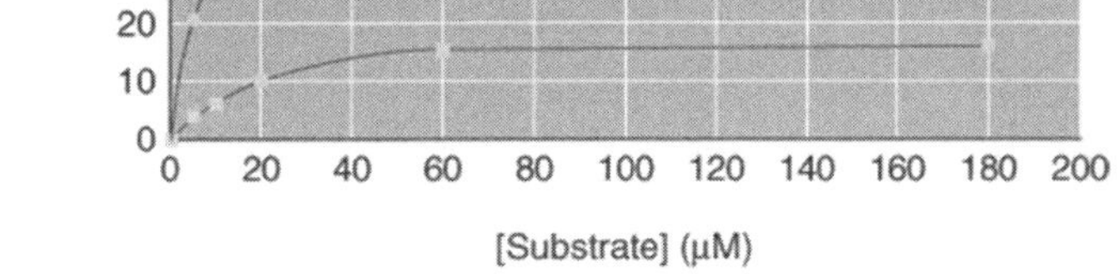

6.61

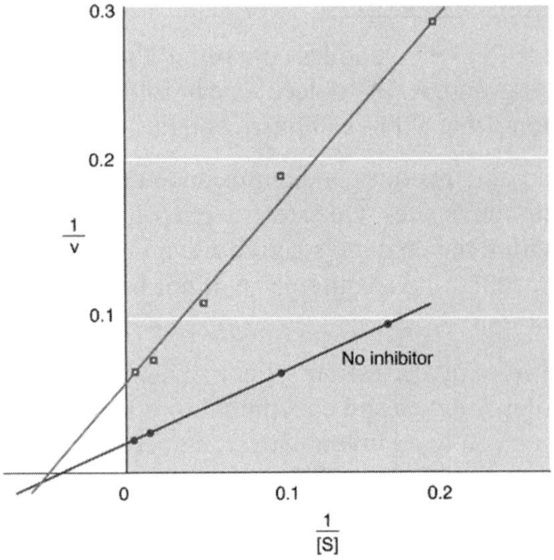

Intercept, horizontal axis = $-\dfrac{1}{K_m}$

Intercept, vertical axis = $\dfrac{1}{V_{max}}$

Slope = $\dfrac{K_m}{V_{max}}$

The type of inhibition being observed is noncompetitive.

6.62 The transition states for each step are as follows:

Step 1

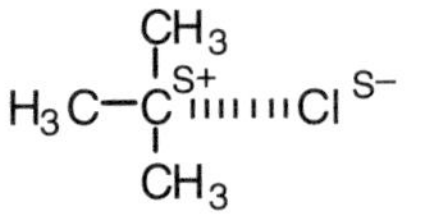

Step 2

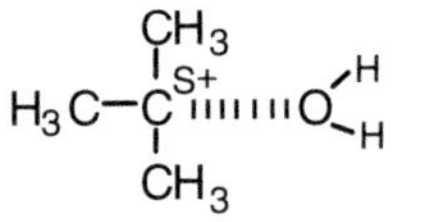

6.64 The inhibition is competitive.

6.65 V_{max}=10 μm/s K_m=5 μm

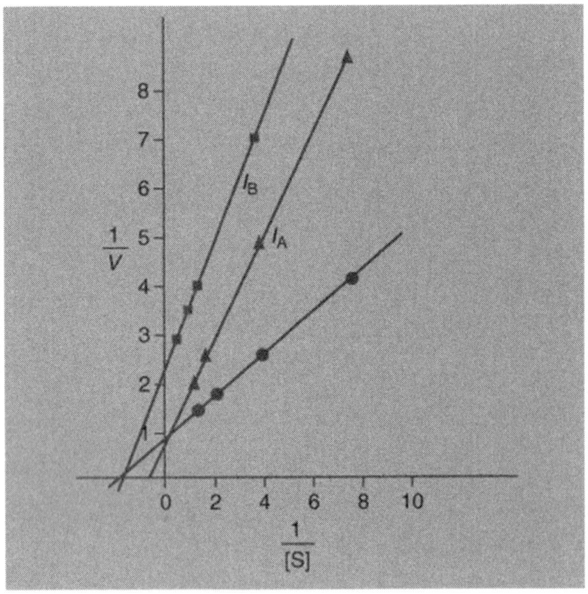

6.67 Catalase has a higher affinity for the substrate and a higher turnover number than does the carbonic anhydrase.

6.68 This enzyme has a greater tendency to dehydrate malate to fumarate than the reverse reaction. In order for the citric acid cycle to operate with the enzyme production of malate, malate concentrations must be very low.

6.70 The rate is first order in each reactant. The overall rate expression is: Rate = k[A][B]. The overall reaction is second order.

6.71 Crowding will tend to raise activity coefficients. If the substances are not allowed to diffuse from the reaction center the concentration in terms of moles per liter can be quite low but the local concentration can be considerably higher.

6.73 This serine residue is part of a serine triad. This makes the serine OH much more reactive. The diisopropylfluorophosphate fits into the active site next to the serine and reacts rapidly.

6.74 Alcohol dehydrogenase selectively metabolizes ethyl alcohol in preference to ethylene glycol. Treatment of the patient with ethyl alcohol will block the conversion of ethylene glycol to toxic metabolites. The ethylene glycol can then be excreted.

6.76 See Figure 6.18.

6.77 The catalytic OH group would be a stronger nucleophile. In aqueous solution the OH would be solvated by water that must be lost before the OH can act as a nucleophile. This would require energy. In a water-free pocket of the enzyme, this desolvation is not necessary, so less energy is required and the OH is more reactive.

6.79 Electrons from the carboxylate anion attack the proton of the histidine releasing the electrons of the histidine making the system more basic

6.80 The K_m value for acetaldehyde indicates that the acetaldehyde binds strongly to ADH as it is converted to ethanol (which binds less strongly). Ethanol is then easily released from the active site allowing the binding of another acetaldehyde.

7 Carbohydrates

Brief Outline of Key Terms and Concepts

7.1 MONOSACCHARIDES

aldoses, ketoses; Fischer projections

MONOSACCHARIDE STEREOISOMERS

D- vs. L-sugars; enantiomers
diastereomer
epimer

CYCLIC STRUCTURE OF MONOSACCHARIDES

HEMIACETALS, HEMIKETALS
HAWORTH STRUCTURES
CONFORMATIONAL STRUCTURES (puckered rings, chair structures)
MUTAROTATION
- anomer; anomeric carbon
- α- vs. β-carbon; relative stability of α- vs. β-anomers; equilibrium mixtures in water

REACTIONS OF MONOSACCHARIDES

OXIDATION
- aldonic acid, aldaric acid
- uronic acid
- lactone
- reducing sugars

REDUCTION; alditols
ISOMERIZATION
- enediol
- epimerization

ESTERIFICATION
- includes phosphoesters

GLYCOSIDE FORMATION
- glycoside; glycosidic linkage
- acetal; ketal
- disaccharide; polysaccharide

GLYCOSYLATION REACTIONS

IMPORTANT MONOSACCHARIDES

GLUCOSE FRUCTOSE
GALACTOSE

MONOSACCHARIDE DERIVATIVES

URONIC ACIDS AMINO SUGARS
DEOXY SUGARS

7.2 DISACCHARIDES

MALTOSE CELLOBIOSE
LACTOSE SUCROSE

7.3 POLYSACCHARIDES

and oligosaccharides

Functions: energy storage
structural materials.

HOMOGLYCANS

STARCH: AMYLOSE, AMYLOPECTIN
GLYCOGEN
CELLULOSE
CHITIN

HETEROGLYCANS

N-GLYCANS, *O*-GLYCANS
GLYCOSAMINOGLYCANS (GAG)

7.4 GLYCOCONJUGATES

PROTEOGLYCANS

GLYCOPROTEINS; GLYCOPROTEIN FUNCTIONS

GLYCOCONJUGATES are molecules in which carbohydrates are linked covalently to proteins or lipids.

PROTEOGLYCANS are composed of relatively large amounts of carbohydrate (GAG units) covalently linked to small polypeptide components.

GLYCOPROTEINS are proteins linked to carbohydrate through N- or O-linkages.

7.5 THE SUGAR CODE

LECTINS: TRANSLATORS OF THE SUGAR CODE

THE GLYCOME

glycoforms, glycomics
microheterogeneity

FUNCTIONS OF CARBOHYDRATES:

energy sources
structural elements
precursors in the synthesis of biomolecules
biological information storage

7.1 MONOSACCHARIDES

Name by type (aldose, ketose), number of carbons (pentose, hexose), or both (aldohexose, ketohexose).

Simplest sugars are trioses: glyceraldehyde, dihydroxyacetone.

FISCHER PROJECTIONS in organic chemistry are always shown as a cross, with the carbon understood to be at the center. When the carbon atom was labeled in the center, it was understood that that was not a Fischer projection but a line-bond structure that implied no stereochemistry whatsoever. In your text, however (as in many biochemistry texts), it is understood that any straight-chain carbohydrate drawn vertically (with the most oxidized carbon nearest the top) is a Fischer projection, even when the carbons are shown.

MONOSACCHARIDE STEREOISOMERS

HOW MANY STEREOISOMERS ARE POSSIBLE?

Because carbohydrates have many chiral centers, there are many stereoisomers. Identifying which carbons are chiral is the first step in determining the number of stereoisomers. Recall that for a carbon to be chiral, it must have *four different atoms or groups* attached to it.

The number of possible stereoisomers is related to the number of chiral carbons:

number of stereoisomers = 2^n n = number of chiral carbons

Why is this true? *Each* chiral carbon can have two different arrangements of groups (that is, they can be mirror images of each other).[1] Glucose is an example of an aldohexose, which has four chiral carbons.

LET'S LOOK MORE CLOSELY AT GLUCOSE (AN ALDOHEXOSE):

1
2
3
4
5
6

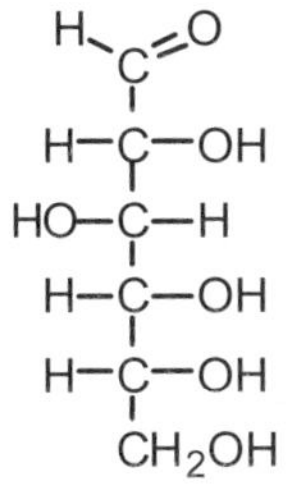

Carbon 1 has only three substituents attached to it, and so it is *not* chiral.

Carbon 2 has four different groups attached to it, so it *is* chiral. Carbons 3, 4, and 5 are also chiral.

Carbon 6 has two hydrogens attached to it, and so it is *not* chiral.

Glucose, therefore, has four chiral carbons.

Number of isomers possible (including glucose) = $2^4 = 2 \times 2 \times 2 \times 2 = \mathbf{16}$

Half of these isomers are enantiomers (mirror images) and are classified as D- or L-. So, of the 16 possible aldohexose isomers, eight are D-isomers and eight are L-isomers. For the structures of all eight D-isomers, see Figure 7.3 in your text for Fischer projections and the end of this Study Guide chapter for the Haworth structures.

---

[1] Organic chemists would differentiate between the two possible arrangements of groups by labeling each chiral carbon atom as *R* or *S*.

Enantiomers

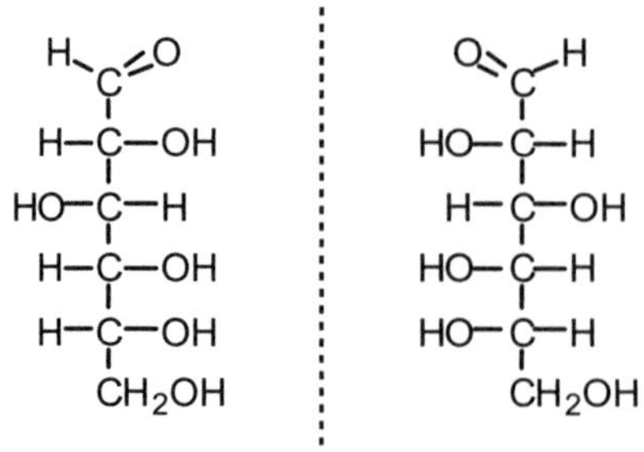

D-Glucose L-Glucose

The remaining isomers are diastereomers. Diastereomers are stereoisomers that are not mirror images of each other.

LET'S LOOK AT FRUCTOSE (A KETOSE) AS ANOTHER EXAMPLE:

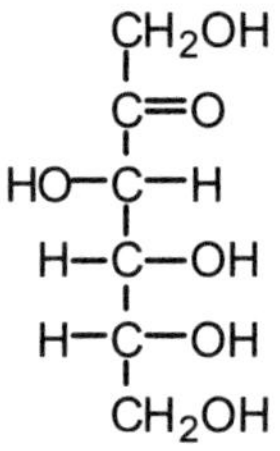

The first carbon has two hydrogens attached to it and is therefore *not* chiral.

The second carbon has only three groups attached to it and is therefore *not* chiral.

The third, fourth, and fifth carbons *are* chiral.

The sixth carbon has two hydrogens attached to it and is therefore *not* chiral.

Thus, fructose has three chiral carbons, so it has $2^3 = 2 \times 2 \times 2 = 8$ possible isomers (4 D- and 4 L-isomers)

D- VS. L-ENANTIOMERS are determined by comparing the Fischer projections to D- and L-glyceraldehyde. Compare the chiral carbons furthest from the carbonyl carbons. Most biologically active sugars are D-sugars.

ENANTIOMERS vs. DIASTEREOMERS vs. EPIMERS

Epimers are diastereomers that differ in configuration at only one chiral center.

Example: Of the following three monosaccharides, which are epimers of each other? D-glucose, D-allose, D-altrose

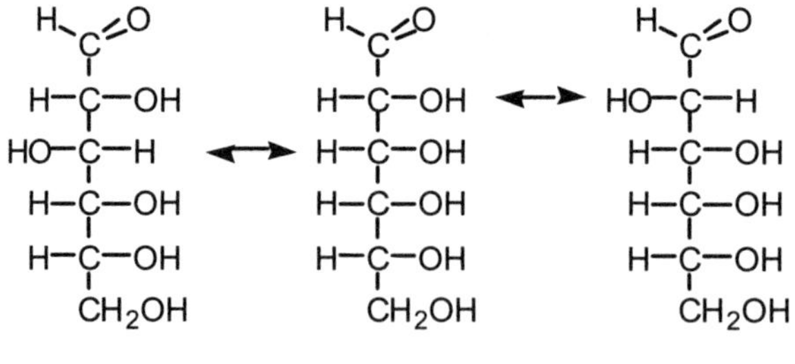

D-Glucose D-Allose D-Altrose

D-Glucose and D-allose are epimers of each other because they differ by only one chiral carbon (at carbon 3). D-allose and D-altrose are also epimers of each other since they differ only at carbon 2. However, D-glucose and D-altrose are diastereomers and *not* epimers of each other because they are different at *both* carbons 2 and 3.

CYCLIC STRUCTURE OF MONOSACCHARIDES

Reversible reaction between an –OH and the aldehyde or ketone on the same molecule; formation of cyclic hemiacetals or hemiketals

Example: C-5 of glucose bonds with C-1, the aldehyde carbon. Approach can be from one side of the aldehyde or the other, so C-1 could either be "up" or "down." C-1 is the anomeric carbon atom.

MUTAROTATION—When α- and β-monosaccharides dissolve in water, they interconvert, forming an equilibrium mixture of open-chain, α- and β- pyranose and furanose forms. The more stable anomer is the most abundant in solution. Note that only the open-chain form can undergo redox reactions.

ANOMERS OF D-SUGARS*: α VS. β

β-D-Glucose

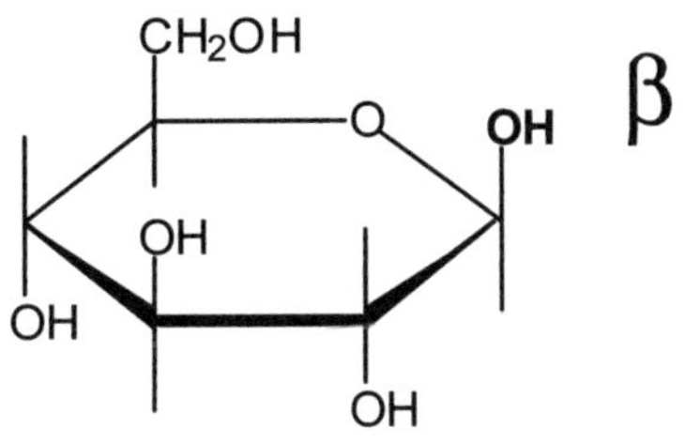

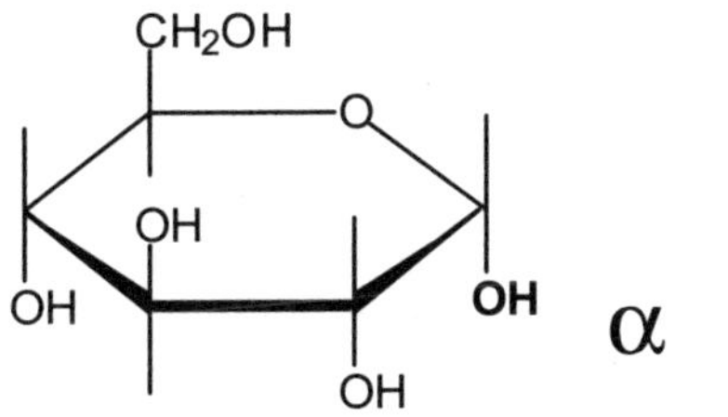

α-D-Glucose

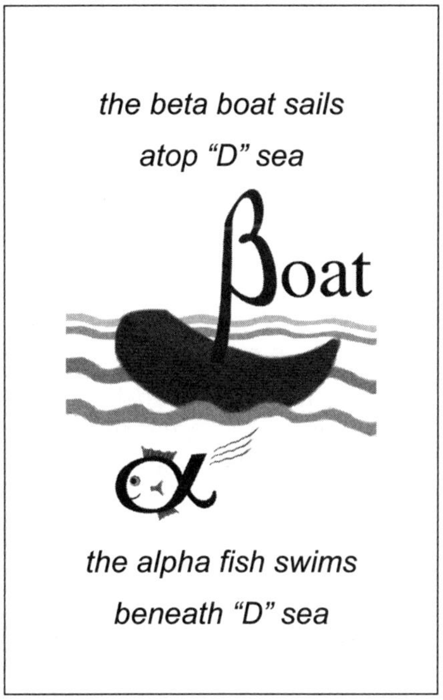
the beta boat sails
atop "D" sea
βoat
the alpha fish swims
beneath "D" sea

*L-sugars are opposite: α is "up" and β is "down"

HAWORTH STRUCTURES OF PYRANOSE AND FURANOSE (CYCLIC) FORMS; DRAWING HAWORTH STRUCTURES FROM FISCHER PROJECTIONS

Haworth structures show the cyclic (pyranose or furanose) forms of a sugar. Note that for the sake of simplicity, biochemists typically don't include H labels in Haworth projections. This convention lets us focus on the location of the OH groups with less clutter.

1. Draw the 5- or 6-membered ring with the oxygen atom in its correct position.
2. Number the carbons in both the Fischer projection and the Haworth structure.
3. First, draw in the atoms attached to the anomeric carbon (that is the first carbon in aldoses and the second carbon in ketoses). For D-sugars, α is "down" and β is "up."*
4. The last carbon is always "up" in D-sugars. Draw that one next.

5. Imagine the Fischer projection "falling over" to the right. The OH groups that were on the left in the Fischer projection will be "up" in the Haworth structure. Remember that the OH on the carbon farthest from the C=O reacted to form the ring, so that carbon will not have an OH attached to it. In other words, the oxygen within the ring had been a hydroxyl oxygen, and the anomeric OH group had been the carbonyl oxygen.
6. Double-check your work by comparing the numbered carbons on both structures.
7. **PRACTICE**. Using the Fischer projections in Figure 7.3 on p. 242, practice drawing Haworth projections of the aldohexoses. The correct Haworth projections for all of the aldohexoses are at the end of this Study Guide chapter (just before the Solutions to Review Questions).

CONFORMATIONAL (CHAIR) STRUCTURES: THE ORGANIC CHEMISTRY CONNECTION

In organic, you learned that the most stable conformation of a six-membered ring is the chair conformation, and the equatorial positions are more stable than the axial positions. It should come as no surprise, then, that the monosaccharide with all of its substituents in the equatorial positions is β-D-glucose, the most common monosaccharide for both energy storage and as a building block for structural materials, namely, cellulose.

β-D-glucose

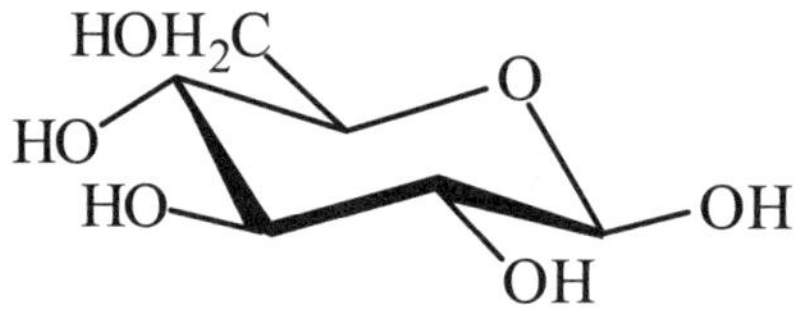

In this structure, the axial positions were omitted for clarity.

Remember that the axial positions were always "up" or "down" and the equatorial positions were labeled relative to the axial positions. In D-hexoses, carbon 6 is always drawn "up" in the Haworth projection. So, to draw the Haworth projection of β-D-glucose, first draw the 6-membered ring skeleton with the oxygen in the proper position. Then, add carbon 6 with its OH. The rest of the OH groups can then be drawn in following the up-down-up-down-up pattern of the substituents in their equatorial positions. (Compare this chair with the Haworth projections.)

REACTIONS OF MONOSACCHARIDES

OXIDATION OF ALDOSES TO:

ALDONIC ACID: aldehyde to carboxylic acid

URONIC ACID: terminal –CH_2OH to carboxylic acid

ALDARIC ACID: BOTH aldehyde and terminal –CH_2OH to carboxylic acids
(Note: To help remember this: "I'll dare Rick to oxidize BOTH ends!")

LACTONES can be formed from aldonic and uronic acids.

REDUCING SUGAR—can be oxidized by Benedict's reagent (a weak oxidizing agent) only if it can exist in its open-chain (aldehyde) form.

REDUCTION of aldehyde and ketone groups to alcohols = **ALDITOLS**

ISOMERIZATION via an enediol intermediate
Aldose-ketose interconversion
Epimerization

ESTERIFICATION forms of phosphate and sulfate esters

Phosphate esters are important in metabolism; they convert –OH, a terrible leaving group, into a good leaving group for nucleophilic substitution

Sulfate esters are in proteoglycan components of connective tissue; they bind large amounts of water and small ions due to the many negative charges on sulfates and form sulfate bridges between carbohydrate chains

GLYCOSIDE FORMATION: cyclic hemiacetal to acetal; cyclic hemiketal to ketal

Glycosidic linkage, glycoside (fructoside vs. glucoside, furanoside vs. pyranoside)

The glycosidic linkage locks the ring in its cyclic form, so it cannot oxidize or mutarotate. (So, β-methyl glucoside is not a reducing sugar.)

AGLYCONES are noncarbohydrate components of glycosides

Formation of **DISACCHARIDES, POLYSACCHARIDES**

IMPORTANT MONOSACCHARIDES

GLUCOSE—primary fuel for living cells, preferred energy source of brain cells and cells with few or no mitochondria, building block for cellulose, starch, glycogen

FRUCTOSE—a ketohexose isomer of glucose; fruit sugar; used as a sweetening agent (twice as sweet as sucrose)

GALACTOSE—an epimer of glucose; needed to synthesize lactose, glycolipids, certain phospholipids, proteoglycans, glycoproteins; can be made from glucose

In galactosemia, an enzyme needed to metabolize galactose is missing.

MONOSACCHARIDE DERIVATIVES

URONIC ACIDS (terminal $-CH_2OH$ group is oxidized to a carboxylic acid)

D-glucuronic acid—combines with molecules to improve their water solubility (and help to remove waste products from the body)

L-iduronic acid = epimer of D-glucuronic acid; both are in components of connective tissues

AMINO SUGARS: AMINO GROUP REPLACES AN –OH GROUP (USUALLY ON CARBON 2)

Common in complex carbohydrate molecules that are attached to cellular proteins and lipids

Most common amino sugars (in animals): D-glucosamine, D-galactosamine

Acetylated amino sugars: *N*-acetylglucosamine, *N*-acetylneuraminic acid, sialic acids

DEOXYSUGARS: –H REPLACES AN –OH

Fucose—formed from D-mannose; found in glycoproteins (example: ABO blood group determinates on the surface of red blood cells)

2-Deoxy-D-ribose is the pentose sugar component in DNA.

7.2 DISACCHARIDES

DESIGNATION OF GLYCOSIDIC LINKAGES

Identify anomeric hydroxyl group (α or β) and the carbons that are linked. Example: α(1,4) means that carbon 1 in the α position of one monosaccharide is linked to carbon 4 of another monosaccharide.

DIGESTION

Digestion occurs in the small intestine. If enzymes are deficient, then in the large intestine, they draw in water (causing diarrhea) and/or are fermented by bacteria, producing gas (causing bloating and cramps). Example: lactose intolerance caused by reduced lactase synthesis after childhood

REDUCING SUGARS

If one of the rings can convert to its open-chain form to regenerate the aldehyde, then the sugar is a reducing sugar. Lactose, maltose, and cellobiose are reducing sugars.

LACTOSE: galactose β(1,4) glucose, found in milk[2]

MALTOSE: glucose α(1,4) glucose, degradation product of starch, malt sugar

CELLOBIOSE: glucose β(1,4) glucose, degradation product of cellulose

SUCROSE: glucose α,β(1,2) fructose, energy source produced in plants. Since the glycosidic bond links both anomeric carbons, sucrose is a nonreducing sugar.

7.3 POLYSACCHARIDES: ENERGY STORAGE OR STRUCTURAL MATERIALS

Hundreds to thousands of sugar units; linear (unbranched) or branched

HOMOGLYCANS—MADE FROM ONLY ONE TYPE OF MONOSACCHARIDE

STARCH and **GLYCOGEN**: compact structures are ideal for energy storage.

STARCH: CARBOHYDRATE STORAGE IN PLANTS

AMYLOSE forms long, tight left-handed helices, contains several thousand glucose residues, and contains only one reducing end.

AMYLOPECTIN contains a few thousand to 106 glucose residues, and its branches prevent helix formation

Iodine test: iodine inserts into helices of amylose and results in blue color

Digestion of starch (initial products are maltose, maltotriose, α-limit dextrins)

1. Mouth: salivary enzyme α-amylase
2. Small intestine: pancreatic α-amylase, other enzymes to convert to glucose
3. Glucose: from small intestine to bloodstream, to liver, then to rest of body

GLYCOGEN: CARBOHYDRATE STORAGE IN VERTEBRATES

—More branches than amylopectin: as many as 1 every 4 at the core of the molecule and 1 every 8-12 in the outer regions

—The ends of these many branches are nonreducing—since enzymes work on nonreducing ends, this structure provides great access to enzymes so that glucose can be rapidly mobilized from glycogen.

CELLULOSE: THE MOST ABUNDANT ORGANIC SUBSTANCE ON EARTH

Great strength comes from hydrogen bonding between extended linear cellulose chains.

---

[2] Lactose intolerance is rather normal among adults in the general population. Adults who drink a lot of milk (especially in the United States) are the exception.

Microfibrils = parallel pairs of cellulose molecules (with up to 12,000 glucose residues) held together by hydrogen bonding; bundles of microfibrils contain about 40 pairs

Digestion: only by microorganisms that have cellulase; animals such as termites and cows have these microorganisms in their digestive tracts; if not digested, it is still important as dietary fiber

CHITIN: *N*-ACETYL GLUCOSAMINE

Strong hydrogen bonds between chitin chains; chitin forms microfibrils

Types of chitin structures:

- α-chitin: antiparallel bundles, most stable and common
- β-chitin: parallel bundles
- γ-chitin: mixture of parallel and antiparallel
- β- and γ- are more flexible than α-chitin

SUMMARY OF HOMOGLYCANS

HOMOGLYCAN (OVERALL SHAPE)		SUGAR UNITS, GLYCOSIDIC BONDS	FUNCTION	FOUND IN
GLYCOGEN (helical, branched)		α-D-glucose α(1,4) with α(1,6) branches	Energy storage	Vertebrates (mostly in liver and muscle cells)
STARCH	**AMYLOSE** (left-handed helices)	α-D-glucose α(1,4), unbranched	Energy storage	Plants
	AMYLOPECTIN (branched every 20-25 residues)	α-D-glucose α(1,4) with α(1,6) branches		
CELLULOSE (extended linear chains with hydrogen bonding between chains)		β-D-glucose β(1,4), unbranched	Structural, forms microfibrils and bundles	Plants (both primary and secondary cell walls)
CHITIN		*N*-acetyl glucosamine β-(1,4), unbranched	Structural	Arthropod exoskeletons, fungi cell walls

HETEROGLYCANS – MADE FROM TWO OR MORE TYPES OF MONOSACCHARIDES

MAJOR CLASSES:

N-LINKED: attached to polypeptides by an *N*-glycosidic bond with the side chain amide group of Asn
3 types of Asn-linked oligosaccharides: high mannose, hybrid, complex

O-LINKED: attached to polypeptides by the –OH of Ser or Thr or the –OH of membrane lipids

GLYCOSAMINOGLYCANS (GAGS)—PRINCIPAL COMPONENTS OF PROTEOGLYCANS

Linear; composed of disaccharide repeat units that contain a hexuronic acid (except keratan sulfate—contains galactose)

Classified according to sugar residues, glycosidic linkages, presence and location of sulfate groups

FIVE CLASSES: hyaluronic acid, chondroitin sulfate, dermatan sulfate, heparin and heparan sulfate, keratan sulfate

GAG chains have *many* negative charges that repel each other and attract large volumes of water, making its final volume about 1000 times larger!

MUREIN (PEPTIDOGLYCAN)

Major structural component of bacterial cell walls, supplies strength and rigidity

Contains *N*-acetyl glucosamine (NAG), *N*-acetyl muramic acid (NAM), and several different amino acids

Three basic components of murein (peptidoglycan)

1. backbone: NAG-NAM disaccharide repeat units linked by β(1,4) glycosidic bonds
2. parallel tetrapeptide chains; each chain attached to *N*-acetyl muramic acid
3. peptide cross-bridges link the tetrapeptide chains of adjacent molecules

7.4 GLYCOCONJUGATES: COVALENT LINKAGES OF CARBOHYDRATES TO PROTEINS AND LIPIDS

GLYCOLIPIDS = Oligosaccharide-containing lipids located on the outer surface of plasma membranes (see Chapter 11).

PROTEOGLYCANS = GAG CHAINS LINKED TO CORE PROTEINS (see Figure 7.37 on p. 260)

N- and O-glycosidic linkages, *very* high carbohydrate content

Located in extracellular matrix (intercellular material) of tissues

Large numbers of GAGs trap large volumes of water and cations

Contribute support and elasticity to tissues

Example: cartilage—strength, flexibility, resilience; part of meshwork (with matrix proteins such as collagen, fibronectin, laminin) that supports multicellular tissues

Genetic diseases associated with proteoglycan metabolism: mucopolysaccharidoses (example: Hurler's syndrome—dermatan sulfate accumulates)

GLYCOPROTEINS = PROTEINS COVALENTLY LINKED TO CARBOHYDRATE

N-glycosidic linkage to asparagine or *O*-glycosidic linkage to serine or threonine

Carbohydrate content varies (1% to more than 85% by weight)

ASPARAGINE-LINKED CARBOHYDRATE

N-glycosidic linkage between *N*-acetylglucosamine (GlcNAc) and asparagine

Core is constructed on a membrane-bound lipid molecule

HIGH-MANNOSE TYPE	GlcNAc and mannose
COMPLEX TYPE	GlcNAc and mannose (may contain fucose, galactose, sialic acid)
HYBRID TYPE	features of both high-mannose and complex types

MUCIN-TYPE CARBOHYDRATE

O-glycosidic linkage, most common is between *N*-acetylgalactosamine (GalNAc) and the hydroxyl group of Ser or Thr

Carbohydrate components vary in size and structure (examples: Gal-β(1,3)-GalNAc, disaccharide found in antifreeze glycoprotein of Antarctic fish; complex oligosaccharides of blood groups such as the ABO system).

GLYCOPROTEIN FUNCTIONS AND THE ROLE OF THE CARBOHYDRATE COMPONENT

EXAMPLES OF GLYCOPROTEIN FUNCTIONS:

- Metal-transport proteins transferrin and ceruloplasmin
- Blood-clotting factors
- Complement (proteins involved in cell destruction during immune reactions)
- Hormones (example: follicle-stimulating hormone (FSH)—stimulates development of eggs and sperm)
- Enzymes (example: ribonuclease (RNase) degrades ribonucleic acid)
- Integral membrane proteins {examples: Na^+-K^+-ATPase—ion pump in the plasma membrane of animal cells; major histocompatibility antigens (cell surface markers used to cross-match organ donors and recipients)}

GLYCAN COMPONENT STABILIZES PROTEIN MOLECULES:

Protects from denaturation (example: bovine RNase A vs. RNase B)

Increases resistance to proteolysis (carbohydrates on the protein's surface may shield the peptide chains from proteolytic enzymes)

AFFECTS BIOLOGICAL FUNCTION

Saliva: high viscosity of salivary mucins is due to high content of sialic acid residues

Antifreeze glycoproteins in Antarctic fish retard the growth of ice crystals by hydrogen bonding with water molecules

7.5 THE SUGAR CODE

CONTRIBUTIONS OF CARBOHYDRATES TO INFORMATION TRANSFER:

- posttranslational modification of proteins via glycosylation = after a protein is synthesized (translation), its structure is modified with the addition of carbohydrates
- oligosaccharides have many more structural possibilities than peptides in that oligosaccharides may be branched, may be linked via an α- or β-glycosidic bond, or may be linked at different positions [e.g., a (1 → 3) vs. (1 → 4)]
- relatively inflexible chains allow for more specificity in ligand binding

Lectins: Translators of the Sugar Code

LECTINS are carbohydrate binding proteins (CBPs) that:

—are not antibodies and have no enzymatic activity
—consist of two or four subunits
—possess recognition domains that bind to specific carbohydrate groups via hydrogen bonds, van der Waals forces, and hydrophobic interactions

Examples of biological processes—cell-cell interactions—that involve lectin binding:

INFECTIONS BY MICROORGANISMS
Bacterial lectins attach the bacteria cell to the host cell by binding to oligosaccharides on the cell's surface (example: gastritis and stomach ulcers)

MECHANISMS OF MANY TOXINS
Lectin-ligand binding initiates endocytosis into the host cell (example: cholera)

PHYSIOLOGICAL PROCESSES SUCH AS LEUKOCYTE ROLLING
SELECTINS are a family of lectins that act as cell adhesion molecules that slow the movement of neutrophils through blood vessels
INTEGRIN is a lectin that cause neutrophils to stop rolling (so they can migrate to the location where they are needed)

The Glycome = the total set of sugars and glycans that a cell or organism produces

Unlike proteins, there is no template for glycan synthesis—they are produced step-by-step on an assembly line in the ER and Golgi complex.

MICROHETEROGENICITY is the production of **GLYCOFORMS**—slightly different forms of glycan components of glycoproteins—possibly for cell- or tissue-specific purposes

Complex Recognition Phenomena

CELL-MOLECULE INTERACTIONS
Insulin receptor—binds to insulin and facilitates transport of glucose into cells, in part by recruiting glucose transporters to the plasma membrane

Glucose transporter—transports glucose into cells

CELL-VIRUS INTERACTIONS
gp120 (target cell binding glycoprotein of HIV) attaches to CD4 receptor on surface of several human cell types (removing oligosaccharides from gp120 reduces its binding to CD4 receptor)

CELL-CELL INTERACTIONS
Cell structure glycoproteins—components of glycocalyx (cell coat), important in cellular adhesion; **CELL ADHESION MOLECULES (CAMs)**

<u>HAWORTH PROJECTIONS FOR α-D-ALDOHEXOSES *(PYRANOSE FORMS)*</u>

α-D-Allose 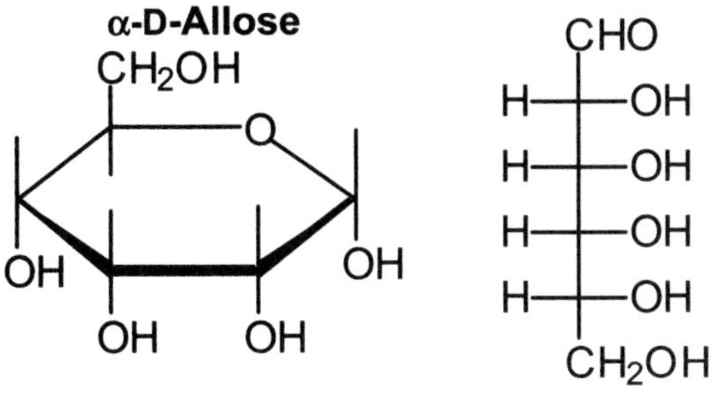	**α-D- Gulose** 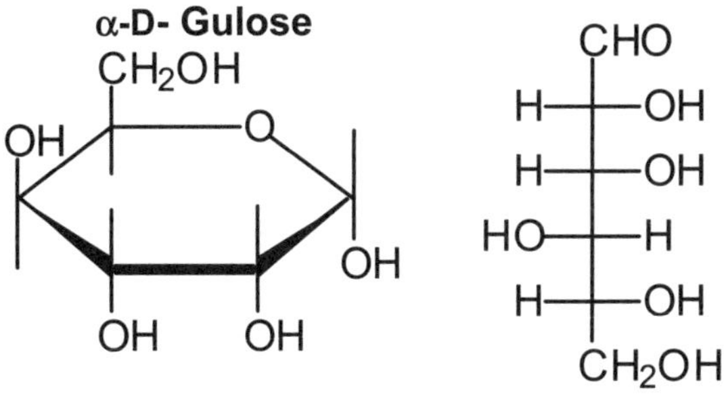
α-D-Altrose 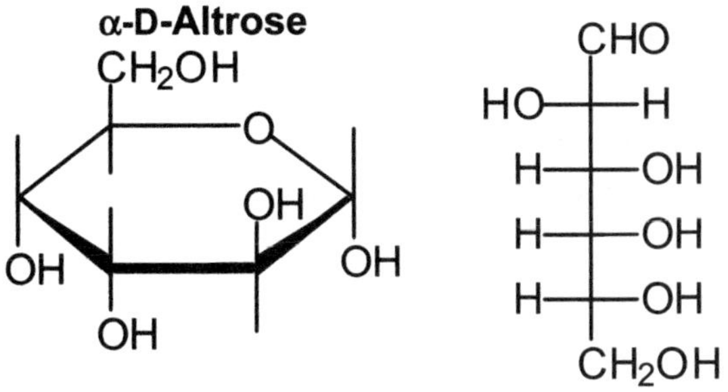	**α-D-Idose** 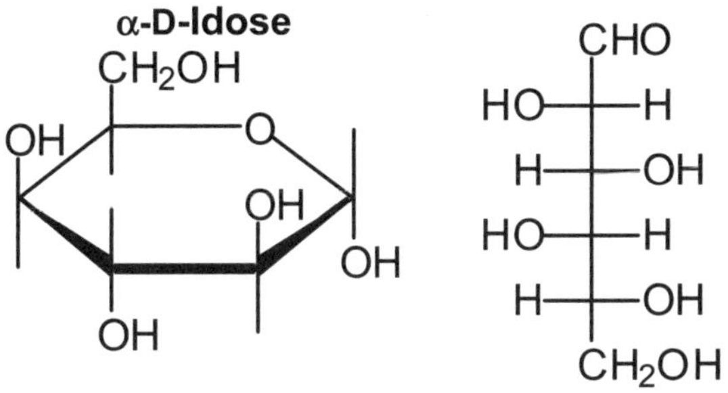
α-D-Glucose 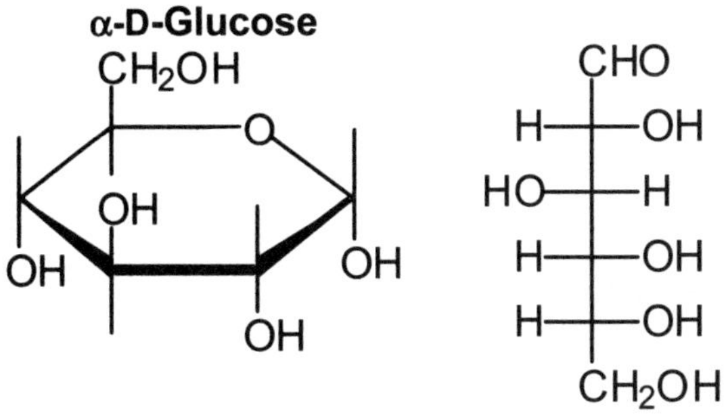	**α-D-Galactose** 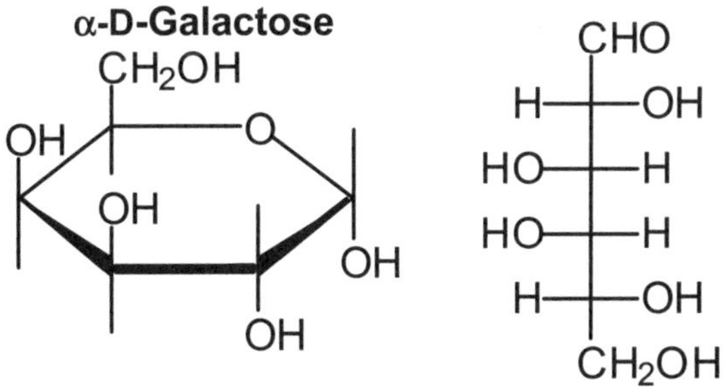
α-D-Mannose 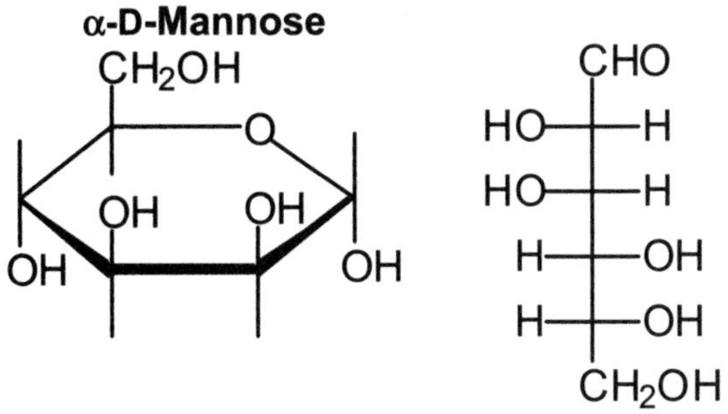	**α-D-Talose** 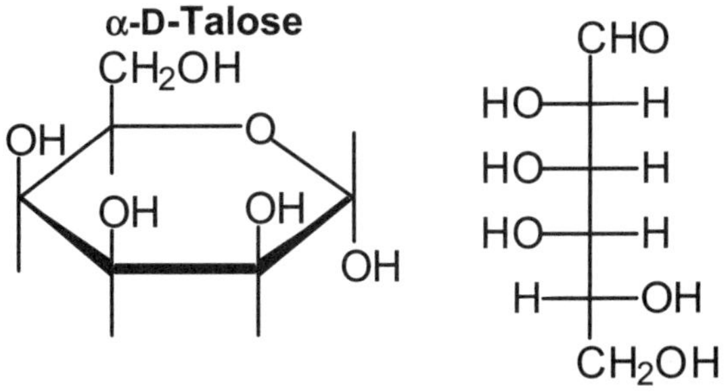

AFTER STUDYING THIS CHAPTER, YOU SHOULD BE ABLE TO:

- Convert a Fischer projection to a Haworth structure or draw a Haworth structure given only the name of a compound.
- Identify pairs of compounds as epimers, anomers, enantiomers, diastereomers, or an aldose-ketose pair. Identify an anomer as α or β.
- Distinguish between reducing and nonreducing sugars.
- Determine the number of possible stereoisomers of a given sugar.
- Name a sugar given its structure. Include α or β and appropriate designations for any glycosidic linkages present.
- Draw the structure of a di-, tri-, or polysaccharide, given the names of the component mono- (or di-)saccharides and the type(s) of glycosidic linkages present.
- How do small differences in carbohydrate structure account for large differences in function? (For example, compare cellulose and glycogen.)
- Demonstrate an understanding of the role of glycans and glycosylation in information storage and transfer. Describe the advantages of sugars that make them well suited for encoding information.

Chapter 7: Solutions to Review Questions

7.1 a. monosaccharide – a polyhydroxy aldehyde or ketone containing at least three carbon atoms

b. aldose – a monosaccharide with an aldehyde functional group

c. ketose – a monosaccharide with a ketone functional group

d. diasterisomer – a stereoisomer that is not an enantiomer (mirror-image isomer)

e. epimer – a molecule that differs from the configuration of another by one asymmetric center.

7.2 a. hemiacetal – one of the family of organic molecules with the general formula

RCH(OR)OH that is formed by the reaction of one molecule of alcohol with an aldehyde

b. hemiketal – one of the family of organic molecules with the general formula RRC(OR)OH that is formed by the reaction of a molecule of alcohol with a ketone

c. anomer – one of two possible diasterisomers that may form during the cyclization reaction of a hemiacetal or hemiketal

d. furan – an organic molecule with a five-sided ring containing one oxygen atom; furanoses are named in reference to furan

e. pyran – an organic molecule with a six-sided ring containing one oxygen atom; pyranoses are named in reference to pyran

7.4 a. alditol – a sugar alcohol; the product of the reduction of the aldehyde or ketone group of a monosaccharide

b. enediol – the intermediate formed during the isomerization reactions of monosaccharides; contains a double bond with a hydroxyl group on each carbon of the double bond

c. epimerization – the reversible interconversion of epimers

d. acetal – one of a family of organic compounds with the general formula $RCH(OR')_2$; formed from the reaction of a hemiacetal with an alcohol

e. ketal – one of a family of organic compounds with the general formula $RRC(OR)_2$; formed from the reaction of a hemiketal with an alcohol

7.5 a. glycosidic link – an acetal linkage formed between two monosaccharides linked through at least one anomeric carbon of the monosaccharides

b. glycoside – the acetal form of a sugar

c. disaccharide – a glycoside composed of two monosaccharide residues

d. oligosaccharide – an intermediate-sized carbohydrate composed of 2 to 10 monosaccharides

e. polysaccharide – a linear or branched polymer of monosaccharides linked by glycosidic bonds

7.7 a. deoxysugar – a monosaccharide in which an H has replaced an OH group

b. galactosemia – a genetic disorder in which an enzyme required to metabolize galactose is missing; symptoms include liver damage, cataracts, and severe mental retardation

c. cellobiose – the disaccharide product of cellulose degradation; two molecules of glucose linked by β(1,4) glycosidic bond

d. glycan – a polymer of monosaccharides; a polysaccharide

e. chitin – the principal structural component of the exoskeletons of arthropods and the cell walls of many fungi; a homoglycan composed of N-acetylglucosamine residues

7.8 a. homoglycan – high-molecular-weight carbohydrate polymers that contain only one type of monosaccharide

b. heteroglycan – high-molecular-weight carbohydrate polymer that contains more than one kind of monosaccharide.

c. amylose – a type of plant starch; an unbranched polymer of D-glucose residues linked with α(1,4) glycosidic linkages

d. amlopectin – a type of plant starch; a branched polymer containing α(1,4) and α(1,6)-glycosidic linkages

e. enterocyte – cell walls that line the small intestine; absorbs digested nutrients such as monosaccharides

7.10 a. glycoconjugate – a molecule that possesses covalently bound carbohydrate components (e.g., glycoproteins and glycolipids)

b. glycolipid – a glycosphingolipid; a molecule in which a monosaccharide, disaccharide, or oligiosaccharide is attached to a ceramide through an O-glycosidic linkage

c. proteoglycan – a large molecule containing large numbers of glycosaminoglycan chains linked to a core protein molecule

d. glycoprotein – a conjugated protein in which carbohydrate molecules are covalently bound to amino acid side chains

e. sugar code – the vast increase in the information coding capacity of proteins with the covalent attachment of carbohydrate groups.

7.11 a. lectin – a carbohydrate binding protein

b. glycoform – one of several slightly different forms of a glycan component of a glycoprotein

c. microheterogeneity – variations in the glycan components of each type of glycoprotein

d. glycome – the total set of sugars an glycans that a cell or organism produces

e. glycomics – the investigation of the structural and functional properties of all the carbohydrate molecules produced by organisms

7.13 In D-family sugars, the OH on the chiral carbon farthest from the carbonyl group is on the right side in a Fischer projection formula. So both (+)-glucose and (−)-fructose are D-sugars despite their rotation of plane-polarized light in opposite directions.

7.14 a. Ribonuclease B is an example of a glycoprotein.

b. Each proteoglycan contains glycosaminoglycans such as chondroitin sulfate and dermatan sulfate which are linked to a core protein via glycosidic linkages.

c. Lactose is an example of a disaccharide.

d. Heparin is a glycosaminoglycans

7.16 a. Nonreducing, b. Nonreducing, c. Reducing, d. Nonreducing, e. Reducing

7.17 Starch and glycogen are both homoglycans containing glucose monomers linked by α-(1,4) glycosidic bonds with branch points connected by α-(1,6) glycosidic bonds. Glycogen, however, is much more highly branched than starch. Cellulose is a linear polymer of glucose linked by β-(1,4) glycosidic bonds.

7.19 a. Raffinose consists of α-D-galactose with an α(1,6) linkage to glucose, which has an α,β(1,2) linkage to fructose. Its systematic name is: galactopyranosyl-α-(1,6)-glucopyranosyl-α,β(1,2)-fructofuranose.

b. Raffinose is a nonreducing sugar.

c. Raffinose is not capable of Mutarotation

7.20 a. Glycogen stores glucose.

b. Glycosaminoglycans are components of proteoglycans.

c. Glycoconjugates may serve as membrane receptors.

d. Proteoglycans provide strength, support, and elasticity to tissue.

e. Hormones such as FSH and enzymes such as RNase are glycoproteins.

f. Polysaccharides, or glycans, play important roles in the storage of carbohydrate (starch and glycogen) and the structure of plants (cellulose).

7.22 In glycoproteins carbohydrate moieties are most frequently linked to the amide nitrogen of asparagine and the hydroxyl oxygen of serine and threonine residues.

7.23 Chondroitin sulfate and proteoglycans are extensively negatively charged at physiological pH and as such are spread out, binding large amounts of water. The interwoven chains block the passage of large molecules. Smaller molecules can pass between the chains.

7.25 Proteoglycans are extremely large molecules that contain a large number of glycosaminoglycan chains linked to a core protein. They are found primarily in extracellular fluids where their high carbohydrate content allows them to bind large amounts of water. Glycoproteins are conjugated proteins in which the prosthetic groups are carbohydrate molecules. The carbohydrate groups stabilize the molecule through hydrogen bonding, protecting the molecule from denaturation, or shielding the protein from hydrolysis. The carbohydrate groups on the glycoproteins on the surface of cells play an important role in a variety of recognition phenomena.

7.26 Numerous carbohydrate groups protect glycoproteins from denaturation because they are hydrophilic groups that surround the protein and protect against protease-catalyzed peptide bond cleavage.

CHAPTER 7: SOLUTIONS TO FILL-IN-THE-BLANK QUESTIONS

7.28 Epimers

7.29 Aldoses

7.31 Diastereoisomers

7.32 Cellulose

7.34 Schiff

7.35 Epimers

7.37 Heteroglycans

CHAPTER 7: SOLUTIONS TO SHORT-ANSWER QUESTIONS

7.38 All simple aldoses have a chiral center but dihydroxyacetone, the simplest ketose, is symmetrical and lacks a chiral center.

7.40 Because the test for reducing sugars requires an aldehyde group, all aldoses are reducing sugars because they contain the required aldehyde. Ketoses give a positive test because they are converted to aldoses in the alkaline-reducing sugar test solution.

7.41 Proteoglycans contain large numbers of negatively charged GAG chains that allow the absorption of exceptionally large amounts of incompressible water molecules that impart resilience to cartilage. The rope-like properties of collagen molecules are responsible in large part for the tensile strength of cartilage.

CHAPTER 7: SOLUTIONS TO THOUGHT QUESTIONS

7.43

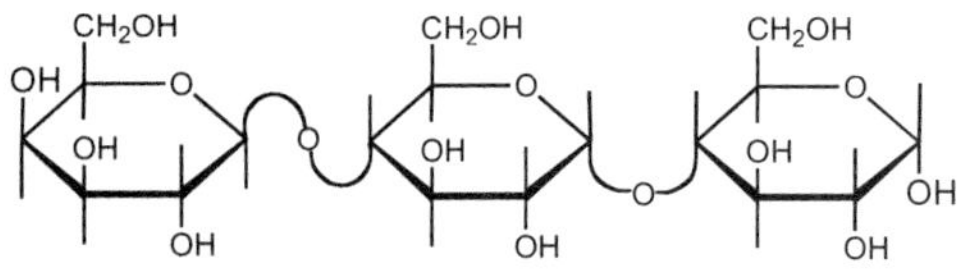

7.44 When steroids are conjugated with a uronic acid, the OH groups of the uronic acid form hydrogen bonds with the water. This structural feature increases the solubility of the conjugated steroid molecule.

7.46 Pathogenic organisms bind to the milk oligosaccharides instead of the oligosaccharides on the surface of the infant's intestinal cells, thus preventing infections.

7.47 Seeds are the dispersal units of seed plants. Plants produce fruit that entice animals (the dispersal mechanism) to eat it by synthesizing sugars such as fructose. Grain-producing plants use wind dispersal as a means of spreading seeds. Consequently, carbohydrate to be used as an energy source during germination is in the form of starch.

7.49 A bitter taste is sufficiently unpleasant that the animal will immediately make the connection between the type of plant and the unpleasant sensation. As a result the animal will stop eating and then avoid the offending plant in the future. If the toxin's taste is neutral or sweet sufficient time may elapse between consuming the toxin and the onset of its effects. The animal may continue to eat the plant, therby defeating the toxins purpose.

7.50 The water that is absorbed in large quantities by proteoglycans is incompressible. Therefore, tissues that contain proteoglycans in large amounts receive some protection against mechanical stress, i.e., the tissue resists deformation when pressure is applied.

7.52 The properties of amylose and cellulose differ because of the structure of the glysosidic linkage in each molecule. The α(1,4) linkage of amylose can form helices tht are water soluble because of extensive hydrogen bonding between water and the hydroxyl groups of glucose residues. In contrast, the β(1,4) linkage of cellulose causes it to form an extended, stiff and rod-like structure. In this conformation the hydroxyl groups of the glucose residues form hydrogen bonds with parallel chains, thus holding them together so as to form water insoluble microfibrils.

7.53 Mannuronic acid has four chiral centers, so it has a maximum number of isomers equal to 2^4 or 16.

7.55 In an individual with an AB blood type, one half of the antigens would be linked with galactose and the other half with N-acetyl-galactoseamine.

7.56 Phosphate esters can form at positions 2-6 of an aldohexose because all these carbons bear alcoholic OH groups. In contrast, a phosphate at the anomeric carbon would be a mixed anhydride.

7.58 The information provided reveals that the fructose molecule has a five-membered ring since the hydroxyl at carbon six is methylated.

7.59 The two diasterisomeric products of the reduction of fructose are:

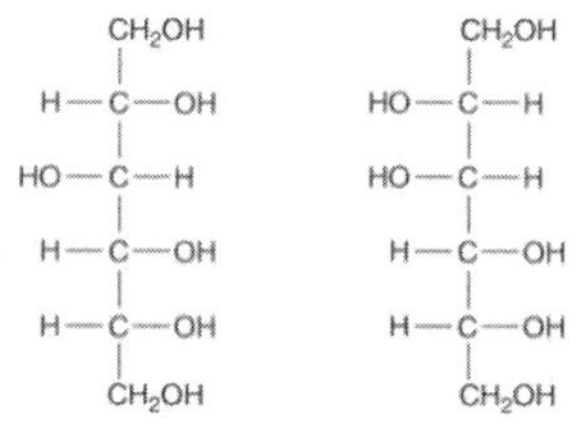

7.61 In order for the sugar to undergo mutarotation there must be a hemiacetal or hemiketal as part of the structure. Sucrose's anomeric carbons are linked in a full acetal.

7.62 The structure of the oligosaccharide is as follows:

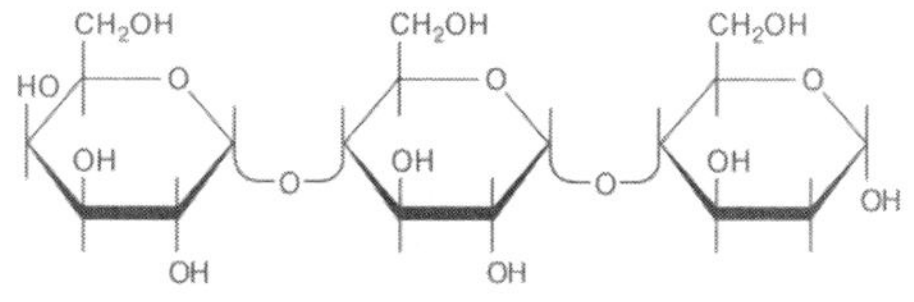

7.64 Three sugars are possible. Talose is the sugar produced by the epimerization of galactose at C2. The C4 epimer is glucose, and the C3 epimer is galactose. (Refer to Figure 7.3 to view the structures.)

7.65 A free sugar has many OH groups that hydrogen bond and raise its boiling point above its decomposition temperature. Conversion of the sugar to a volatile derivative lowers its boiling point below its decomposition point and analysis by GC or GC/MS can be accomplished easily.

7.67 Moisture loss is prevented because of hydrogen bonding between the OH groups of sorbitol and water.

7.68 If glyceraldehyde forms a four-membered ring, the bond angles create significant strain. As a result the four membered ring does not form. The larger five-membered ring formed by the tetroses is relatively strain free and forms easily.

7.70 The test for a reducing sugar requires an alkaline medium. Under these conditions, fructose converts to an aldose and gives a positive test.

8 Carbohydrate Metabolism

Brief Outline of Key Terms and Concepts

8.1 GLYCOLYSIS

aerobic respiration
aerobic vs. anaerobic organisms

THE REACTIONS OF THE GLYCOLYTIC PATHWAY

convert glucose to two pyruvate molecules. A small amount of energy is captured in two ATP and two NADH.

GLUCOSE SENSOR
ALDOL CLEAVAGE
PHOSPHOANHYDRIDE BOND
SUBSTRATE-LEVEL PHOSPHORYLATION
TAUTOMERS; TAUTOMERIZATION

THE FATES OF PYRUVATE

In the presence of oxygen the cells of aerobic organisms convert pyruvate into CO_2 and H_2O

CITRIC ACID CYCLE
AMPHIBOLIC PATHWAY
ELECTRON TRANSPORT SYSTEM
DECARBOXYLATION
FERMENTATION
LACTIC ACID FERMENTATION

THE ENERGETICS OF GLYCOLYSIS

Enzymes of irreversible reactions:
hexokinase, PFK-1, pyruvate kinase

REGULATION OF GLYCOLYSIS

Allosteric effectors:
AMP, ATP, citrate, acetyl-CoA,
fructose-2,6-bisphosphate
fructose-1,6-bisphosphate
Hormones: glucagon, insulin
cAMP—a second messenger

8.2 GLUCONEOGENESIS is the synthesis of new glucose molecules from noncarbohydrate precursors.

– Occurs primarily in the liver.

GLUCONEOGENESIS REACTIONS are the reverse of glycolysis except for three reactions that bypass irreversible glycolysis reactions.

BIOTIN   MALATE SHUTTLE
SUBSTRATE CYCLE FLUX CONTROL

GLUCONEOGENESIS SUBSTRATES

CORI CYCLE   GLUCOSE-ALANINE CYCLE
GLUCOGENIC AMINO ACIDS

GLUCONEOGENESIS REGULATION

Substrate availability, allosteric effectors, hormones
Allosteric effectors: ATP, AMP, fructose-2, 6-bisphosphate, acetyl-CoA
Hormones: insulin, glucagon, cortisol

8.3 THE PENTOSE PHOSPHATE PATHWAY

produces NADPH, ribose-5-phosphate, fructose-6-phosphate and glyceraldehyde-3-phosphate.

Hexose monophosphate shunt
Antioxidant
TPP (thiamine pyrophosphate)

8.4 METABOLISM OF OTHER IMPORTANT SUGARS

FRUCTOSE METABOLISM

bypasses two regulatory enzymes:
HEXOKINASE AND PFK-1

8.5 GLYCOGEN METABOLISM

GLYCOGENESIS

GLYCOGEN SYNTHASE
GLYCOGEN BRANCHING ENZYME
GLYCOGENIN

GLYCOGENOLYSIS requires glycogen phosphorylase and debranching enzyme.
Limit dextrin

REGULATION OF GLYCOGEN METABOLISM

Several allosteric regulators; hormones:
GLUCAGON, INSULIN, and EPINEPHRINE
covalent modification:
forms of glycogen phosphorylase:
phosphorylase a—active form
phosphorylase b—inactive form

TERMS FROM IN-CHAPTER QUESTIONS:

glucose tolerance
malignant hyperthermia
von Gierke's disease
hepatomegaly
Pasteur effect
hypoglycemia
Cori's disease

Study Hints/Strategy

Studying metabolism brings together what you have learned about structures of biomolecules and how energy flows in living systems.

An important tip in studying metabolism is to key in on the ultimate goal of the pathway and how each reaction brings you closer to the desired end product.

Begin by visualizing the structure of the starting material and follow the changes that occur until you get to the final product. This simplifies remembering the sequence or order of reactions. Focus on the structure of the intermediates. At each step, note any changes that occur to the molecule to understand the kind of reaction that is occurring. Remember that each enzyme name clues the type of reaction it catalyzes. For example, a kinase transfers phosphoryl groups.

Know yourself and your learning styles, and tap into your strengths. If you learn best by reading, create lists to study from. Visual learners will do well to create diagrams of the pathways. Those who learn best by hearing will want to find a study partner to talk through the pathways. Kinesthetic learners learn by doing, and practicing writing out the pathways might be the best bet. Most of us are a combination of the four. Try each style of learning.—You might be surprised by which method works the best. Studying metabolism lends itself to all four learning styles, so it is a great subject to dive into. (It also gives meaning to everything we have learned so far!)

Example: Glycolysis cleaves glucose to form two 3-carbon molecules. In this pathway, a kinase adds a *second* phosphoryl group to fructose-6-phosphate. *Why*? After cleavage, both halves will have a phosphoryl group that traps each half in the cell. Later, those phosphates will be transferred to ADP to form ATP.

Pathway	***General function***	***Each pathway is activated when:***
Glycolysis	glucose → 2 pyruvate makes 2 ATP, 2 NADH	. . . energy is needed: either anaerobic (pyruvate forms lactate) or aerobic (pyruvate enters the citric acid cycle)
Gluconeogenesis	pyruvate → glucose requires 4 ATP, 2 GTP	. . . the liver needs to raise blood sugar levels and liver glycogen is depleted
Glycogenolysis	glycogen → glucose	. . . muscles need glucose for energy or the liver needs to raise blood sugar levels
Glycogenesis	glucose → glycogen requires 1 UTP/glucose	. . . glucose is in excess
Pentose phosphate pathway	glucose-6-P → ribose-5-P + other sugars makes 2 NADPH	. . . NADPH is needed for lipid synthesis . . . ribose is needed for nucleotide synthesis

[1] Visual, aural, read/write, and kinesthetic modes of learning are described on the VARK web page, http://vark-learn.com. VARK is a questionnaire that can be used to discover your learning preferences. If this link is outdated, try a general search on VARK at www.google.com. Copyright for VARK is held by Neil D. Fleming, Christchurch, New Zealand, and Charles C. Bonwell, Green Mountain, Colorado, USA.

8.1 GLYCOLYSIS: THE OLDEST AND SIMPLEST WAY TO PRODUCE ENERGY

- Glycolysis provides energy (2 ATP), intermediates (2 pyruvate), and reducing power (2 NADH).
- It splits one glucose (with six carbons) into two pyruvates (with three carbons each).
- It does not need O_2 (it's anaerobic).
- It is **AMPHIBOLIC**, i.e., it functions in both anabolic and catabolic processes.
- It is activated when the cell needs energy and inhibited when there is plenty of ATP.

THE OVERALL REACTION OF GLYCOLYSIS: **ΔG = –103.8 KJ/MOL**

D-Glucose + 2ADP + 2P_i + 2NAD^+ → 2Pyruvate + 2ATP + 2NADH + 2H^+ + 2H_2O

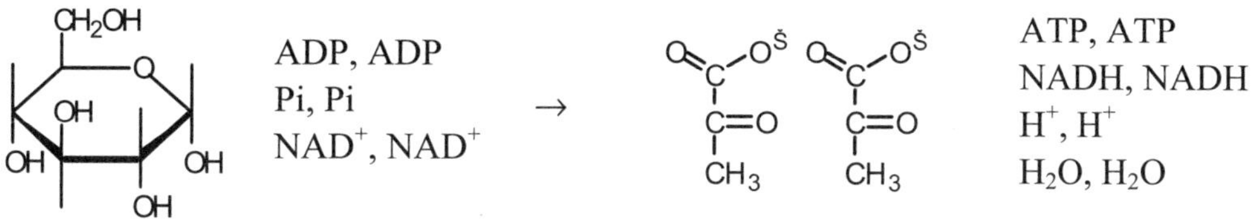

THE REACTIONS OF THE GLYCOLYTIC PATHWAY

STAGE ONE:

Glucose is phosphorylated and cleaved to form 2 glyceraldehyde-3-phosphate.

STAGE TWO:

Energy recapture stage (oxidation-reduction and phosphorylation)

(SEE THE FOLLOWING TWO PAGES FOR REACTION DETAILS)

THE FATES OF PYRUVATE

IF O_2 IS PRESENT:

pyruvate → acetyl-CoA → citric acid cycle (2CO_2, NADH, $FADH_2$)
NADH and $FADH_2$ → may transfer its electrons to form H_2O via the electron transport system (ETS), and the energy released is used to make ATP

IF O_2 IS ABSENT:

The ETS and the citric acid cycle stop. The cell must rely on glycolysis to get the energy (ATP) it needs. To continue ATP synthesis glycolysis needs a steady supply of NAD^+, which must be regenerated from NADH. Otherwise, the cell's NAD^+ supply would become depleted and glycolysis would stop.

IN MUSCLE CELLS AND SOME BACTERIA: HOMOLACTIC FERMENTATION

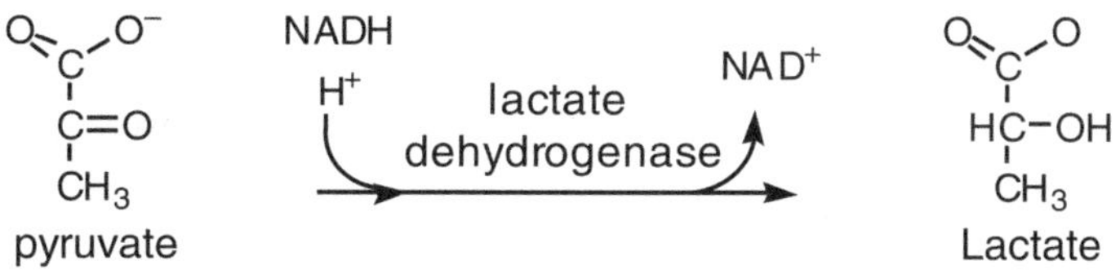

IN YEAST CELLS: ALCOHOLIC FERMENTATION

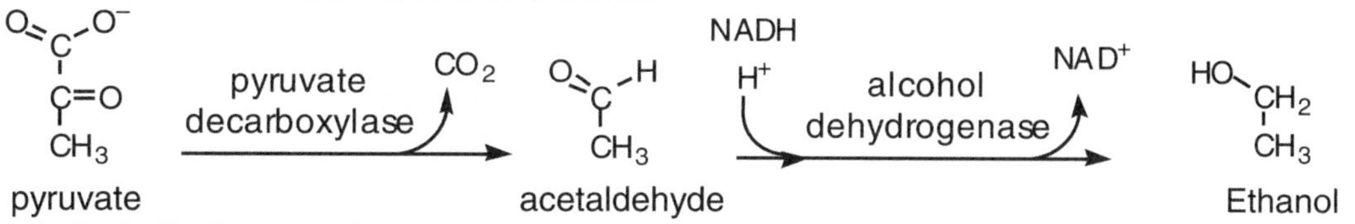

(Alcoholic fermentation is a decarboxylation followed by a reduction.)

GLYCOLYSIS

STAGE ONE: GLUCOSE IS PHOSPHORYLATED AND CLEAVED TO FORM TWO G-3-P

Glycolysis occurs only in the cytoplasm. The three irreversible reactions—hexokinase, PFK-1, and pyruvate kinase are highly regulated.

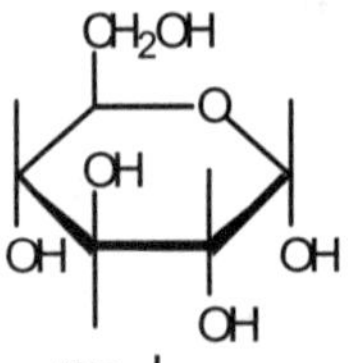

glucose

1. **Hexokinase**, Mg^{2+} *or* **Glucokinase**(GK) (primarily in liver cells)

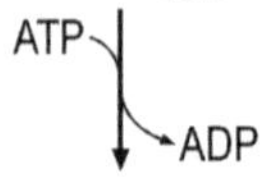

Now this molecule is charged so that it cannot pass through the cell membrane
Investment of one ATP

IRREVERSIBLE!

Inhibited by ATP and glucose-6-phosphate (except for glucokinase)

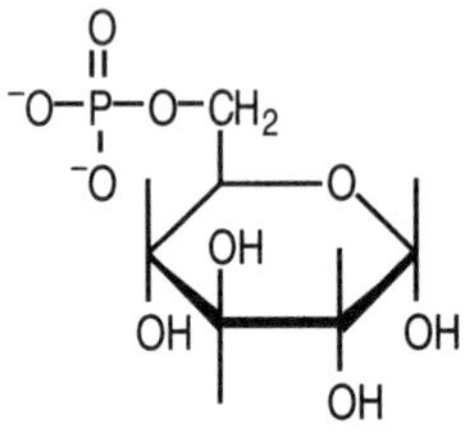

glucose-6-phosphate

2. **Phosphoglucose isomerase**

↓↑

Now C-1 can be phosphorylated
Enediol intermediate

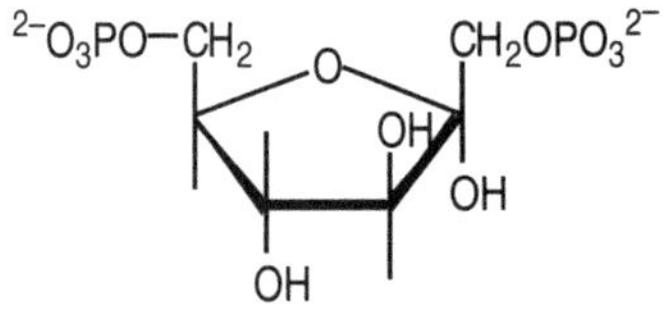

fructose-6-phosphate

3. **Phosphofructokinase-1**, Mg^{2+} (PFK-1)

Now both halves will have a phosphate after cleavage
Phosphates will make ATP later

IRREVERSIBLE!

Activated by fructose-2,6-bisphosphate
Inhibited by ATP and citrate

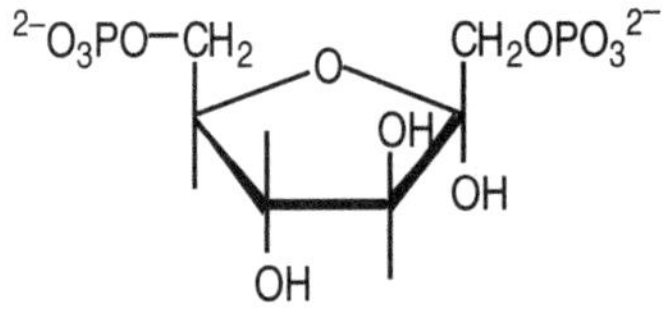

fructose-1,6-bisphosphate

4. **Aldolase**

↓↑

Aldol cleavage
Reverse aldol condensation

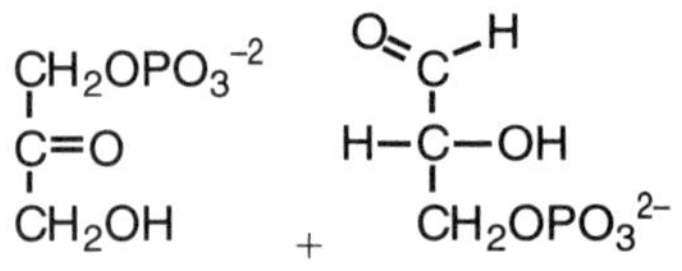

dihydroxyacetone phosphate (DHAP)

glyceraldehyde-3-phosphate

5. **Triose phosphate isomerase**

Now the DHAP half will not be wasted

GLYCOLYSIS

STAGE TWO: ENERGY RECAPTURE STAGE (OXIDATION-REDUCTION-PHOSPHORYLATION)

*From this point on, there are **two** of each molecule below.*

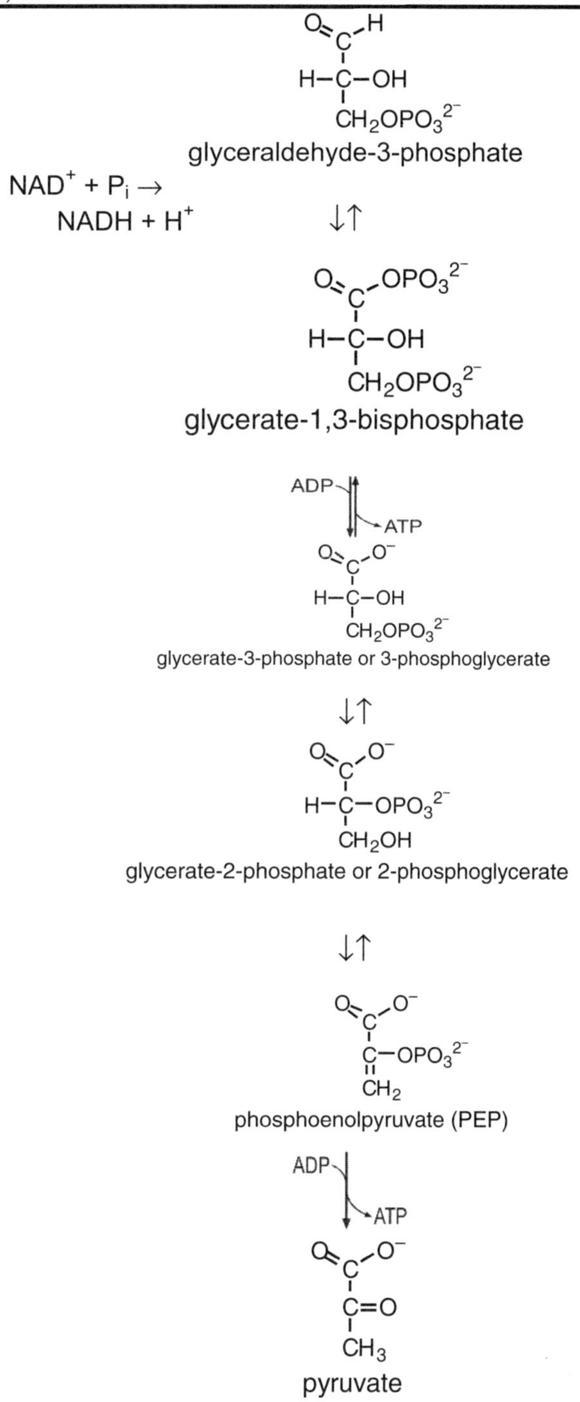

6. **Glyceraldehyde-3-phosphate dehydrogenase**

Mechanism: Figure 8.5 (text)
Oxidation of aldehyde with reduction of NAD$^+$ to NADH
Phosphorylation without ATP
Product has a high phosphoryl group transfer potential

7. **Phosphoglycerate kinase**, Mg^{2+}

Recovers earlier ATP investment
Substrate-level phosphorylation: transfer of phosphoryl group to ADP

8. **Phosphoglycerate mutase**

Isomerization sets up PEP formation
Adds and removes phosphate

9. **Enolase**

High phosphoryl group transfer potential because without the phosphate, it is an enol
Dehydration reaction

10. **Pyruvate kinase**, Mg^{2+}

Enol formed tautomerizes to the stable keto form
Net two ATP formed

IRREVERSIBLE!

What happens next depends upon: *whether oxygen is present,*
the type of cell, and
the energy needs of the cell.

THE ENERGETICS OF GLYCOLYSIS

The three irreversible reactions have a significantly negative ΔG value. The other reactions have a ΔG value close to zero and are reversible.

REGULATION OF GLYCOLYSIS

To regulate glycolysis, regulate the enzymes that catalyze the three reversible reactions:
- hexokinase or glucokinase (GK)
- phosphofructokinase-1 (PFK-1)
- pyruvate kinase

For further details, see the comparison of glycolysis regulation and gluconeogenesis regulation later in this chapter.

8.2 GLUCONEOGENESIS: THE GENESIS OF NEW GLUCOSE

- is an *anabolic* pathway used to synthesize glucose from certain amino acids or lactate
- requires energy: the hydrolysis of 4 ATP and 2 GTP
- is the reverse of glycolysis except for reactions that bypass the three irreversible reactions of glycolysis
- is important because the brain and red blood cells need a steady supply of glucose (use glucose as their primary energy source).
- occurs primarily in the liver after glycogen is depleted

THE OVERALL REACTION OF GLUCONEOGENESIS: PYRUVIC ACID TO GLUCOSE

$$2C_3H_4O_3 + 4ATP + 2GTP + 2NADH + 2H^+ + 6H_2O \rightarrow C_6H_{12}O_6 + 4ADP + 2GDP + 2NAD^+ + 6HPO_4^{2-} + 6H^+$$

GLUCONEOGENESIS REACTIONS (SEE PP. 290-294 FOR REACTION DETAILS)

GLUCONEOGENESIS SUBSTRATES

GLUCOGENIC AMINO ACIDS

ALANINE CYCLE:

In exercising muscle, alanine transaminase transfers $-NH_3^+$ from glutamate to pyruvate:

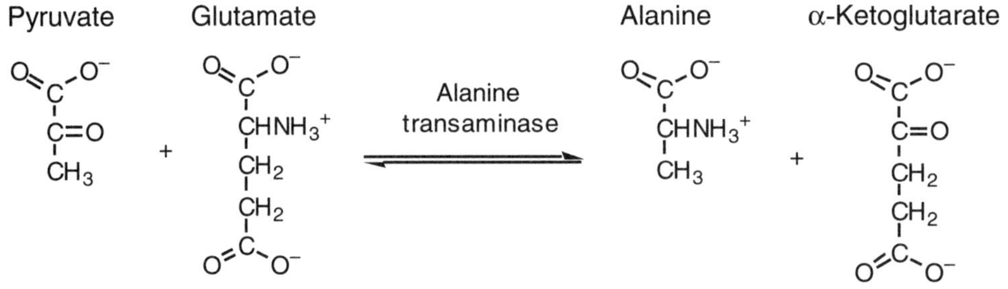

Alanine is transported to the liver, where it is converted back to pyruvate again via alanine transaminase in the reverse reaction.

(For lactate and glycerol, see the page that follows the gluconeogenesis reactions.)

REACTIONS OF GLUCONEOGENESIS THAT ARE NOT THE REVERSE OF GLYCOLYSIS *(AND SO USE DIFFERENT ENZYMES)*

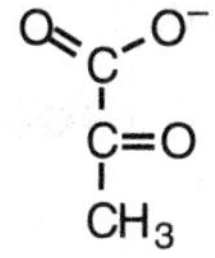

pyruvate + CO_2

Pyruvate carboxylase (biotin)

$ATP + H_2O \rightarrow ADP + P_i + H^+$ ↓

Mitochondria
Biotin is a coenzyme, CO_2 carrier
Certain amino acids can be used to make OAA.

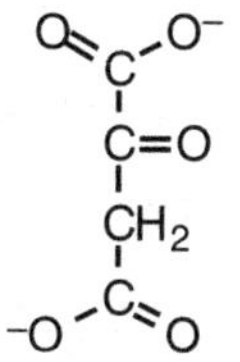

oxaloacetate (OAA)

PEP carboxykinase

$GTP \rightarrow GDP$ ↓

Mitochondria (some species), but OAA cannot cross the membrane, so the malate shuttle is used:
OAA converts to malate, which is transported across the membrane and then converted back to OAA

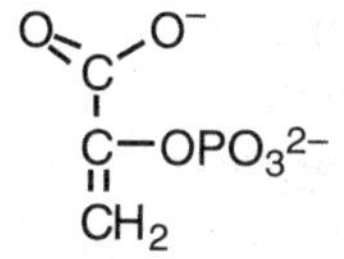

PEP + CO_2

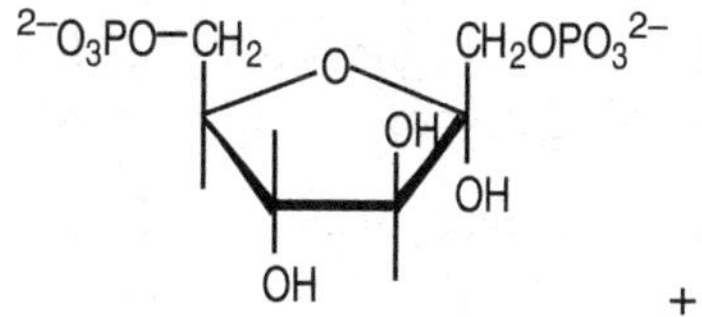

fructose-1,6-bisphosphate + H_2O

Fructose-1,6-bisphosphatase

↓

Note that ATP is not regenerated
Allosteric regulation:
Activated by citrate
Inhibited by fructose-2,6-bisphosphate and AMP

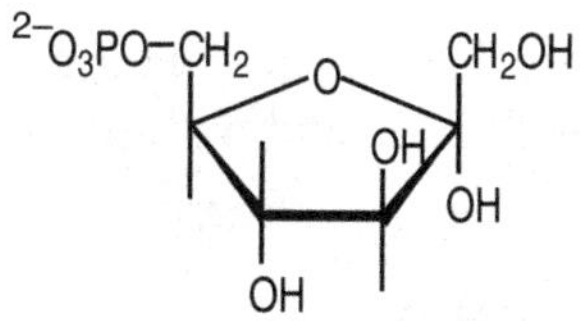

fructose-6-phosphate + P_i

Glucose-6-phosphatase

Only in the liver and the kidney; takes place in the endoplasmic reticulum
Glucose produced in the liver is released into the blood

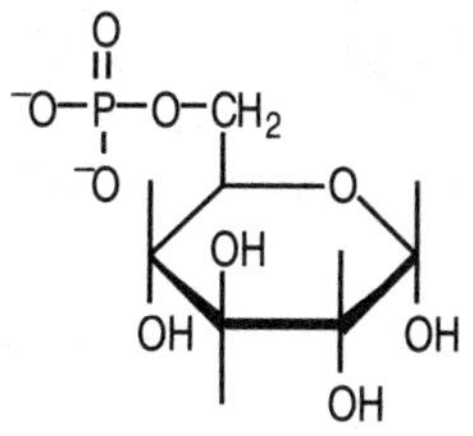

glucose-6-phosphate + H_2O ⟶

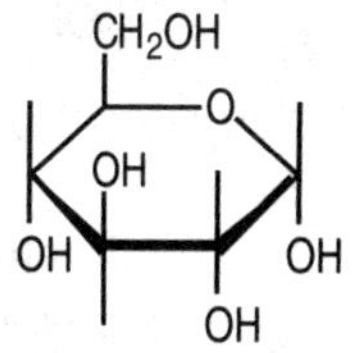

glucose + P_i

LACTATE: CORI CYCLE:

Exercising muscle: .... glucose to pyruvate to lactate (glycolysis/lactic acid fermentation); lactate enters bloodstream

Transport to liver: .... lactate to pyruvate to glucose (gluconeogenesis)

GLYCEROL → GLYCEROL-3-PHOSPHATE → DIHYDROXYACETONE PHOSPHATE (DHAP)

This occurs only in the liver, when cytoplasmic [NAD^+] is high; uses an ATP but creates an NADH (which can produce ATP via the electron transport system). Glycerol is a product of fat catabolism.

SUMMARY OF REGULATION OF GLYCOLYSIS AND GLUCONEOGENESIS

COMPARTMENTATION:

Glycolysis occurs only in the cytoplasm, and several gluconeogenesis reactions occur in the mitochondria. Glycerol kinase is found only in the liver.

REGULATION BASED UPON THE NEEDS OF THE CELL:

Keep in mind the ultimate goal of each pathway, and their regulation will make sense. For example, glycolysis produces energy, reducing power, and intermediates. So, when the cell needs energy (when reserves are low), glycolysis is activated and gluconeogenesis is inhibited. When the cell has enough energy, glycolysis is inhibited and gluconeogenesis is activated. *Substrate cycle:* paired reactions that are coordinately regulated

THE LIVER MODULATES BLOOD SUGAR LEVELS

When blood sugar is low, the liver releases glucose into the bloodstream, first by degrading glycogen and then by gluconeogenesis (in response to glucagon).

When blood sugar is high, insulin inhibits gluconeogenesis and activates glycolysis in the liver. Also, hexokinase D in the liver is not inhibited by glucose-6-phosphate, so it lets the liver remove glucose from the blood to store as glycogen.

REGULATION BY COVALENT MODIFICATION

PFK-2 and fructose-2,6-bisphosphatase is actually one enzyme that is bifunctional: without its phosphate, it is PFK-2, and with its phosphate, it is fructose-2,6-bisphosphatase.

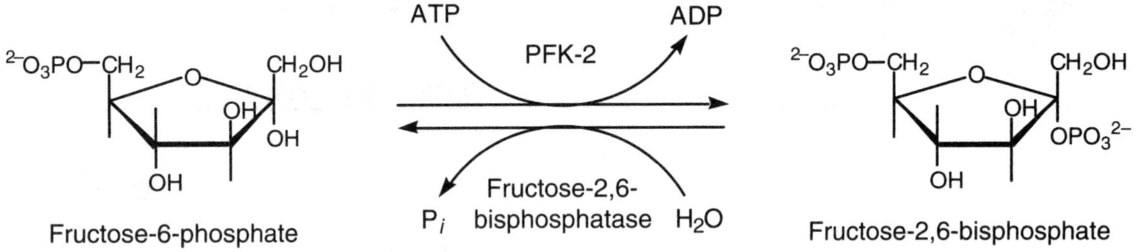

When blood glucose is high, insulin activates PFK-2, which increases fructose-2,6-bisphosphate levels, which:

- activates PFK-1 (glycolysis) in the liver
- inhibits fructose-2,6-bisphosphatase (inhibits gluconeogenesis)

Glycolysis and Gluconeogenesis Regulation

When Energy Is Needed Activate Glycolysis

Glucose-6-phosphate

↓

2 Pyruvate

To Store Energy or to Raise Blood Glucose Activate Gluconeogenesis

2 Pyruvate (also: lactate, glycerol, some amino acids)

↓

Glucose-6-phosphate

PFK-1 activated by:
- AMP
- Fructose-2,6-bisphosphate

Pyruvate kinase activated by:
- AMP
- Fructose-1,6-bisphosphate (feed-forward control)

Insulin stimulates the synthesis of:
- glucokinase
- PFK-1
- PFK-2

Pyruvate carboxylase activated by:
- Acetyl-CoA (high during starvation; product of fatty acid catabolism)

Fructose 1,6-bisphosphatase activated by:
- ATP

Glucagon stimulates the synthesis of:
- PEP carboxykinase
- fructose-1,6-bisphosphatase
- glucose-6-phosphatase

Inhibit Gluconeogenesis

Pyruvate carboxylase inhibited by:
- Acetyl-CoA

Fructose-1,6-bisphosphatase inhibited by:
- AMP
- Fructose-2,6-bisphosphate

(Fructose-2,6-bisphosphate indicates high levels of glucose.)

Inhibit Glycolysis

Hexokinase inhibited by:
- ATP
- Glucose-6-phosphate

PFK-1 inhibited by:
- ATP
- Citrate

Pyruvate kinase inhibited by:
- ATP
- Acetyl-CoA

8.3 THE PENTOSE PHOSPHATE PATHWAY

ULTIMATE GOALS:

1. to make NADPH (reducing power) for syntheses (for example fatty acids) and to help prevent oxidative damage (it is a great reducing agent)
2. to make sugar intermediates, especially ribose-5-phosphate, a component of nucleotides and nucleic acids

LOCATION: Cytoplasm; especially in cells that synthesize lipids and cells that are at high risk for oxidative damage
In plants, during the dark reactions of photosynthesis

OXIDATIVE PHASE: **ALL THREE REACTIONS ARE IRREVERSIBLE.**

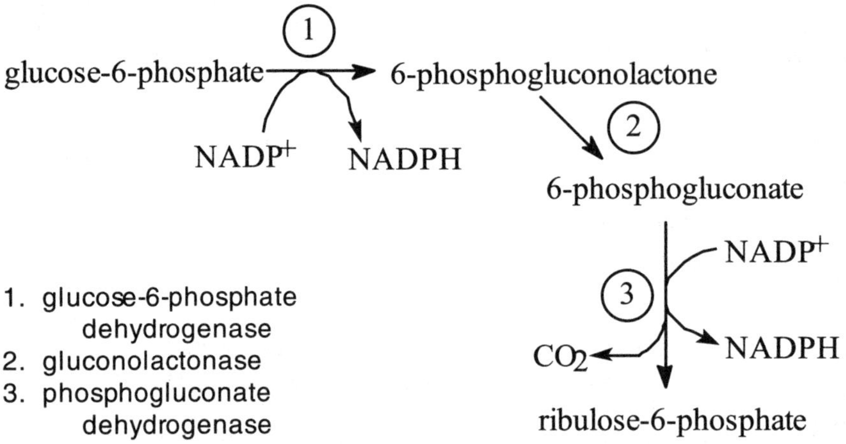

1. glucose-6-phosphate dehydrogenase
2. gluconolactonase
3. phosphogluconate dehydrogenase

REGULATION: Glucose-6-phosphate dehydrogenase (G6PD) is inhibited by NADPH and activated by glucose-6-phosphate and GSSG (oxidized glutathione), which indicates oxidative damage and a need for NADPH.

A high-carbohydrate diet triggers the synthesis of the enzymes G6PD and phosphogluconate dehydrogenase.

NONOXIDATIVE PHASE: ALL REACTIONS ARE REVERSIBLE.

What happens to ribulose-5-phosphate depends on the metabolic needs of the cells:

1. If the cell needs ribose-5-phosphate for nucleotide biosynthesis, ribose-5-phosphate isomerase will isomerize ribulose-5-phosphate.
2. If the cell only needs reducing power (NADPH), then the carbons of ribulose-5-phosphate will be recycled and converted into fructose-6-phosphate and glyceraldehyde-3-phosphate, intermediates of glycolysis. To completely recycle the ribulose-5-phosphate carbons via glycolysis, at least three molecules of ribulose-5-phosphate are needed initially. See the reactions outlined below.
3. If the cell needs ribose-5-phosphate but does not need reducing power (NADPH), the cell can also use the reverse reactions from the nonoxidative phase, starting with two fructose-6-phosphates and one glyceraldehyde-3-phosphate to make 3 ribose-5-phosphates.

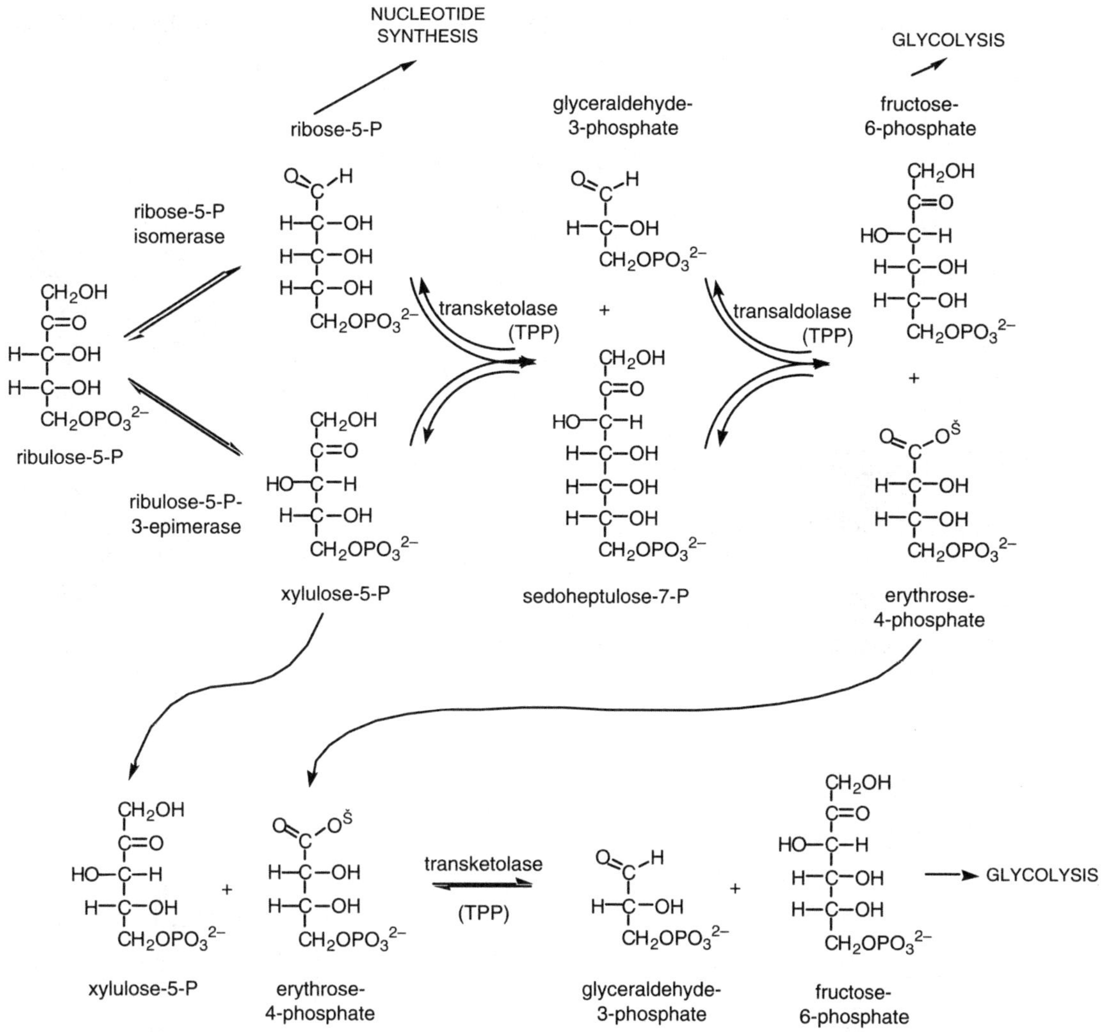

TRANSKETOLASE: TRANSFERS 2 CARBONS FROM A KETO SUGAR TO AN ALDO SUGAR

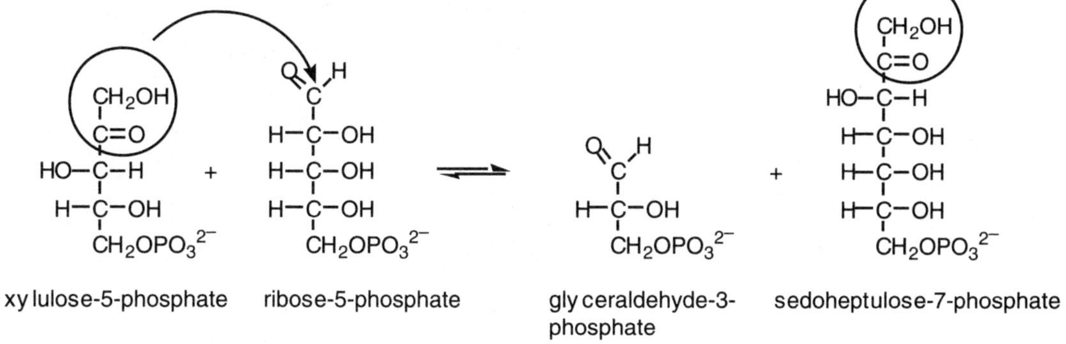

TRANSALDOLASE: TRANSFERS 3 CARBONS FROM A KETO SUGAR TO AN ALDO SUGAR

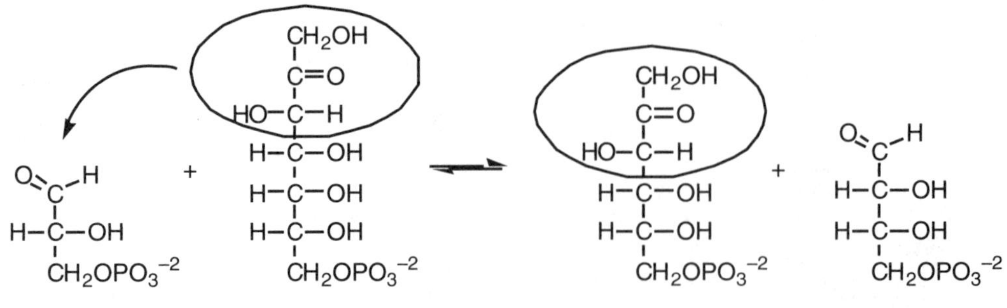

8.4 METABOLISM OF OTHER IMPORTANT SUGARS

FRUCTOSE METABOLISM

LIVER: Four enzymes convert fructose to two G-3-P molecules, bypassing the regulatory enzymes hexokinase and PFK-1.

GALACTOSE, MANNOSE: Several reactions convert galactose to UDP-glucose. Hexokinase phosphorylates mannose to mannose-6-phosphate, which is isomerized to fructose-6-phosphate.

8.5 GLYCOGEN METABOLISM

Most glycogen is found in the muscle and liver (≈10% liver mass and ≈1% muscle mass). The actual amount of glycogen depends on the nutritional state of the organism.

Remember that glycogen is made of α(1,4)-linked glucose with α(1,6)-linked branches. Special enzymes are needed to form and degrade the branches. Glycogen degradation is not simply a reverse of glycogenesis.

The ends of all of those branches are nonreducing ends. Enzymes for both glycogenesis and glycogenolysis work only on the nonreducing ends. One huge glycogen molecule can be coated with working enzymes to release glucose very quickly when needed.

The ultimate goal of **GLYCOGENESIS** depends on where it takes place:

MUSCLE: stores excess glucose as glycogen: **ENERGY STORAGE.**

LIVER: removes glucose from the blood in response to insulin, which signals high blood glucose levels.

Glycogen synthase, and thus glycogenesis, is activated by glucose-6-phosphate, an indicator of excess glucose. High levels of ATP (as well as glucose-6-phosphate) inhibit the glycogenolysis by inhibiting glycogen phosphorylase.

The ultimate goal of **GLYCOGENOLYSIS** also depends on where it takes place:

MUSCLE: degrades glycogen to produce glucose-6-phosphate for **ENERGY** (→ glycolysis).

LIVER: degrades glycogen to produce free glucose for **EXPORT** (to other cells), so glucose-6-phosphate is dephosphorylated and sent to the blood.

Glucagon signals **LOW BLOOD GLUCOSE,** and epinephrine signals an immediate need for energy. Glycogen phosphorylase, and thus glycogenolysis, is activated by AMP, an indicator that energy is needed.

GLYCOGENESIS: GENESIS OF GLYCOGEN
BEGINNING WITH GLUCOSE-6-PHOSPHATE

Other pathways to glycogen exist, namely glucose to C3 molecules to liver glycogen.

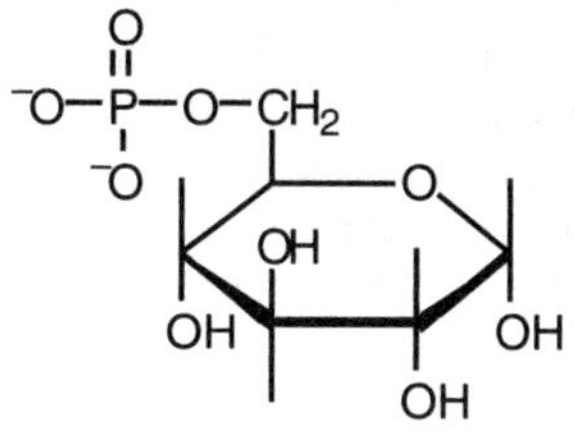

glucose-6-phosphate

↓

Phosphoglucomutase

Intermediate is glucose-1,6-bisphosphate
Mechanism involves a phosphoryl group attached to a Ser residue on the enzyme

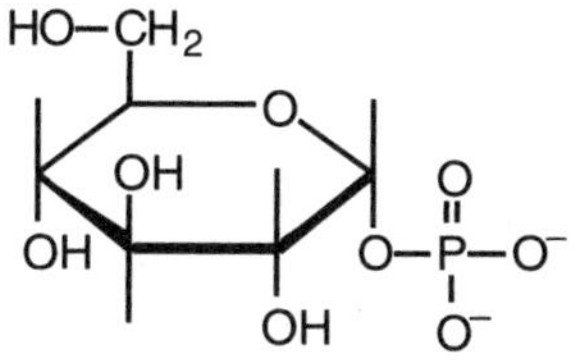

glucose-1-phosphate

UDP-glucose pyrophosphorylase

UDP is a great leaving group
UDP-glucose is held more securely (than glucose alone) in the active site

UTP→PPi ↓↑

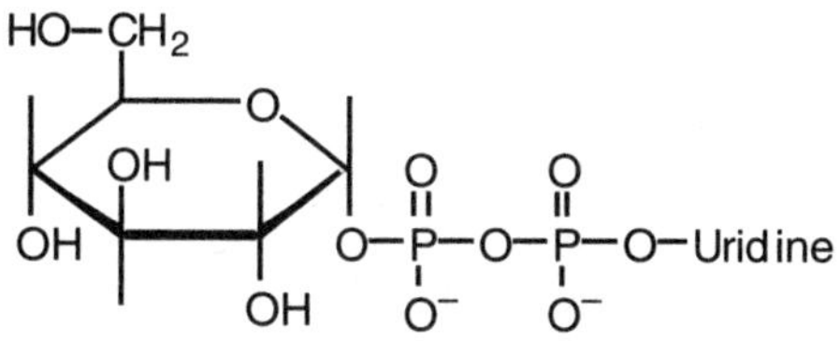

UDP-glucose

Irreversible PPi hydrolysis drives the previous reaction forward

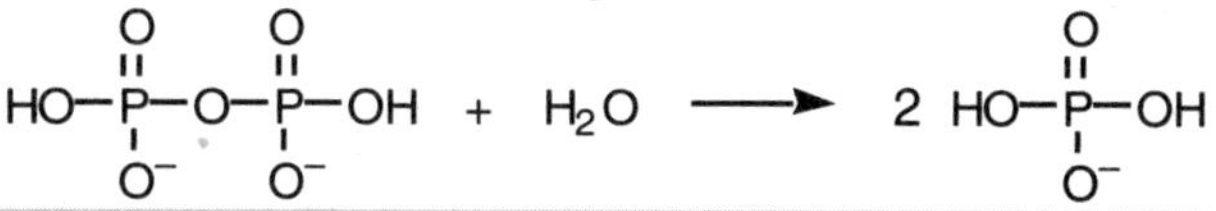

glycogen (n glucose residues) + UDP-glucose —glycogen synthase→ glycogen (n+1 residues)

↓

UDP + glycogen$_{(n\,+\,1\ \text{residues})}$

Glycogen synthase

Creates α(1,4) glycosidic bonds
Needs glycogenin ("primer" protein) or an existing glycogen chain
Transfers glucose from UDP to glycogen chain

Branching enzyme

creates α(1,6)-linkages (branches)

GLYCOGENOLYSIS

Glycogen phosphorylase

Cleaves α(1,4) linkages; stops 4 glucose units before a branch

Limit dextrin is a glycogen molecule that is been degraded to its branch points

glycogen (*n* residues) + Pi (HPO_4^{2-})

↓

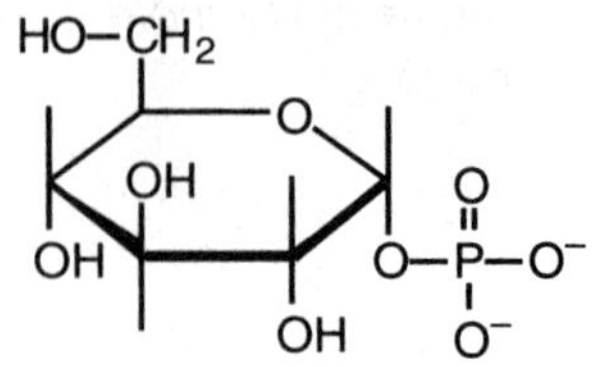

glycogen (*n* - 1 residues) + glucose-1-phosphate

↓↑

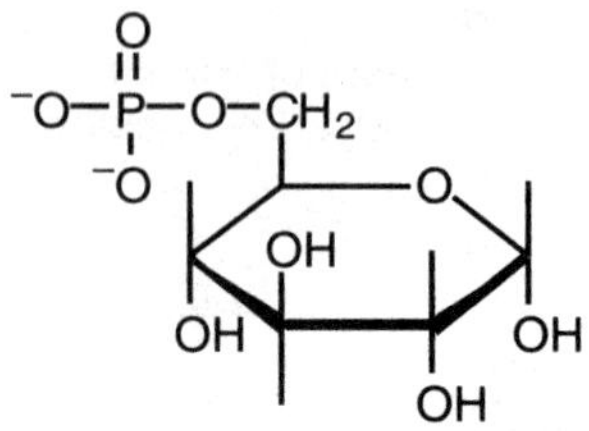

glucose-6-phosphate

Phosphoglucomutase

Debranching enzyme (amylo-α(1,6)-glucosidase)

Transfers the outer three of the four glucose residues attached to a branch point to a nearby nonreducing end.

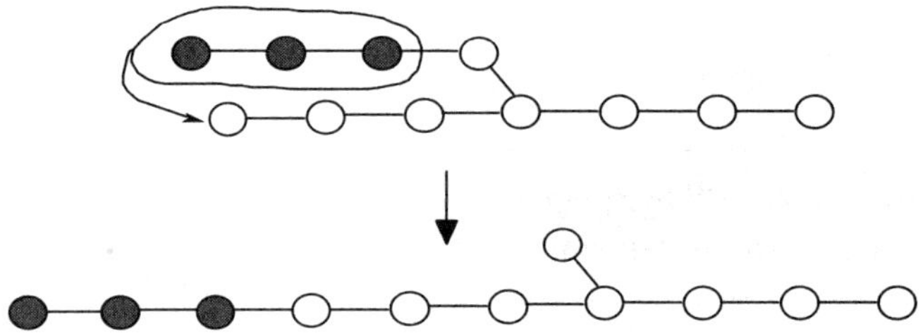

Debranching enzyme (amylo-α(1,6)-glucosidase)

Hydrolysis of α(1,6)-linkages

Removes single glucose residue at each branch point

The unbranched polymer produced by 1,6-glucosidase is then degraded by glycogen phosphorylase.

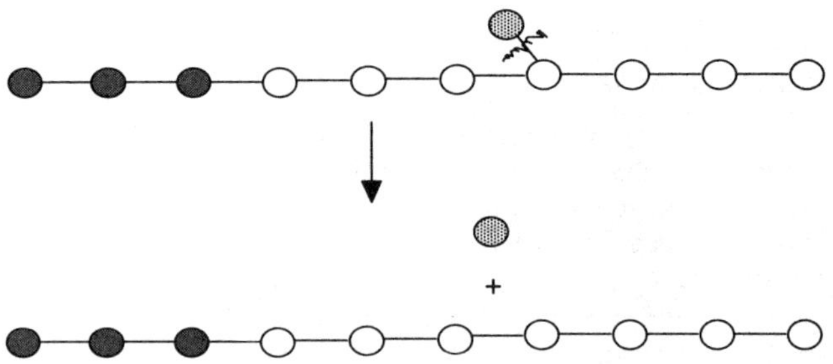

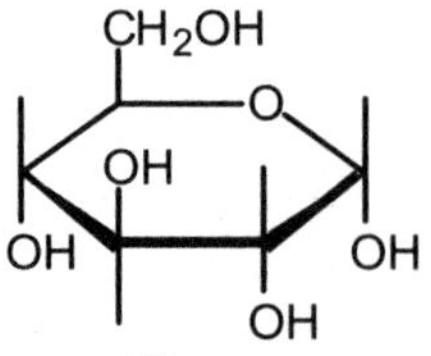

Glucose

Regulation of Glycogen Metabolism

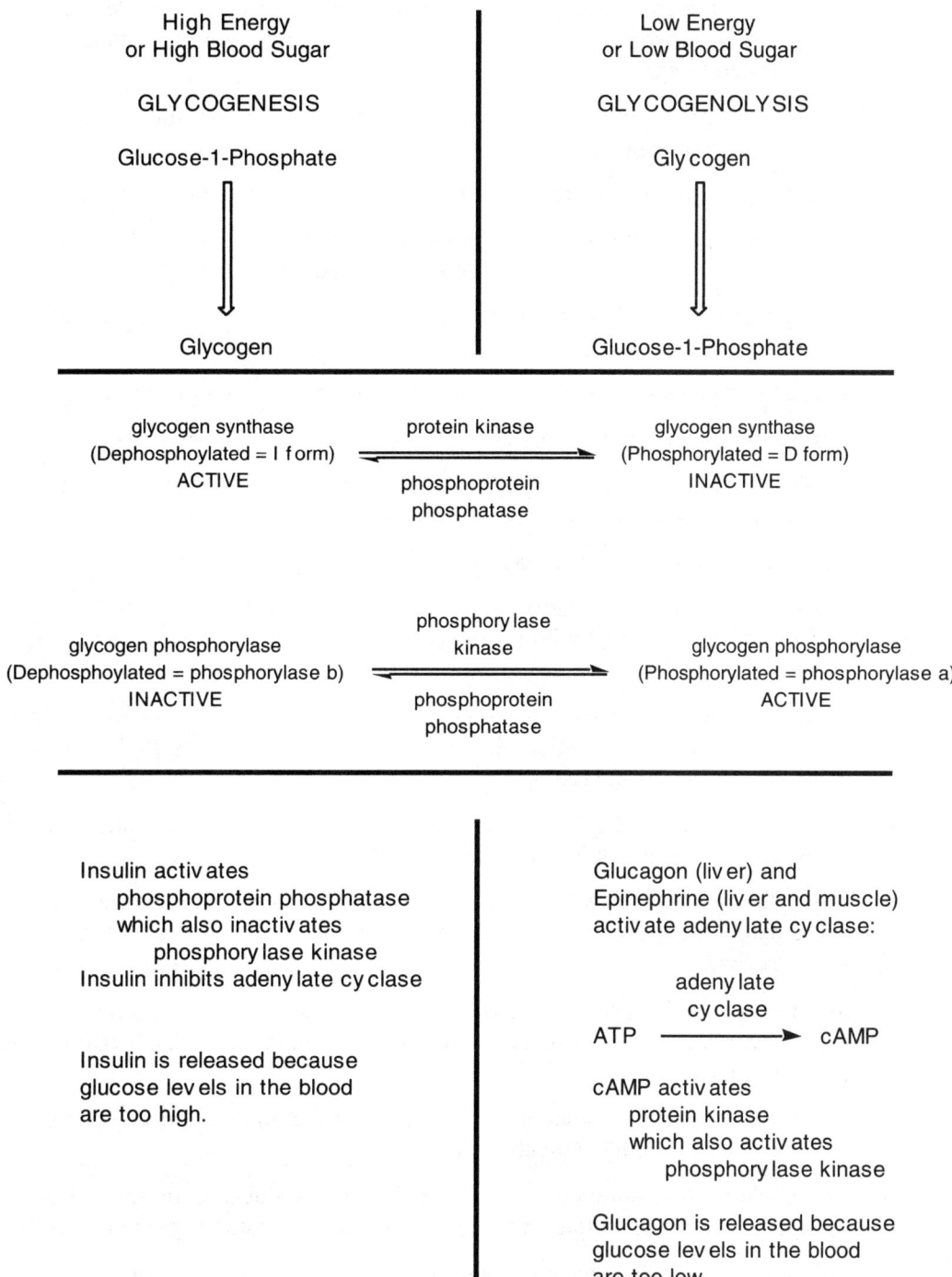

High Energy or High Blood Sugar	Low Energy or Low Blood Sugar
GLYCOGENESIS	GLYCOGENOLYSIS
Glucose-1-Phosphate ⇩ Glycogen	Glycogen ⇩ Glucose-1-Phosphate

glycogen synthase (Dephosphoylated = I form) ACTIVE ⇌ glycogen synthase (Phosphorylated = D form) INACTIVE

- forward: protein kinase
- reverse: phosphoprotein phosphatase

glycogen phosphorylase (Dephosphoylated = phosphorylase b) INACTIVE ⇌ glycogen phosphorylase (Phosphorylated = phosphorylase a) ACTIVE

- forward: phosphorylase kinase
- reverse: phosphoprotein phosphatase

Insulin activates
phosphoprotein phosphatase
which also inactivates
phosphorylase kinase
Insulin inhibits adenylate cyclase

Insulin is released because glucose levels in the blood are too high.

Glucagon (liver) and Epinephrine (liver and muscle) activate adenylate cyclase:

ATP —(adenylate cyclase)→ cAMP

cAMP activates
protein kinase
which also activates
phosphorylase kinase

Glucagon is released because glucose levels in the blood are too low.

Epinephrine is released in response to stress: Fight or flight! Need energy *FAST*!

Note: This page is a different version of Figure 8.22, p. 311 in your text.

CHAPTER 8: SOLUTIONS TO REVIEW QUESTIONS

8.1 a. glycolysis – the enzymatic pathway that converts a glucose molecule into two molecules of pyruvate; the anaerobic process generates energy in the form of two ATP molecules and two NADH molecules

b. pentose phosphate pathway – a biochemical pathway that produces NADPH, ribose and several other sugars

c. gluconeogenesis – the synthesis of glucose from noncarbohydrate molecules

d. glycogenolysis – a biochemical pathway that removes glucose molecules from glycogen polymers when blood glucose levels are low

e. glycogenesis – a biochemical pathway that adds glucose to growing glycogen polymers when blood glucose levels are high

8.2 a. anaerobic organism – an organism that does not use oxygen to generate energy

b. aerobic organism – an organism that uses oxygen to generate energy

c. aerobic respiration – the metabolic process in which oxygen is used as a terminal electron acceptor to generate energy from food molecules

d. aldol cleavage – the reverse reaction of the aldol condensation

e. substrate-level phosphorylation – the synthesis of ATP from ADP by phosphorylation coupled with the exergonic breakdown of a high-energy organic substrate molecule

8.4 a. Crabtree effect – the physiological capacity of *Saccaromyces cervisiae* cells to ferment sugar to produce ethanol, which kills their competitors, and then use the ethanol as an energy source

b. transcription factor – a protein that regulates or initiates the synthesis of specific mRNAs by binding to DNA sequence called response elements

c. response element – a DNA sequence within the promoter of genes that are coordinately regulated; transcription is triggered when a specific hormone receptor complex binds

d. insulin – a peptide hormone released from pancreatic β-cells; among its many efforts are the promotion of glucose uptake into the cells of certain target organs (muscle and adipose tissue)

e. malate shuttle – oxaloacetate is transferred from the mitrochondrial matrix to the cytoplasm by reversible conversion to malate

8.5 a. Cori cycle – a metabolic process in which lactate produced in tissue such as skeletal muscle is transferred to liver where it becomes a substrate for gluconeogenesis

b. glucagon – a peptide hormone released from pancreatic α-cells; among its effects are increasing the level of glucose in blood via the breakdown of liver glycogen

c. glucose-alanine cycle – a method of recycling α-keto acids between muscle and liver and for transporting ammonia to the liver

d. hypoglycemia – blood glucose levels that are lower than normal

e. antioxidant – a substance that prevents the oxidation of other molecules

8.7 a. Insulin – a hormone that stimulates glycogenesis and inhibits glycogenolysis.

b. Glucagon – a hormone that stimulates glycogenolysis and inhibits glycogenesis.

c. Fructose-2,6-bisphosphate – an effector molecule that activates PFK- 1 and stimulates glycolysis.

d. UDP-glucose – an activated form of glucose; a substrate for glycogen synthesis that is held more securely in the active site than glucose alone

e. cAMP – cyclic AMP – a second messenger molecule produced from ATP in response to glucagon or epinephrine

f. GSSG – the oxidized form of glutathione; an activator of glucose-6-phosphate dehydrogenase, which catalyzes a key regulatory step in the pentose phosphate pathway

g. NADPH – an important reducing agent; the reduced form of $NADP^+$, required for reductive processes (e.g. lipid biosynthesis) and antioxidant mechanisms (NADPH is a powerful antioxidant).

8.8 a. Aldolase – the reversible conversion of fructose-1,6-bisphosphate to DHAP and glyceraldehyde-3phosphate

b. Enolase – the reversible conversion of glycerate-2-phosphate to phosphoenolpyruvate

c. Hexokinase – the phosphorylation of glucose to yield glucose-6-phosphate

d. Amylo-a(1,6)-glucosidase – hydrolysis of the α(1,6)glycosidic bonds of glycogen

e. Phosphoglucomutase – the reversible conversion of glucose-1-phosphate and glucose-6-phosphate

8.10 a. ATP consumption – reaction 1(glucose-6-phosphate synthesis) and reaction 3 (fructose-1,6-bisphosphate synthesis)

b. ATP synthesis – reactions 7(glycerate-3-phosphate synthesis) and 10 (pyruvate synthesis)

c. NADH synthesis – reaction 6 (glycerate-1,3-bisphosphate synthesis)

8.11 The reactions catalyzed by hexokinases I, II, and III, PFK-1 and pyruvate kinase are allosterically regulated.

8.13 In the Pasteur effect, oxygen inhibits glucose consumption. In other words glucose is oxidized more rapidly in aerobic cells that are deprived of oxygen compared to when oxygen is available. In the Crabtree effect, which is observed in *S, cerevisiae* ,glycolysis is unaffected by oxygen. Instead, excess pyruvate is converted to ethanol, which is then excreted into the environment where it kills microbial competitors. In addition glucose represses aerobic metabolism, the opposite of what occurs in the Pasteur effect.

8.14 Glycogenis is a primer protein that is required for glycogen synthesis. Each glycogen chain is synthesized by glycogen synthesis by extending from the preexisting tetrasaccharide, which is linked to a specific tyrosine residue in glycogenin.

8.16 a. gluconeogenesis – cytoplasm and mitochondria

b. glycolysis - only in the cytoplasm

c. pentose phosphate pathway - cytoplasm

8.17 In glycolysis, the entry level substrates are sugars and the product is pyruvate. The main purposes of glycolysis are to provide the cell with energy and/or several metabolic intermediates. The substrates for gluconeogenesis are pyruvate, lactate, glycerol, and several amino or α-keto acids. Gluconeogenesis provides the body with glucose when blood glucose levels are low.

8.19 Such organisms utilize acetaldehyde as a hydrogen acceptor in order to regenerate NAD^+. Ethanol is the product of the reduction of acetaldehyde.

8.20 Under anaerobic conditions, pyruvate is reduced to lactate to regenerate NAD^+.

8.22 Glycolysis occurs in two stages. In stage I, glucose is phosphorylated and cleaved to two molecules of glyceraldehyde-3-phosphate. During this stage, two ATP molecules are consumed. In stage 2, each glyceraldehyde-3-phosphate is converted to pyruvate, a process in which four ATP and two NADH are produced.

8.23 Only (e), AMP, inhibits gluconeogenesis. Lactate, ATP, pyruvate, glycerol, and acetyl-CoA all stimulate gluconeogenesis.

8.25 Futile cycles are prevented by having the forward and reverse reactions catalyzed by different enzymes, both of which are independently regulated.

8.27

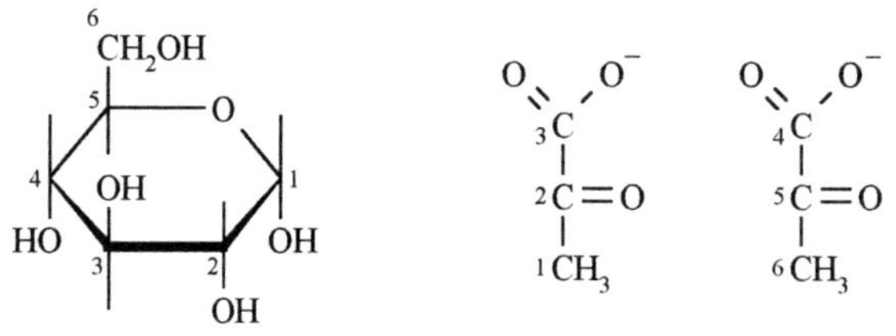

8.28 Refer to Figures 8.3 and 8.4, stages 1 and 2 reactions of glycolysis, where glucose is converted into pyruvate. Pyruvate may then be converted to ethanol by alcoholic fermentation:

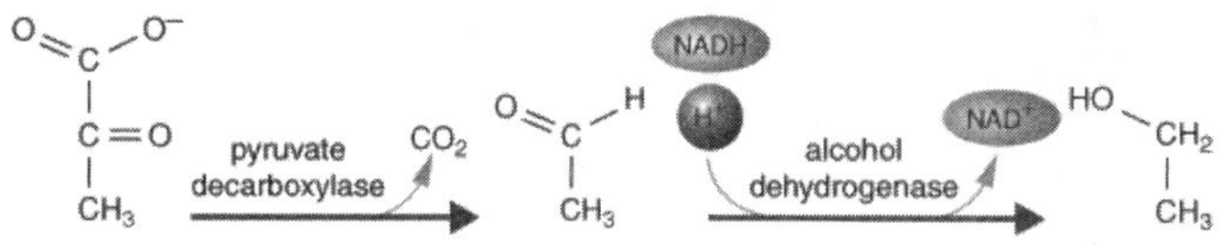

8.29

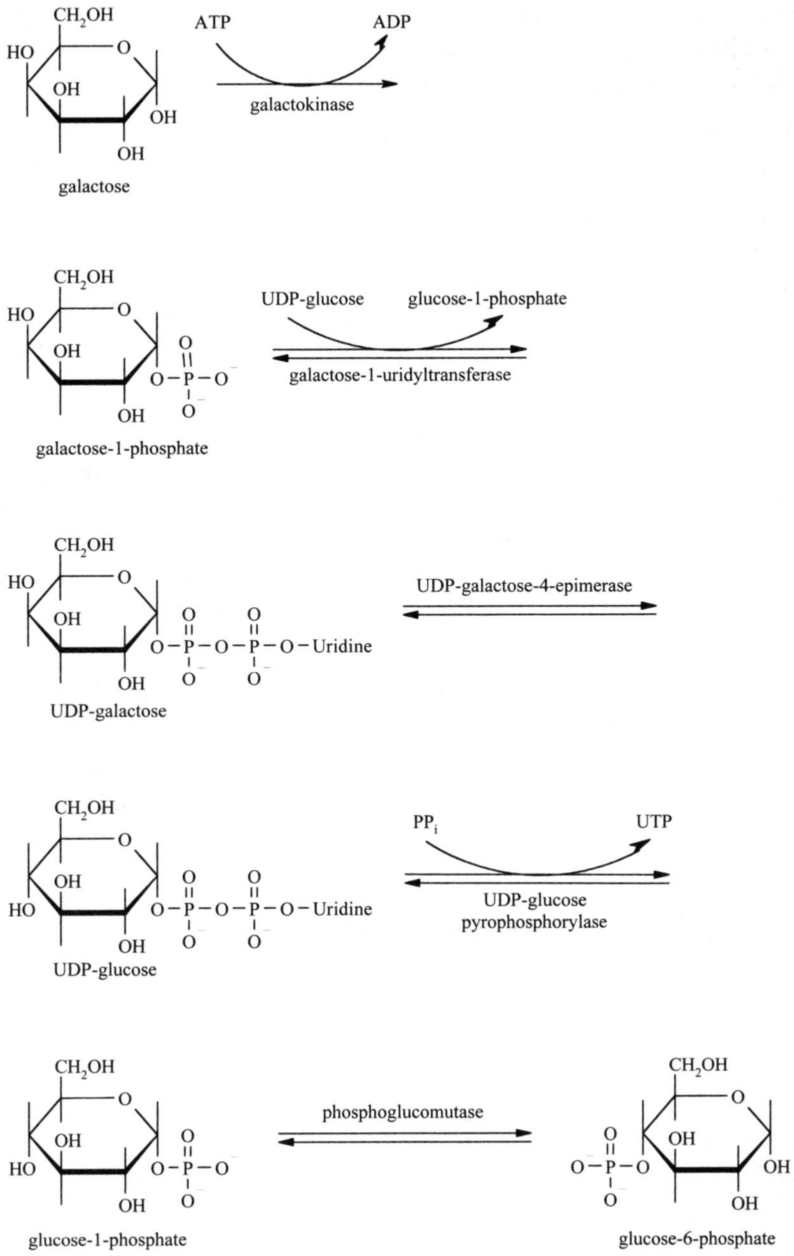
ATP
ADP
galactokinase
galactose
UDP-glucose
glucose-1-phosphate
galactose-1-uridyltransferase
galactose-1-phosphate
UDP-galactose-4-epimerase
UDP-galactose
PP$_i$
UTP
UDP-glucose pyrophosphorylase
UDP-glucose
phosphoglucomutase
glucose-1-phosphate
glucose-6-phosphate

8.31 Severe hypoglycemia is so dangerous because brain cells (and red blood cells) rely solely on glucose for their energy needs. Severe hypoglycemia causes fainting, and requires *immediate* medical attention. When the condition is prolonged and/or recurrent, it may also result in long-term brain impairment or dysfunction.

8.32 ATP hydrolysis at the beginning of glycolysis drives the pathway forward so that ATP is produced under varying conditions of substrate availability and product concentrations.

8.34 In the Cori cycle, lactate produced under anaerobic conditions in exercising muscle is passed to the liver, where it is converted back to pyruvate and then to glucose, which passes back to the muscle. In the muscle, glucose undergoes glycolysis to form pyruvate, which reacts with NADH to form lactate and NAD^+. The physiological function of the Cori cycle is to allow glycolysis to continue by regenerating NAD^+ so that under anaerobic conditions, exercising muscle can produce ATP for energy.

8.35 The general effect of insulin is to lower blood glucose levels by initiating a process that leads to the inhibition of glycogenolysis and the activation of glycogenesis. Glucagon has the opposite effect: it raises blood glucose levels by initiating a process that leads to the activation of glycogenolysis and the inhibition of glycogenesis. Refer to Figure 8.21 for the effects of insulin and glucagon on the activation or inhibition of specific enzymes, including adenylate cyclase, which catalyzes the synthesis the secondary messenger cAMP.

8.37 Insulin and glucagon are produced by cells in the pancreas. Epinephrine and cortisol are produced by cells in the adrenal glands in response to stress.

8.38 In muscle, glycogen is synthesized to store glucose for energy, or metabolized when energy is needed. In the liver, glycogenesis and glycogenolysis are regulated to maintain blood glucose levels.

CHAPTER 8: SOLUTIONS TO FILL-IN-THE-BLANK QUESTIONS

8.40 Tautomers

8.41 Amphibolic

8.43 Crabtree

8.44 Pasteur

8.46 Malate

8.47 Cori

8.49 UDP-glucose

CHAPTER 8: SOLUTIONS TO SHORT-ANSWER QUESTIONS

8.50 The glycolytic reaction in which oxidation occurs is the conversion of glyceralde-hyde-3-phosphate to glycerate-1,3-bisphosphate.

8.52 A diauxic shift is a significant change in gene expression that occurs in response to a change in an organism's environment. A diauxic shift occurs in yeast when ethanol has killed their competitors, glucose levels are depleted, and O2 is available. Under these conditions, the glucose-induced repression of aerobic metabolism ends. As the levels of the enzymes required for the aerobic oxidation of acetyl-CoA (the product of ethanol oxidation) increase, the yeast cells use their "waste product" to generate energy.

8.53 Assuming that pyruvate (a three-carbon compound) is used solely in the resynthesis of glucose (a six-carbon compound), 6 ATPs would be required. The conversion of two molecules of pyruvate and 2 CO2 to oxaloacetate requires 2 ATP (1 ATP for each molecule of pyruvate). The subsequent decarboxylation of oxaloacetate to yield two molecules of phosphoenolpyruvate and CO2 requires 2 ATP equivalents in the form of 2 GTP. The last ATP-requiring reaction in gluconeogenesis is the phosphorylation of two molecules of glycerate-3-phosphate at the expense of 2 ATP to yield two molecules of glycerate-1,3-bisphosphate.

CHAPTER 8: SOLUTIONS TO THOUGHT QUESTIONS

8.55

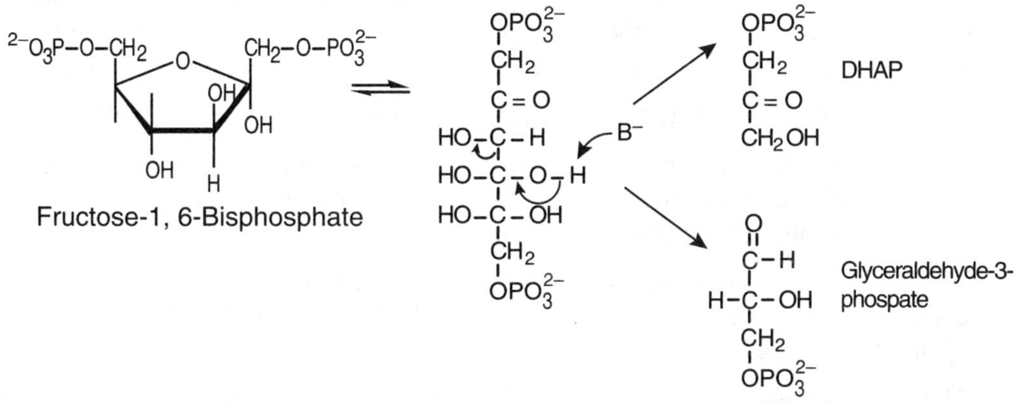

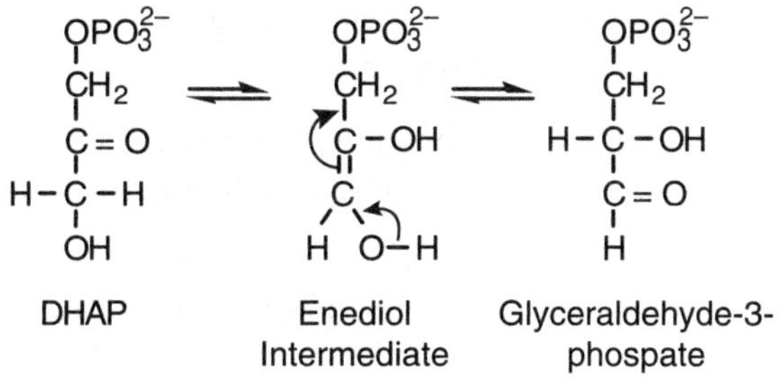

8.56 Hexokinases I, II and III which occur in liver and other body tissues, have high affinities for glucose and are inhibited by the glucose-6-phosphate. As a result when blood glucose levels are high the body's cells do not phosphorylate more glucose molecules than are required to meet their immediate needs. In the liver hexokinase IV (glucokinase), which has a relatively low affinity for glucose and is not inhibited by glucose-6-phosphate, diverts glucose into glycogen synthesis. As a result of the kinetic properties of hexokinase IV after a carbohydrate meal, the liver does not begin glycogen synthesis until other tissues have satisfied their glucose requirements.

8.57 Glucokinase (GK) acts as a glucose sensor because of its relatively low affinity for glucose. Since it does not usually work at high velocity GK is sensitive to small changes in blood glucose levels. In each cell type in which GK occurs it is linked to a signal transduction pathway. For example, in pancreatic β-cells, GK-catalyzed glucose phosphorylation, which occurs when blood glucose levels are high, triggers insulin release.

8.58 In such an individual, following a carbohydrate meal blood glucose levels would be higher than normal. Recall that the kinetic properties of hexokinase D allow the liver to remove excess glucose from blood. Skeletal muscle would accumulate some additional glycogen, but most excess glucose would be used to synthesize triacylglycerol in adipocytes, a process that is promoted by insulin. A significant amount of liver glycogen is synthesized from glucose produced by gluconeogenesis.

8.59 In the synthesis of new glycogen molecules, a primer protein called glycogenin is used to initiate glycogen formation. Glucose is transferred from UDP-glucose to a specific tyrosine residue of the glycogenin. This glucose then serves as the starting point for a new growing glycogen molecule.

8.61 Phosphoenolpyruvate has a high phosphoryl group transfer potential because the transfer of the enol phosphate to another molecule produces a vinyl alcohol. The vinyl alcohol tautomerizes rapidly to the keto-form making the transfer almost irreversible.

8.62 Two common oxidizing agents in anaerobic metabolism are NAD^+ and $NADP^+$.

8.64 Ethanol is the most reduced molecule and acetate is the most oxidized. The degree of oxidation of an organic molecule can be correlated with its oxygen content, that is, acetate has more oxygen than does ethanol.

8.65 When the cell membrane is compromised, the contents of the glycosome and the general cytoplasm mix and glycolysis proceeds unchecked. All the ATP in the cell is then used up and the cell dies.

8.67 The OH group itself is not very easily displaced. However conversion of an OH group to a phosphate ester is very easily accomplished. These esters are good leaving groups and when displaced have the same effect as if an OH had reacted.

8.68 When a phosphate group is transferred from the phosphoenolpyruvate to another group, an enol alcohol remains. This compound rapidly tautomerizes to a ketone thus driving the phosphoryl transfer reaction to completion.

8.70 Once muscle protein has been degraded to amino acids, a large percentage of them are converted to either oxaloacetate or pyruvate. Both of these molecules are substrates for the gluconeogenesis pathway.

8.71 If either NADPH or NADH were the sole hydrogen carrier it would be impossible to determine the need for carbon units in either the pentose phosphate pathway or glycolysis. A decrease in the concentration of either reducing agent activates enzymes that shunt carbon into its respective pathway.

8.73 The products of ethanol oxidation in the liver are acetaldehyde (a toxic molecule) and NADH. When excess NADH molecules are produced (above that required to generate energy) fat synthesis increases. Because the livers of children are not fully developed, the accumulation of fat easily damages them. Liver function can thereby be compromised.

9 Aerobic Metabolism I: The Citric Acid Cycle

Brief Outline of Key Terms and Concepts

Overview

- OBLIGATE ANAEROBE; AEROTOLERANT ANAEROBE
- FACULTATIVE ANAEROBE
- OBLIGATE AEROBE

9.1 OXIDATION-REDUCTION REACTIONS

- CONJUGATE REDOX PAIR
- STANDARD REDUCTION POTENTIAL ($E°'$)

REDOX COENZYMES

- NICOTINIC ACID (VITAMIN)
- NAD(P)—NICOTINAMIDE ADENINE DINUCLEOTIDE (PHOSPHATE)
- ELECTRON ACCEPTOR; HYDRIDE TRANSFER
- RIBOFLAVIN (VITAMIN B_2)
- FLAVIN MONONUCLEOTIDE
- FLAVIN ADENINE DINUCLEOTIDE
- FLAVOPROTEINS

AEROBIC METABOLISM

In living organisms both energy-capturing and energy-releasing processes consist primarily of redox reactions.

In redox reactions electrons move between an electron donor and an electron acceptor.

In many reactions both electrons and protons are transferred.

In biological systems most redox reactions involve hydride ion transfer ($NADH/NAD^+$) or hydrogen atom transfer ($FADH_2/FAD$).

9.2 CITRIC ACID CYCLE

CONVERSION OF PYRUVATE TO ACETYL-COA

by the enzymes in the pyruvate dehydrogenase complex requires the coenzymes TPP, FAD, NAD^+, COENZYME A, and LIPOIC ACID. TPP is THIAMINE PYROPHOSPHATE.

REACTIONS OF THE CITRIC ACID CYCLE

The **CITRIC ACID CYCLE** begins with the condensation of a molecule of acetyl-CoA with oxaloacetate to form citrate, which is eventually reconverted to oxaloacetate. In one turn of the cycle, 2 CO_2, 3 NADH, 1 $FADH_2$, and 1 GTP are produced.

FATE OF CARBON ATOMS IN THE CITRIC ACID CYCLE

THE AMPHIBOLIC CITRIC ACID CYCLE

An **AMPHIBOLIC PATHWAY** plays a role in both anabolism and catabolism. The citric acid cycle intermediates used in anabolic processes are replenished by several **ANAPLEROTIC REACTIONS.**

CITRIC ACID CYCLE REGULATION

The citric acid cycle is closely regulated, thus ensuring that the cell's energy and biosynthetic needs are met.

ALLOSTERIC EFFECTORS AND SUBSTRATE AVAILABILITY primarily regulate the enzymes citrate synthase, isocitrate dehydrogenase, α-ketoglutarate dehydrogenase, pyruvate dehydrogenase, and pyruvate carboxylase.

The Citric Acid Cycle and Human Disease

Several rare human diseases have been attributed to deficits in citric acid cycle enzymes.

THE GLYOXYLATE CYCLE

Organisms in which the glyoxylate cycle occurs can use two-carbon molecules to sustain growth.

In plants, the glyoxylate cycle is in organelles called glyoxysomes.

As biomolecules are oxidized to CO_2, they transfer electrons to NAD^+ (and FAD), which is then reduced to NADH and $FADH_2$. In turn, the electrons are transferred to the electron transport chain and ultimately to oxygen, which is reduced to H_2O. The energy released during this process pumps H^+ ions to the other side of the membrane. When they naturally flow back in, hydrogen ions are (typically) forced to pass through special channels that synthesize ATP. So, ATP captures energy that results from the transfer of electrons from fuel molecules to (ultimately) oxygen.

9.1 Oxidation-Reduction Reactions

(Review oxidation-reduction reactions in Chapter 1 (p. 18) of your text.)

$\Delta G^{\circ\prime} = -nF\Delta E^{\circ\prime}$	where $\Delta G^{\circ\prime}$ =	the standard free energy
	n =	number of electrons transferred
	F =	Faraday's constant, 96.5 kJ/V·mol)
	$\Delta E^{\circ\prime}$ =	the difference in reduction potential between the electron donor and the electron acceptor under standard conditions

For redox reactions to be spontaneous (have a negative $\Delta G^{\circ\prime}$), they must have a positive $\Delta E^{\circ\prime}$. (Refer to the equation above; n and F are always positive, so for $\Delta G^{\circ\prime}$ to be negative, $\Delta E^{\circ\prime}$ must be positive.)

Electrons spontaneously transfer from the more negative $E^{\circ\prime}$ to the more positive $E^{\circ\prime}$. So, **the electron donor will have the more negative $E^{\circ\prime}$.**

When combining two half reactions, make sure that the overall equation is balanced in the number of electrons transferred.[1] When you reverse a half-reaction, also reverse the sign of its $E^{\circ\prime}$. To calculate $\Delta E^{\circ\prime}$, add the two values of $E^{\circ\prime}$ (including the sign-change of the reaction that you reversed).[2]

Use one of the redox coenzymes, NAD^+/NADH, and the reduction of acetaldehyde as an example. The two half reactions with their $\Delta E^{\circ\prime}$ values (as listed in Table 9.1, p. 321) of your text) are as follows:

			$E^{\circ\prime}$
$NAD^+ + H^+ + 2e^-$	$\rightarrow$	NADH	–0.32 V
Acetaldehyde + $2H^+ + 2e^-$	$\rightarrow$	ethanol	–0.20 V

For the $\Delta E^{\circ\prime}$ to be positive and for the electrons to balance on both sides of the overall equation, we must reverse the first reaction:

NADH	$\rightarrow$	$NAD^+ + H^+ + 2e^-$	+0.32 V
Acetaldehyde + $2H^+ + 2e^-$	$\rightarrow$	ethanol	–0.20 V
Acetaldehyde + H^+ + NADH	$\rightarrow$	ethanol + NAD^+	+0.12 V[3]

To identify oxidizing agents and reducing agents:

1. Locate each CONJUGATE REDOX PAIR. It often helps to draw the molecular structures.

   In the equation above, the two conjugate redox pairs are acetaldehyde/ethanol and NAD^+/NADH.

2. In each conjugate redox pair, one member will be its oxidized form, and the other will be its reduced form. Label them. Remember that the oxidized form will have more oxygens

[1] Remember: In calculating $\Delta E^{\circ\prime}$, do not multiply the value of $E^{\circ\prime}$ by the number of electrons transferred or by the coefficient in the balanced equation. This is different from ΔG° and ΔH°calculations.

[2] This method avoids the confusion of which $E^{\circ\prime}$ to subtract from. If you feel more comfortable with a different method, by all means use it.

[3] We have already seen this reaction as the second step of homolactic fermentation, catalyzed by alcohol dehydrogenase. (See Question 9.2, page 324, of your text.)

(and/or fewer H's and a higher oxidation state), and the reduced form will have fewer oxygens (and/or more H's and a lower oxidation state).

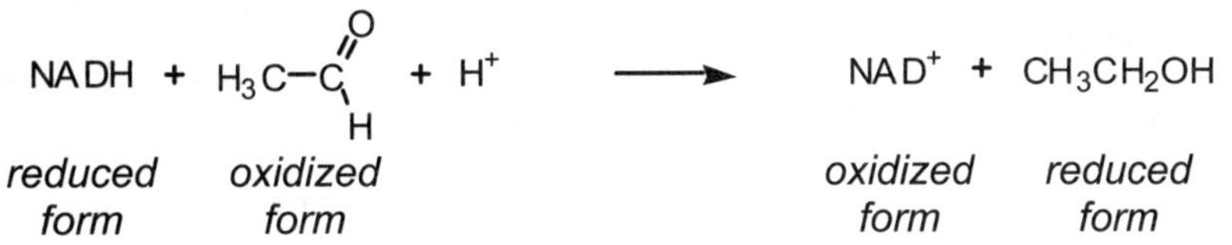

3. Check your labels using the balanced equation (above). Each side of the equation should have one reduced form and one oxidized form.
4. Of the two reactants, the reduced form is the reducing agent since it gives away its extra electrons, reducing the other reactant (and becoming oxidized in the process). Conversely, the oxidized form is the oxidizing agent since it takes electrons from the other reactant, oxidizing it (and becoming reduced in the process).

REDOX COENZYMES

NICOTINIC ACID

NAD(P), **NICOTINAMIDE ADENINE DINUCLEOTIDE (PHOSPHATE)**, is an electron carrier; see "The Organic Connection" in Chapter 6, p. 00, of this Study Guide for more information on NAD^+/NADH.

RIBOFLAVIN (vitamin B_2)

RIBOFLAVIN is the vitamin precursor for **FLAVIN MONONUCLEOTIDE (FMN)** and **FLAVIN ADENINE DINUCLEOTIDE (FAD)**, prosthetic groups for the **FLAVOPROTEINS**.

FMN has a three-ring isoalloxazine group that stabilizes a radical intermediate and can transfer electrons one at a time. The structure of FAD is FMN with another phosphate, another sugar, and an adenine and can also transfer one electron at a time.

FMN— – FLAVIN MONONUCLEOTIDE

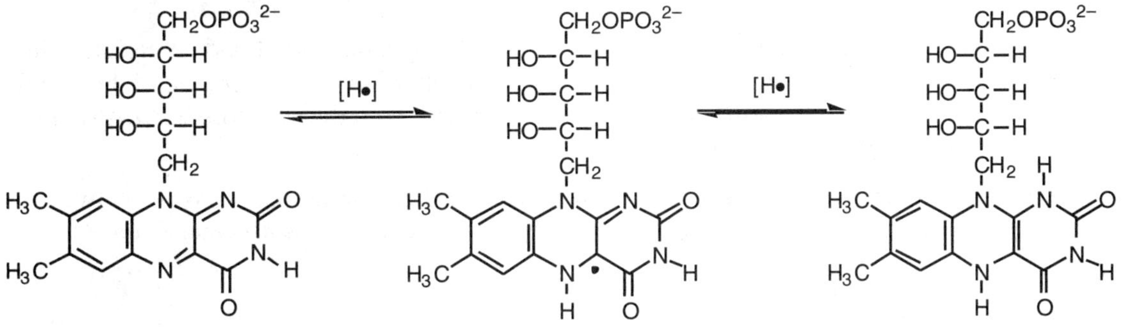

FMN (oxidized or quinone form) | FMNH• (radical or semiquinone form) | $FMNH_2$ (reduced or hydroquinone form)

9.2 CITRIC ACID CYCLE

THE GOALS OF THE CITRIC ACID CYCLE ARE TO:

1. Convert fuel molecules to **ENERGY** by producing reducing power (3 NADH and 1 $FADH_2$ per acetyl-CoA), which can enter the electron transport chain and provide the energy to synthesize ATP. One GTP (which can easily be converted to ATP) is produced directly, although this is a small energy bonus compared to the number of ATP that are ultimately produced from the electron transport chain.

2. Produce metabolic **INTERMEDIATES.** The carbon precursors for lipids, sugars, amino acids, and nucleic acids are all derived from citric acid cycle intermediates.

OVERALL REACTION

$$\text{Acetyl-CoA} + 3NAD^+ + FAD + GDP + P_i + 2H_2O \rightarrow 2CO_2 + 3NADH + FADH_2 + CoASH + GTP + 3H^+$$

In eukaryotes, the citric acid cycle takes place in the mitochondria. Remember that glycolysis occur in the cytoplasm.

COENZYMES

THIAMINE PYROPHOSPHATE (TPP)—decarboxylates and transfers aldehyde groups; used by pyruvate dehydrogenase (E_1 of the pyruvate dehydrogenase complex) and the α-ketoglutarate dehydrogenase complex

LIPOIC ACID—carries hydrogens or acetyl groups; used by dihydrolipoyl transacetylase (E_2 of the pyruvate dehydrogenase complex) and dihydrolipoyl transsuccinylase of the α-ketoglutarate dehydrogenase complex

FAD is used by dihydrolipoyl dehydrogenase (E_3 of the pyruvate dehydrogenase complex and of the α-ketoglutarate dehydrogenase complex) and succinate dehydrogenase.

NAD is used by dihydrolipoyl dehydrogenase (E3 of the pyruvate dehydrogenase complex and of the α-ketoglutarate dehydrogenase complex), isocitrate dehydrogenase, and malate dehydrogenase

COENZYME A (COASH)—carries acetyl groups via a thioester bond (acetyl-CoA); used by dihydrolipoyl transacetylase (E_2 of the pyruvate dehydrogenase complex) and dihydrolipoyl transsuccinylase of the α-ketoglutarate dehydrogenase complex

ACETYL-COA

Acetyl-CoA is an acetyl group bound to coenzyme A by a thioester bond. When studying reaction mechanisms that involve acetyl-CoA, it is helpful to remember that the linkage is a thioester. Compare the structure of acetyl-CoA below with the structure of CoASH in Figure 9.9, p. 328 of your text.

{Note that coenzyme A is abbreviated CoASH. To be consistent, instead of "acetyl-CoA" we should probably say "acetyl-S-CoA" but, acetyl-CoA is just easier and is in common usage.}

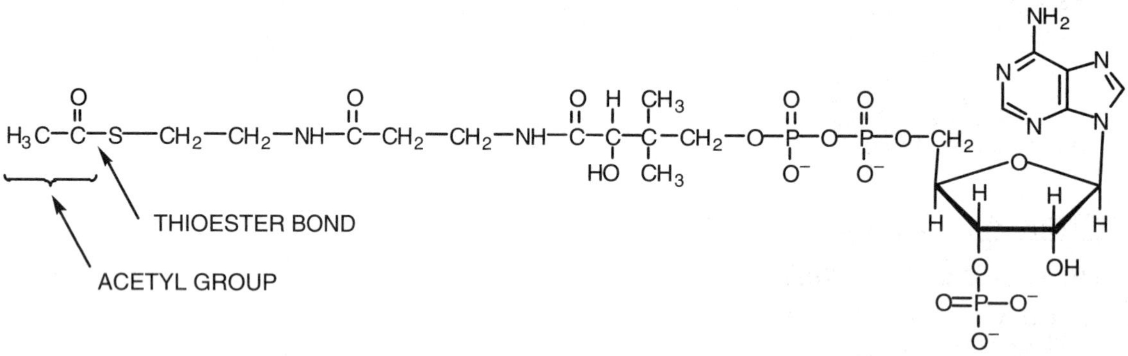

CONVERSION OF PYRUVATE TO ACETYL-COA $\Delta G^{\circ\prime} = -33.5$ KJ/MOL

OVERALL REACTION: AN OXIDATIVE DECARBOXYLATION

pyruvate + NAD^+ + CoASH → acetyl-CoA + NADH + CO_2 + H_2O + H^+

PYRUVATE DEHYDROGENASE COMPLEX = MULTIPLE COPIES OF:

E_1: pyruvate dehydrogenase (or pyruvate decarboxylase); needs TPP

E_2: dihydrolipoyl transacetylase; needs lipoic acid and CoASH

E_3: dihydrolipoyl dehydrogenase; needs FAD and NAD^+

MECHANISM OF THE PYRUVATE DEHYDROGENASE COMPLEX

(This is simplified to emphasize the changes to the acetyl group. For details regarding the linkage to the TPP thiazole ring, see Figure 9.10 of your text.)

E1: Decarboxylation of pyruvate to form hydroxyethyl-TPP (HETPP):

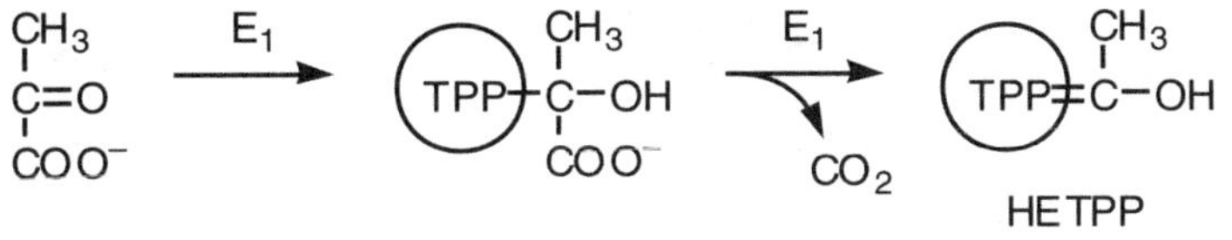

E2 converts the hydroxyethyl group of HETPP to acetyl-CoA:

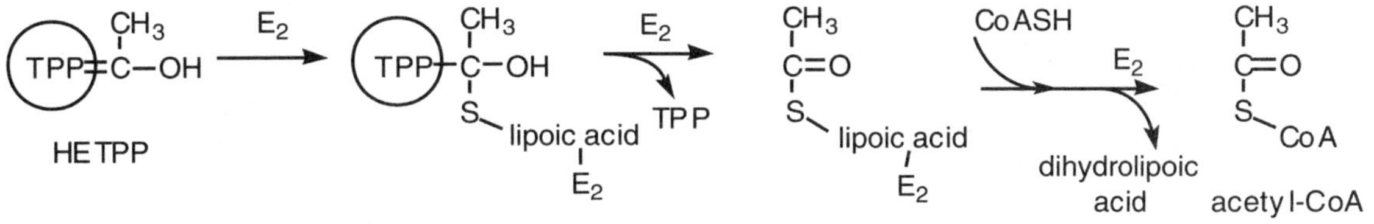

E3 regenerates lipoic acid; FAD is also regenerated spontaneously.

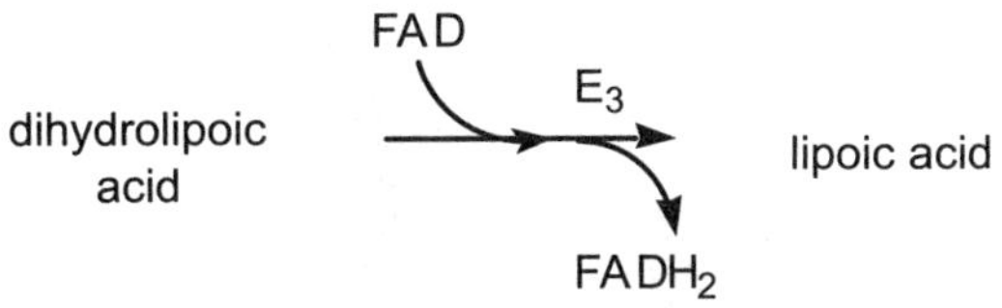

$FADH_2 + NAD^+ \rightarrow FAD + NADH + H^+$

REGULATION:

In general, inhibitors are products and indicators that the cell has plenty of energy (e.g., ATP), while activators are substrates and indicators that the cell needs energy (e.g., AMP)

PRODUCT INHIBITION: acetyl-CoA and NADH are inhibitors.

ALLOSTERIC: ATP inhibits; AMP, NAD^+, CoASH are activators.

COVALENT MODIFICATION: The products acetyl-CoA and NADH activate a kinase, which phosphorylates and inactivates the pyruvate dehydrogenase complex. Substrates pyruvate, CoASH, and NAD^+ inhibit this kinase, removing an inhibitor (that is not quite the same as activating a reaction, but it helps). Low levels of ATP activate a phosphoprotein phosphatase, which dephosphorylates and activates the pyruvate dehydrogenase complex.

REACTIONS OF THE CITRIC ACID CYCLE TAKE PLACE IN THE MITOCHONDRIAL MATRIX

Acetyl-CoA brings in two carbons, and two carbons leave as CO_2. Oxaloacetate is regenerated so the cycle can continue.

THE CITRIC ACID CYCLE

ACETYL-COA BRINGS IN TWO CARBONS; TWO CARBONS LEAVE AS CO_2

Acetyl-CoA enters the cycle.

acetyl-CoA   oxaloacetate

1. Citrate synthase

Aldol condensation, release of CoASH

Large negative ΔG

Note that the tertiary alcohol cannot be oxidized.

→ CoASH

↓

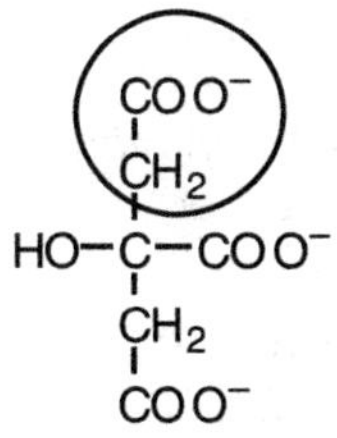

citrate

2. Aconitase

Now the alcohol is secondary and can be oxidized.

Mechanism: dehydrate, then rehydrate. Intermediate is *cis*-aconitate.

This reaction is stereospecific. Only the product with the stereochemistry shown is formed.

↓↑

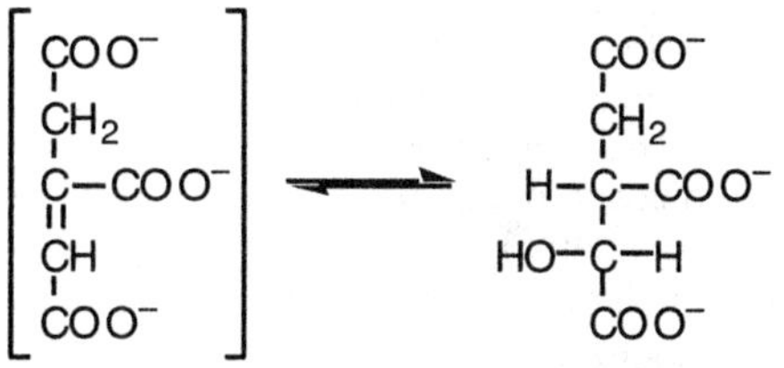

cis-aconitate   isocitrate

↓↑

3. Isocitrate dehydrogenase

Oxidation of isocitrate to form an oxalosuccinate intermediate

Reduction of NAD^+ to NADH

Decarboxylation

NAD^+ → NADH

↓

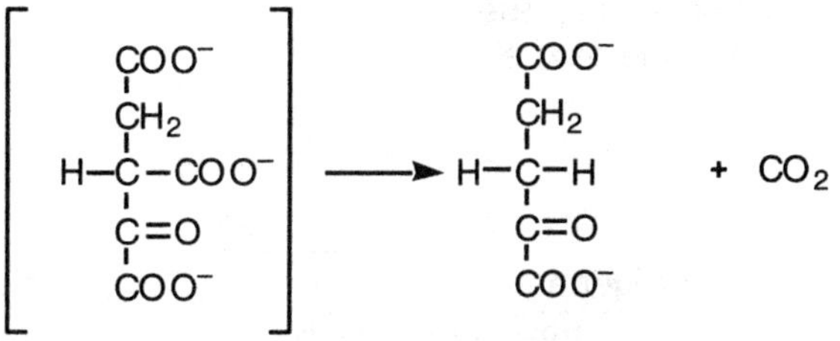

oxalosuccinate   α-ketoglutarate

4. α-Ketoglutarate dehydrogenase complex

Oxidation of α-ketoglutarate

Reduction of NAD^+ to NADH

Decarboxylation

Large negative ΔG

Similar to pyruvate dehydrogenase complex, formation of a thioester with CoASH

Allosterically regulated

+ CoASH + NAD^+

↓

NADH + H^+ + CO_2

↓

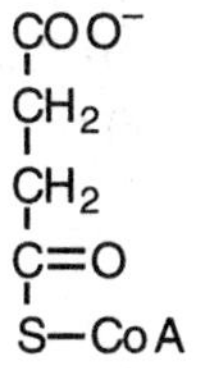

succinyl-CoA

CITRIC ACID CYCLE, CONTINUED . . .

REGENERATION OF OXALOACETATE TO CONTINUE THE CYCLE

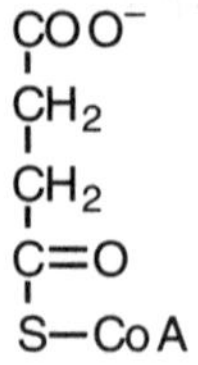

succinyl-CoA

↓↑

5. Succinate thiokinase

$GDP + P_i \rightarrow GTP$

Cleavage of thioester bond, goodbye CoASH

Substrate-level phosphorylation of GDP in mammals (other organisms use ADP)

$GTP + ADP \rightleftharpoons GDP + ATP$

Succinate is symmetrical! Carbons 1 and 4 are now identical, as are carbons 2 and 3.

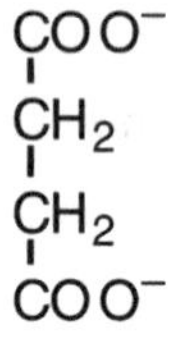

succinate

↓↑

6. Succinate dehydrogenase

$FAD \rightarrow FADH_2$

Oxidation of succinate

Reduction of FAD to $FADH_2$

Bound tightly to inner mitochondrial membrane (This is the only citric acid cycle enzyme not in the mitochondrial matrix.)

Subsequent transfer of electrons from $FADH_2$ to coenzyme Q drives this reaction forward.

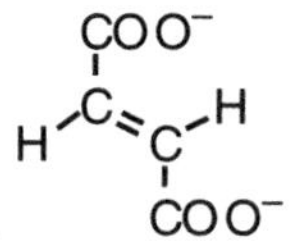

fumarate

↓↑

7. Fumarase (also called fumarate hydratase)

Stereospecific hydration

This reaction sets up the formation of OAA in the next step.

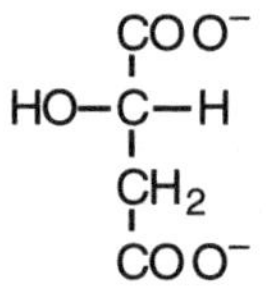

malate

↓↑

Malate dehydrogenase

$NAD^+ \rightarrow NADH + H^+$

Compare the structure of OAA with that of succinate. It took three steps to change a CH_2 to a C=O.

$\Delta G^{\circ\prime} = +29$ kJ/mol, BUT the next reaction with another molecule of acetyl-CoA pulls this reaction forward by removing OAA (so the *actual* ΔG will be negative).

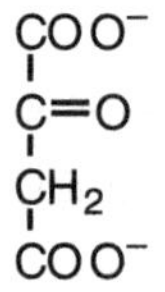

oxaloacetate (OAA)

FATE OF CARBON ATOMS IN THE CITRIC ACID CYCLE

To trace specific (labeled) carbon atoms through the citric acid cycle:

DRAW OUT THE STRUCTURES OF THE CITRIC ACID CYCLE ON ONE PAGE, AS IN FIGURE 9.8 OF YOUR TEXT. CIRCLE THE LABELED CARBON IN EACH STRUCTURE, FOLLOWING IT AROUND THE CITRIC ACID CYCLE. WHEN YOU GET BACK TO CITRATE, SWITCH TO DRAWING A SQUARE AROUND THE LABELED CARBON. FOR EACH TURN OF THE CYCLE, USE A DIFFERENT SHAPE (DIFFERENT COLORED MARKERS ARE EVEN BETTER). THIS IS A NEAT WAY TO SEE WHERE SPECIFIC CARBON ATOMS END UP WITH EACH TURN OF THE CYCLE.

When you get to succinate, you have to choose between carbons 1 and 4 (or between carbons 2 and 3), since succinate is symmetrical, and those carbons are equivalent. Try repeating the exercise above, choosing an equivalent (but different) carbon. It is interesting to see how that will change the number of turns of the cycle it takes to lose the labeled carbon as CO_2.

Try using this method to answer Question 9.5 on page 335 of your text.

THE CITRIC ACID CYCLE IS AMPHIBOLIC (I.E., BOTH CATABOLIC AND ANABOLIC).

In the catabolic process 2 carbons enter (acetyl-CoA) and they leave oxidized (CO_2).

To be anabolic, the citric acid cycle must produce intermediates. If 2 carbons enter and 2 CO_2 leave, how can the citric acid cycle produce intermediates? There are two possible solutions:

1. There must be a way to bypass the CO_2-generating steps of the cycle. This is the glyoxylate cycle, used by plants (and discussed later).
2. There must be additional ways for carbon to enter the cycle (other than via acetyl-CoA). Mammals use this option. **ANAPLEROTIC REACTIONS** are reactions that replenish citric acid cycle intermediates.

ANAPLEROTIC REACTIONS

- pyruvate → oxaloacetate → citric acid cycle (instead of pyruvate → acetyl-CoA)

  Pyruvate carboxylase is activated by acetyl-CoA (which indicates that there is not enough OAA). Excess OAA can be used for gluconeogenesis. (Remember that the first two steps of gluconeogenesis are pyruvate → oxaloacetate → PEP.)
- certain fatty acids (also certain amino acids) → succinyl-CoA
- glutamate → α-ketoglutarate
- aspartate → oxaloacetate

CITRIC ACID CYCLE INTERMEDIATES ARE USED IN THESE BIOSYNTHESES:

Amino acids and proteins (from oxaloacetate and α-ketoglutarate); heme and chlorophyll (from succinyl-CoA)

Glucose (from oxaloacetate)

Fatty acids and cholesterol (from acetyl CoA)

Pyrimidines (from oxaloacetate) and purines (from α-ketoglutarate)

CITRIC ACID CYCLE REGULATION

Remember the ultimate goals of the citric acid cycle, and its regulation will make sense. When the cell needs energy or intermediates, the cycle will be activated. The cycle will be inhibited when the cell has plenty of ATP and its energy needs are being met. It will also be inhibited at low substrate concentrations—including NAD^+, FAD, and ADP.

Low energy	High Energy
low NADH/ NAD^+ ratio[4]	high NADH/ NAD^+ ratio
low ATP/ADP ratio	high ATP/ADP ratio

Metabolic branch points: Look at Figure 9.13 in your text (or at the list above). The four intermediates mentioned are oxaloacetate, acetyl-CoA, α-ketoglutarate, and succinyl-CoA. So, it makes sense that their enzymes are closely regulated. It also makes sense that the two CO_2-generating reactions would be closely regulated.

ISOCITRATE DEHYDROGENASE

isocitrate + NAD^+ → α-ketoglutarate + NADH + CO_2

Activated by high ADP, NAD^+ — Inhibited by ATP, NADH

Isocitrate dehydrogenase is also important for citrate metabolism. Only citrate can penetrate the mitochondrial membrane; acetyl-CoA cannot. When the cell's energy needs are being met, citrate is not needed for the citric acid cycle. Citrate can be used to carry acetyl-CoA from the mitochondria to the cytosol, where it can be used in fatty acid synthesis. Citrate activates the first reaction of fatty acid synthesis, and recall that citrate inhibits glycolysis (at PFK-1). Figure 9.15 in your text is a great summary of citrate metabolism.

α-KETOGLUTARATE DEHYDROGENASE

α-ketoglutarate + NAD^+ → succinyl-CoA + NADH + CO_2 + H^+ activated by NAD^+ and ADP
Inhibited by: NADH and succinly-CoA

Will the citric acid cycle make intermediates or will it make energy? To make intermediates, pyruvate enters the cycle as oxaloacetate (using pyruvate carboxylase). To make energy, pyruvate enters the cycle as acetyl-CoA (using pyruvate dehydrogenase). Obviously, these two enzymes will be closely regulated as well.

Acetyl-CoA activates pyruvate carboxylase and inhibits pyruvate decarboxylase. Why does this make sense? If intermediates are being made, oxaloacetate levels drop and acetyl-CoA will accumulate. Activating pyruvate carboxylase replenishes oxaloacetate; inhibiting pyruvate dehydrogenase prevents more acetyl-CoA from being made.

[4] Watch out for the quintessential trick question, which plays with these ratios. Make sure that this makes sense to you, so that you will immediately see that a ***low*** NADH/NAD^+ ratio and a ***high*** NAD^+/NADH ratio are the same, that they both indicate a need for energy, and that the citric acid cycle will be activated.

THE GLYOXYLATE CYCLE BYPASSES CO_2-GENERATING STEPS
TWO ACETYL-COA REACT TO FORM ONE MALATE

The glyoxylate cycle occurs in plants, some fungi, algae, protozoans, and bacteria.

Glyoxylate enzymes occur in the cytoplasm, except in plants, where they occur in glyoxysomes.

1. A glyoxysome-specific isozyme of citrate synthase

...same as the citric acid cycle...

2. A glyoxysome-specific isozyme of aconitase

...same as the citric acid cycle...

3. Isocitrate lyase

Aldol cleavage

4. Malate synthase

Also requires H_2O

Net:

Four carbons in,
zero carbons out !!

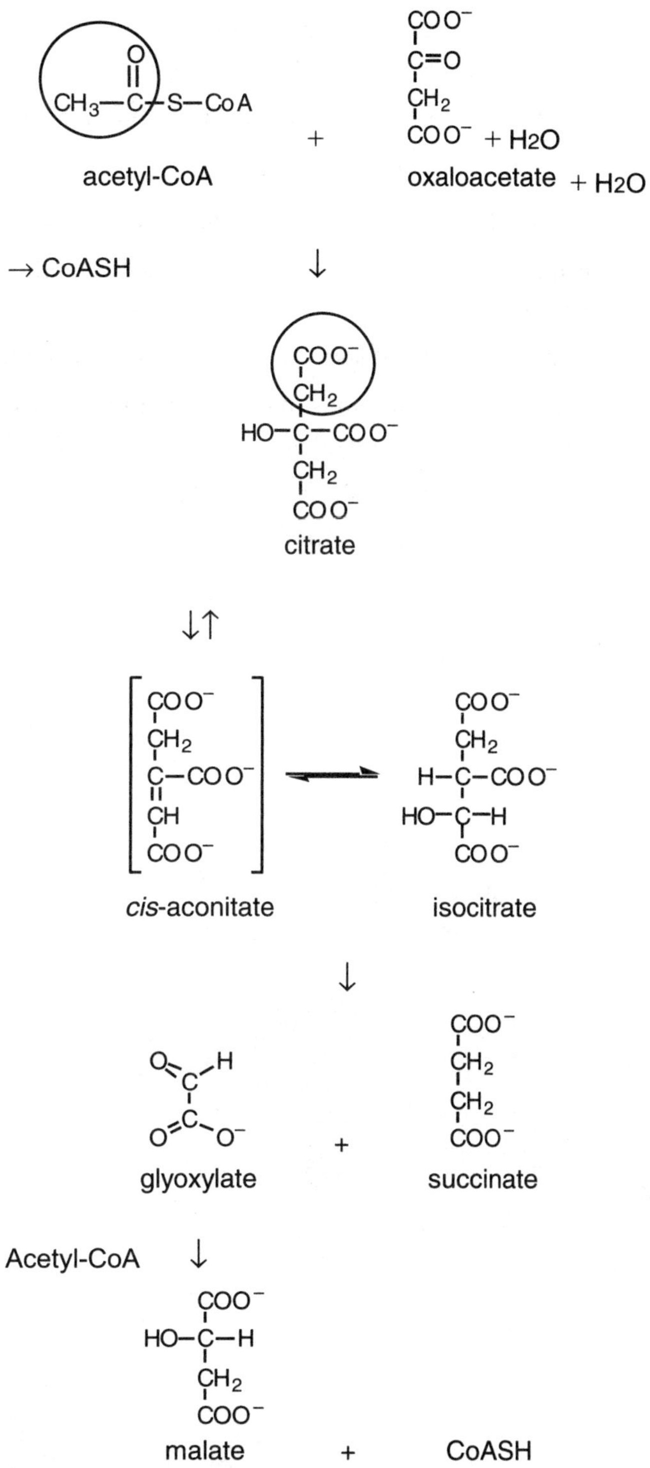

Succinate and malate can each continue around the citric acid cycle to form the intermediate(s) needed

AFTER STUDYING THIS CHAPTER, YOU SHOULD BE ABLE TO:

- Calculate $\Delta E^{\circ\prime}$ for a reaction from half-cell potential ($\Delta E^{\circ\prime}$) data. Determine whether a given redox reaction will proceed in the direction written.
- Identify redox reactions. Identify oxidizing and reducing agents. Identify whether a molecule is oxidized or reduced in a given reaction.
- Calculate $\Delta G^{\circ\prime}$ using the equation $\Delta G^{\circ\prime} = -nF\Delta E^{\circ\prime}$ ($F = 96.5$ kJ/V·mol).
- Identify oxidation states of atoms in molecules.
- Trace specific carbons through the pathways learned to date.
- Determine whether the citric acid cycle will be activated or inhibited, given one or more of the following data:

  high/low levels of specific effectors (such as high ATP levels)

  ratios of concentrations of effectors or indicators of energy levels (such as high NAD^+/NADH or low AMP/ATP)

  specific energy conditions (running vs. resting, for example)

Use this space to note any additional objectives provided by your instructor.

CHAPTER 9: SOLUTIONS TO REVIEW QUESTIONS

9.1 a. obligate anaerobe – an organism that grows only in the absence of oxygen

b. aerotolerant anaerobe – an organism that depends on fermentation for its energy needs and possesses protection from toxic oxygen metabolites (detoxifying enzymes and antioxidant molecules)

c. facultative anaerobe – an organism that possesses the capacity for detoxifying oxygen metabolites; energy is generated using oxygen when available, as an electron acceptor

d. obligate aerobe – an organism that is highly dependent on oxygen for energy protection

e. reactive oxygen species – a reactive derivative of molecular oxygen, including superoxide radicals, hydrogen peroxide, the hydroxyl radical, and singlet oxygen

9.2 a. thiamine pyrophosphate (TPP) – the coenzyme form of thiamine; also called vitamin B_1

b. lipoic acid – a biomolecule that contains a carboxylate group and two sulfhydryl groups that are easily oxidized or reduced; functions as an acyl group carrier in pyruvate dehydrogenase complex and α-ketoglutarate dehydrogenase complex

c. PDHC – pyruvate dehydrogenase complex – the enzyme complex that catalyzes the oxidative decarboxylation of pyruvate to yield acetyl-CoA, carbon dioxide and NADH

d. HETPP – hydroxyethyl-thiamine pyrophosphate – an intermediate in the oxidative decarboxylation of pyruvate to yield acetyl-CoA

e. nucleoside diphosphate kinase – an enzyme that catalyzes the reversible transfer of phosphoryl group between a nucleotide and nucleoside diphosphate

9.4 a. glyoxysome – a peroxisome in germinating seed; glyoxysome enzymes contribute to the conversion of TG reserves to carbohydrate

b. coenzyme A – a carrier of acetyl and acyl groups that contains a terminal reactive thiol

c. $NAD(P)^+$ - nicotinamide adenine dinucleotide phosphate – an oxidized coenzyme form of nicotinic acid in which the N-ribosyl derivative of nicotinamide is linked to adenosine; in $NADP^+$ an additional phosphate is attached to the 2'-OH group of the adenosine sugar; involved in electron transfer in a class of compounds called dehydrogenases

d. FAD – flavin adenine dinucleotide – a prosthetic group consisting of riboflavin, D-ribitol, and adenine that functions in the class of enzymes called flavoproteins

e. FMN – flavin mononucleotide – a prosthetic group consisting of a molecule of riboflavin and D-ribitol phosphate that functions in the class of enzymes called flavoproteins

9.5 Ancient earth possessed an atmosphere that contained methane, ammonia and was devoid of oxygen. With the development of photosynthesis, oxygen was released into the atmosphere. Subsequently, this oxygen reacted with methane to form carbon dioxide and with ammonia to form molecular nitrogen. The continued release of oxygen produced an oxidizing atmosphere consisting of primarily oxygen, nitrogen, and carbon dioxide.

9.7 The citric acid cycle is an important component of aerobic respiration. The NADH and $FADH_2$ produced during oxidation-reduction reactions of the cycle donate electrons to the mitochondrial ETC. Citric acid cycle intermediates are also used as biosynthetic precursors

9.8 Citrate, produced in the mitochondria by the citric acid cycle, is transported across the mitochondrial membrane into the cytoplasm. Once in the cytoplasm, citrate is cleaved by citrate lyase to acetyl-CoA and oxaloacetate. The acetyl-CoA is then used to synthesize fatty acids as well as other biomolecules. The oxaloacetate is reduced to malate and moved across the mitochondrial membrane, where it is reoxidized to oxaloacetate, the citric acid cycle intermediate.

9.10 The glyoxylate cycle is a modified version of the citric acid cycle in which the two reactions that release carbon dioxide are bypassed As a result organisms that have the glyoxylate cycle can grow using two carbon compounds such as acetate and acetyl-CoA.

9.11 Of all the conditions listed only (d) low citrate concentration indicates a low cell energy status and would increase flux through the citric acid cycle:. The remaining conditions indicate that the cell's energy requirements are currently being met, and would reduce flux through the citric acid cycle.

9.13 Isocitrate dehydrogenase requires NAD^+ (or $NADP^+$) and the cofactor Mn^{2+}. The α-ketoglutarate dehydrogenase complex requires TPP (thiamine pyrophosphate),lipoic acid, CoASH (coenzyme A), and NAD^+. Succinate dehydrogenase requires FAD. Malate dehydrogenase requires NAD^+.

9.14 The three mechanisms of control of the irreversible reactions of the citric acid cycle are: substrate availability, product inhibition, and competitive feedback inhibition. Citrate synthase catalyzes reaction 1, and is sensitive to availability of substrates oxaloacetate and acetyl-CoA. In many gram-negative bacteria such as *E. coli*, ATP, NADH, and succinyl-CoA allosterically inhibit citrate synthase. Isocitrate dehydrogenase, which catalyzes the next irreversible reaction (3), is activated by high ADP and NAD^+ and inhibited by ATP and NADH. The third irreversible reaction (4) is catalyzed by α-ketoglutarate dehydrogenase, which is activated by AMP, and inhibited by products succinyl-CoA and NADH.

9.16 The free energy changes are as follows:

a. $1/2O_2 + NADH + H^+ \rightarrow H_2O = NAD^+$

$\Delta E^0 = 0.82V + 0.32V = 1.14V$

$\Delta G^0 = 2(96485 J/molV)(1.14V)$

$=219985$ J

$=220$kJ

b. $S + NADH + H^+ \rightarrow H_2S + NAD^+$

$\Delta E^0 = -0.23V + 0.32V = 0.09V$

$\Delta G^0 = 2(96485 J/molV)(0.09V)$

$= 17{,}367$ J

$= 17$kJ

9.17 The difference in energy yield is

ΔG^0 (O_2 reaction) - ΔG^0(S reaction) =

219985J – 17367J

= 202618J

= 200kJ

9.19 If a small amount of [1-^{14}C]glucose is added to an aerobic yeast culture, the radiolabel will appear in citrate molecules at the carboxyl carbon that is attached to citrate's third carbon atom. In glycolysis, the [1-^{14}C]glucose is split into two molecules of pyruvate, one of which contains the ^{14}C at its carboxyl carbon (C1). If the ^{14}C-labeled pyruvate is converted to acetyl-CoA by pyruvate dehydrogenase, the carboxyl carbon will be released as $^{14}CO_2$. However, if the pyruvate is carboxylated to form oxaloacetate, the ^{14}C will be C1 in oxaloacetate. Upon further reaction with acetyl-CoA, the ^{14}C becomes the carboxyl carbon attached to C3 of citrate.

9.20 Pyruvate dehydrogenase complex is a large multienzyme complex that contains three enzyme activities, each of which is present in multiple copies. E_1 is pyruvate dehydrogenase (or pyruvate decarboxylase) with TPP, E_2 is dihydrolipoyl transacetylase with lipoic acid and CoASH, and E_3 is dihydrolipoyl dehydrogenase, with NAD^+ or FAD. The pyruvate dehydrogenase complex in mammalian cells contains 60 copies of E_2 and 20-30 copies each of E_1 and E_3.

9.22 The roles of genes involved in carcinogenic aerobic glycolysis are as follows. cMyc is a transcription factor that regulates the expression of numerous genes involved in cell division and apoptosis. A tumor with unregulated cMyc diverts glycolytic intermediates into pathways that synthesize molecules required for cell division such as nucleotides (DNA synthesis), amino acids (protein synthesis) and lipids (membrane synthesis). HIF-1 is a transcription factor that ordinarily induces genes that promote survival of cells when oxygen levels are temporarily low. In tumors stabilized HIF-1 reduces oxygen requirements in part by diverting larger than normal amounts of pyruvate away from mitochondrial oxidation. PKB facilitates carcinogenesis by enhancing glucose uptake and glycolysis and diverts glycolytic intermediates into biosynthetic pathways that are required for rapid growth. p53 is a transcription factor that regulates the expression of over 100 genes, some of which regulate energy metabolism. When p53, a major tumor suppressor, is inactivated aerobic glycolysis results, partially because of an increase of glycolytic flux and decreased oxidative phosphorylation.

9.23 Germinating seed uses the glyoxylate cycle to convert acetyl-CoA derived from fatty acids to glucose as follows. Within the glyoxysome the acetyl-CoA product of fatty acid oxidation enters the glyoxylate cycle as it reacts with oxaloacetate to form citrate. After citrate is converted to isocitrate, the latter molecule is split into succinate and the two carbon molecule glyoxylate. Glyoxylate then reacts with a second acetyl-CoA molecule to yield malate and CoASH. Malate is converted to oxaloacetate, which is used to sustain the glyoxylate cycle. Succinate exits the glyoxysome and enters the mitochondrion where it is converted to malate. After transport into the cytoplasm malate is converted to oxaloacetate, which is subsequently converted to phosphoenolpyruvate. Using the enzymes of gluconeogenesis PEP is converted into glucose.

9.25 The net equation for the citric acid cycle is:

$$\text{Acetyl-CoA} + 3\text{NAD}^+ + \text{FAD} + \text{GDP(or ADP)} + \text{P}_i + 2\text{H}_2\text{O} \rightarrow$$

$$2\text{CO}_2 + 3\text{NADH} + \text{FADH}_2 + \text{CoASH} + \text{GTP(or ATP)} + 2\text{H}^+$$

9.26 The steps in the citric acid cycle that are regulated are those catalyzed by citrate synthase, which converts acetyl-CoA and oxaloacetate into citrate; isocitrate dehydrogenase, which converts isocitrate to α-ketoglutarate; and α-ketoglutarate dehydrogenase, which forms succinyl-CoA. These steps represent important metabolic branch points. Acetyl-CoA may be used in biosynthetic pathways such as fatty acid synthesis. Also, acetyl-CoA activates pyruvate carboxylase (and inhibits pyruvate dehydrogenase), which activates the reaction of pyruvate with malate to form oxaloacetate (instead of more acetyl-CoA). Because citrate can penetrate the mitochondrial membrane and be cleaved to form acetyl-CoA and oxaloacetate, it therefore can be used to transport acetyl-CoA out of the mitochondrion for fatty acid synthesis. Citrate also inhibits PFK-1 in glycolysis and citrate synthase in the citric acid; cycle, it also activates the first step in fatty acid synthesis. α-Ketoglutarate plays an important role in amino acid metabolism and other metabolic processes. Succinyl-CoA inhibits and α-ketoglutarate dehydrogenase, and is required in the synthesis of porphyrins.

CHAPTER 9: SOLUTIONS TO FILL-IN-THE-BLANK QUESTIONS

9.28 Amphibolic

9.29 Anaplerotic

9.31 Pyruvate dehydrogenase

9.32 Succinate

9.34 Oxygen-evolving complex

9.35 Reductive

9.37 Reduction

CHAPTER 9: SOLUTIONS TO SHORT-ANSWER QUESTIONS

9.38 Recall that in citrate metabolism when mitochondrial levels of citrate are high, the molecule is transported into the cytoplasm, where it is converted to oxaloacetate and acetyl-CoA, the substrate for fatty acid synthesis. Oxaloacetate can then be reduced by NADH to yield malate. Malic enzyme, which oxidizes malate to form pyruvate, uses NADP+ as the oxidizing agent to yield NADPH.

9.40 Germinating seeds convert the two-carbon acetyl group of acetyl-CoA, the product of fatty acid degradation, into glucose via the glyoxylate cycle. The first two reactions in the glyoxylate cycle are the first two reactions in the citric acid cycle: the formation of citrate from oxaloacetate and acetyl-CoA and the isomerization of citrate to yield isocitrate. These two reactions are glyoxysome-specific variants of the citric acid cycle enzymes. Isocitrate is then split to form succinate and glyoxylate. Succinate is then transported into a mitochondrion, where it enters the citric acid cycle where it is converted to malate. The oxidation of malate yields oxaloacetate, a substrate of the glucose-producing pathway gluconeogenesis.

9.41 The relative activities of the pyruvate dehydrogenase complex (PHDC) and pyruvate carboxylase determine the fate of pyruvate. If PDHC is more active than pyruvate carboxylase, the pyruvate will be used as a precursor of acetyl-CoA, the substrate for the citric acid cycle. If pyruvate carboxylase is more active, then oxaloacetate, the product of the carboxylation of pyruvate, will replenish the intermediates of the citric acid cycle, some of which are used as biosynthetic precursors.

CHAPTER 9: SOLUTIONS TO THOUGHT QUESTIONS

9.43 Fluoroacetate reacts with CoASH to form fluoroacetyl-CoA, which is then converted to fluorocitrate by citrate synthase. Once fluorocitrate enters the active site of aconitase the intermediate 2-fluoro-*cis*-aconitate forms. However the subsequent addition of the hydroxyl group results in the migration of the double bond with the elimination of the electronegative fluorine to give 4-hydroxy-*trans*-aconitate.

9.44

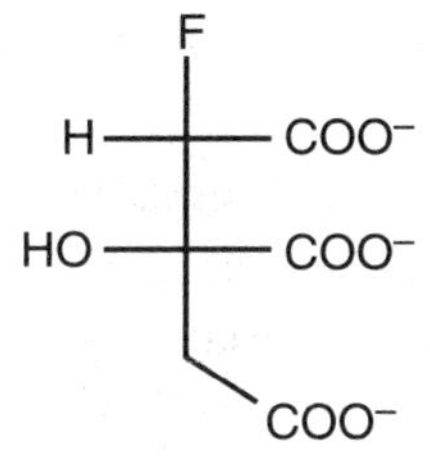

Fluoroacetate

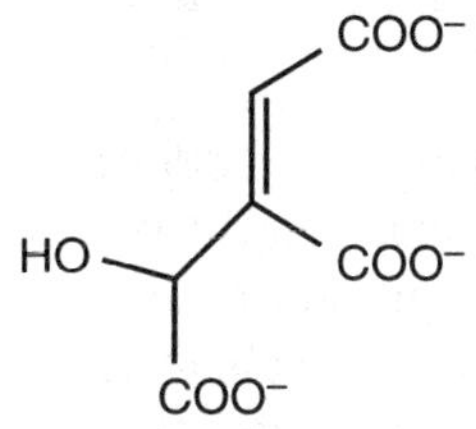

4-Hydroxy-*trans*-aconitate

9.46 After glucose has been absorbed from the intestine, it is transported to the liver, where it is converted to glucose-6-phosphate, which then enters the glycolytic cycle and is eventually transformed into two molecules of pyruvate. Pyruvate is converted to acetyl-CoA, which then can enter the citric acid cycle

9.47 a. $NADH + H^+ + {}^1/_2\, O_2 \rightarrow NAD^+ + H_2O$ $\Delta E_0' = +\ 1.14$ V

$\Delta G^{0\prime} = -nF\Delta E_0'$

$= (-2)(96{,}485 J/V \bullet mol)(+1.14 V)$

$= -\ 219{,}985.8$ J/mol

$= -220$ kJ/mol

b. $Cyt\ c\ (Fe^{2+}) + {}^1/_2\, O_2 \rightarrow Cyt\ c\ (Fe^{3+}) + H_2O$ $\Delta E_0' = 0.58$ V

$\Delta G^{0\prime} = (-2)(96{,}485 J/V \bullet mol)(+0.58 V)$

$= -112$ kJ/mol

9.49 Acetate is converted to acetyl-CoA, which is then converted through the citric acid cycle to oxaloacetate. Via gluconeogenesis, oxaloacetate is converted to glucose. Excess glucose molecules are then stored as glycogen.

9.50 Malonate is a competitive inhibitor of succinate dehydrogenase. It competes reversibly with succinate for the active site in the enzyme. Increasing the concentration of succinate should reverse the effect of the malonate.

9.52 At low oxygen levels the cells switch over to glycolysis as a source of energy. As a result ATP levels fall. Without sufficient ATP, cells cannot maintain appropriate intracellular ion concentrations. Among the consequences of this phenomenon are rising calcium levels and the activation of calcium-dependent enzymes such as the phospholipases, which begin degrading membrane phospholipids. Osmotic pressure also rises causing cells to swell and leak their contents.

9.53 When oxygen levels are reduced products of the citric acid cycle accumulate. Energy generation is primarily by glycolysis. The end product of glycolysis is pyruvate. In order to regenerate the NAD^+ required for the reaction pyruvate is reduced to lactate.

9.54 Inhibition of pyruvate dehydrogenase kinase, an enzyme that contributes to the regulation of pyruvate dehydrogenase, has the effect of increasing the conversion of pyruvate molecules to acetyl-CoA, thereby decreasing lactate levels.

9.55 Pyruvate dehydrogenase catalyzes the conversion of pyruvate to acetyl-CoA, which is then used in the citric acid cycle. Lack of this enzyme would cause pyruvate to accumulate. Excess pyruvate molecules results in high lactate levels. Pyruvate is also converted by transamination to alanine, so the levels of this amino acid would also be expected to rise.

9.56 The acetyl-CoA is used to produce energy, so its carbons are not available for synthesis of oxaloacetate. (Also, acetyl-CoA brings two carbons into the citric acid cycle, and two CO_2 molecules are produced, leaving no net addition of carbons to make oxaloacetate.) The required additional oxaloacetate molecules are synthesized by pyruvate carboxylase from pyruvate and carbon dioxide.

9.58 Common two-carbon fragments that could be present are acetate and acetyl-CoA.

9.59 The half-cells for the reaction are

1/2 02 + 2 H+ + 2 e---> H2O ΔE0 = + 82 V

FADH2 --> FAD + 2H+ + 2 e- ΔE0 = + 22 V

For the formation of water the reaction would be

1/2 O2 + FADH2 --> H2O + FAD

ΔE0 = + 82 V + 0.22 V = 1.04 V

ΔG0 = 2(96485 J/mol.V) (1.04) = 200688 J = 201 kJ

The free energy for 1/2 02 + NADH + H+ --> H2O + NAD+ is 220kJ, which is a higher value than the free energy for the reaction involving FADH2.

10 Aerobic Metabolism II: Electron Transport and Oxidative Phosphorylation

Brief Outline of Key Terms and Concepts

10.1 ELECTRON TRANSPORT

AEROBIC RESPIRATION

ELECTRON TRANSPORT AND ITS COMPONENTS

The **ELECTRON TRANSPORT CHAIN (ETC)** is located in the inner mitochondrial membrane of eukaryotic cells. Electrons are transferred from NADH and $FADH_2$ through the following complexes, ultimately reducing O_2 to H_2O. Free energy released in the ETC pumps H^+ from the matrix into the intermembrane space, creating a proton gradient across the membrane. Electron carriers include IRON-SULFUR CENTERS; NONHEME IRON PROTEINS; COENZYME Q (UQ, UBIQUINONE), and CYTOCHROMES.

COMPLEX I (NADH DEHYDROGENASE COMPLEX)
COMPLEX II (SUCCINATE DEHYDROGENASE COMPLEX)
COMPLEX III (CYTOCHROME B_1 COMPLEX; Q CYCLE)
COMPLEX IV (CYTOCHROME OXIDASE)

Electron Transport: The Fluid-State Model vs. the Solid-State Model

The respirasome is believed to be the functional respiration unit because it is the most active and stable version of the ETC

ELECTRON TRANSPORT INHIBITORS

Antimycin A inhibits cyt b (complex III). Rotenone and amytal inhibit NADH dehydrogenase (complex I). CO, N^{3-}, and CN^- inhibit cytochrome oxidase (complex IV).

10.2 OXIDATIVE PHOSPHORYLATION

THE CHEMIOSMOTIC COUPLING THEORY

The proton (H^+) gradient (ΔpH) created by the ETC is also an electrical potential, Δp, the PROTONMOTIVE FORCE. Protons may only cross the membrane through specific channels (ATP synthase), resulting in the synthesis of ATP.

UNCOUPLER; IONOPHORE

ATP SYNTHESIS

ATP SYNTHASE converts energy from the protonmotive force into rotational energy, which causes changes in conformation: L (loose), T (tight), O (open). ADP and P_i bind, and ATP is released.

CONTROL OF OXIDATIVE PHOSPHORYLATION

ATP synthase is inhibited by high [ATP] and activated by high [ADP]; RESPIRATORY CONTROL is control by [ADP]. The ATP mass action ratio is controlled in part by transport proteins in the inner membrane:

ADP-ATP TRANSLOCATOR
PHOPHATE TRANSLOCASE ($H_2PO_4^-/H^+$ SYMPORTER)

The P/O RATIO reflects the coupling between electron transport and ATP synthesis. The measured maximum P/O ratios for NADH and $FADH_2$ are 2.5 and 1.5, respectively.

THE COMPLETE OXIDATION OF GLUCOSE

GLYCEROL PHOSPHATE SHUTTLE
MALATE-ASPARTATE SHUTTLE

Aerobic oxidation of glucose yields 29.5 to 31 ATP molecules.

UNCOUPLED ELECTRON TRANSPORT

UNCOUPLING PROTEINS (UCP1 or THERMOGENIN; also UCP2, UCP3)

Dissipation of the proton gradient releases energy as heat: NONSHIVERING THERMOGENESIS

10.3 OXYGEN, CELL FUNCTION, AND OXIDATIVE STRESS

RESPIRATORY BURST

REACTIVE OXYGEN SPECIES (ROS)

ROS, unstable RADICALS that are derivatives of O_2, ROS include SUPEROXIDE RADICAL, HYDROXYL RADICAL, AND SINGLET OXYGEN. ROS form as a normal by-product of metabolism, as O_2, a DIRADICAL, accepts electrons one at a time during reduction. Other causes include radiation exposure. REACTIVE NITROGEN SPECIES (RNS) are often classified as ROS because the synthesis of these species is often linked. Examples of RNS: NO (nitric oxide), nitrogen dioxide, and peroxynitrite.

ANTIOXIDANT ENZYME SYSTEMS

The major enzymatic defenses against oxidative stress are SUPEROXIDE DISMUTASE (SOD), GLUTATHIONE PEROXIDASE [part of the GLUTATHIONE-CENTERED SYSTEM (GSH/GSSG)]

PEROXIREDOXIN (PRX), and CATALASE. The TRX-centered system includes TRX (thioredoxin), TR (thioredoxin reductase), and PRX.

ANTIOXIDANT MOLECULES protect cell components from oxidative damage. Prominent antioxidants include GSH and the dietary components α-tocopherol, β-carotene, and ascorbic acid.

10.1 ELECTRON TRANSPORT

Energy. AEROBIC RESPIRATION (using oxygen to generate energy from food) yields much more energy from nutrients than does fermentation. This energy is used to do work such as synthesizing ATP, pumping Ca^{2+} into the mitochondrial matrix, and generating heat in brown adipose tissue. ATP hydrolysis fuels endergonic reactions, runs molecular machines, and in general, acts as the energy "currency" for the cell.

WHY OXYGEN? O_2 PROPERTIES THAT ARE RELEVANT TO AEROBIC METABOLISM:

O_2 is everywhere (almost) and diffuses easily across cell membranes.

O_2 is highly reactive so it accepts electrons readily. (That can also be dangerous because O_2 can form ROS.)

ELECTRON TRANSPORT AND ITS COMPONENTS

ELECTRON DONORS

NADH and $FADH_2$ (from glycolysis, citric acid cycle, and fatty acid oxidation)

INTERMEDIARIES THAT CAN TRANSFER 1 ELECTRON AT A TIME: FMN, FAD, UBIQUINONE (UQ), IRON-SULFUR CLUSTERS, and CYTOCHROMES (contain Fe and/or Cu).

IRON-SULFUR CLUSTERS are held in place by Cys residues in nonheme iron proteins and occur as 2Fe-2S or 4Fe-4S (not counting the cysteinyl sulfurs).

FMN, **FAD**, and **UQ** have resonance-stabilized radicals. (FMN and FAD: see Redox Coenzymes, pp. 322-324 of your text.)

UBIQUINONE (UQ) (COENZYME Q)

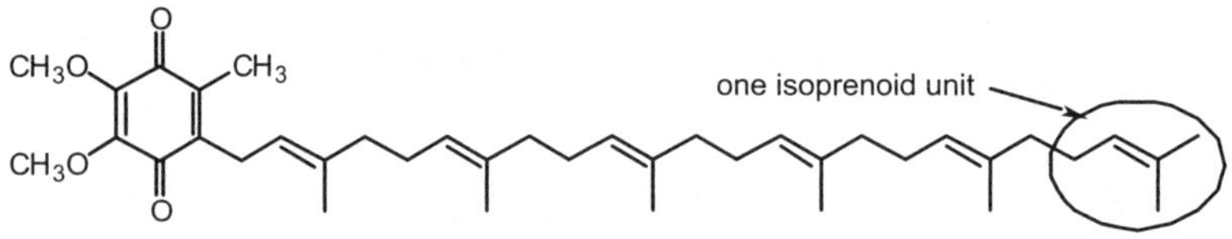

This UQ has 6 isoprenoid units (as some bacteria). Mammalian UQ, with 10 isoprenoid units, is Q_{10}.

The long hydrocarbon tail allows UQ to diffuse within the inner membrane, so it is mobil, i.e., it travels to carry electrons between the donors at complex I or II and the acceptor at complex III. The semiquinone intermediate is a resonance-stabilized radical.

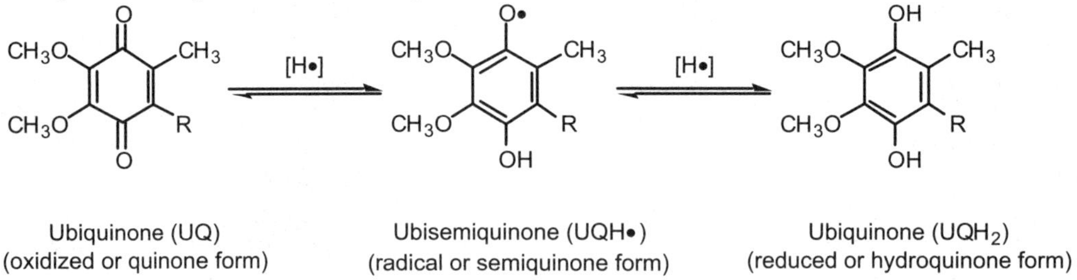

CYTOCHROMES: a series of electron transport proteins that carry an Fe-containing heme prosthetic group (similar to the heme in hemoglobin).

Example: cyt c (Fe^{3+}) + 1e^- → cyt c (Fe^{2+})

oxidized form reduced form

ELECTRON TRANSPORT LOCATION: Eukaryotes: in the inner membrane of the mitochondria

Aerobic prokaryotes: in the plasma membrane

ELECTRON TRANSPORT COMPLEX	PATH OF ELECTRONS FROM e^- DONOR:	PATH OF ELECTRONS TO e^- ACCEPTOR:	RESULT
COMPLEX I NADH DEHYDROGENASE COMPLEX: FMN, ~7 iron-sulfur clusters	from NADH	→FMN →FeS →FeS →UQ→ UQH_2	$4H^+$ transported from the matrix to the intermembrane space new e^- carrier: UQH_2
COMPLEX II SUCCINATE DEHYDROGENASE COMPLEX: Succinate dehydrogenase, FAD (bound), 2 Fe-S proteins	from succinate	→FAD →FeS clusters (several) →UQ → UQH_2	$2e^-$ transferred from succinate (forms fumarate) new e^- carrier: UQH_2
In some cell types: Glycerol-3-phosphate dehydrogenase (on the intermembrane-space-side surface of the inner membrane)	from cytoplasmic NADH	→DHAP →FAD →FeS clusters (several) →UQ → UQH_2	$2e^-$ transferred from cytoplasmic NADH new e^- carrier: UQH_2
Acyl-CoA dehydrogenase (matrix side of inner membrane) (first step of fatty acid oxidation)	from fatty acyl-CoA	→FAD →FeS clusters (several) →UQ → UQH_2	2 e^- transferred from fatty acyl-CoA new e^- carrier: UQH_2
COMPLEX III (Q CYCLE) CYTOCHROME bc₁ COMPLEX 1 Fe-S center cyt b_L cyt b_H cyt c_1 cyt c (loosely attached to intermembrane-space side)	from UQH_2 (space-side) UQH• (space-side, from UQH_2 above) from UQH_2 (space-side)	$1e^-$ → Fe-S → cyt c_1 → cyt c $1e^-$→ cyt b_L → cyt b_H → UQ → UQ• matrix-side $1e^-$→ cyt b_L → cyt b_H → UQ• matrix-side→ UQH_2	$UQH_2 + 2cyt\ c_{ox}(Fe^{3+}) + 2H^+$ matrix → UQ + $2cyt\ c_{red}(Fe^{2+}) + 4H^+$ cytosol $1e^-$ transferred from UQH_2 new e^- carrier: cyt c $1e^-$ transferred from UQH• (space-side) new e^- carrier: UQH• (matrix-side) $1e^-$ transferred from UQH_2 new e^- carrier: UQH_2 Transport of $4H^+$ to intermembrane space
COMPLEX IV CYTOCHROME OXIDASE cyt a, cyt a_3	from cyt c $1e^-$ at a time	→Cu_A → cyt a → cyt a_3	$O_2 + 4H^+ + 4e^- \rightarrow 2H_2O$ $2H^+$ (per cyt c)

2Cu: Cu_A near cyt a; Cu_B near cyt a_3		→ Cu_B → O_2→ H_2O	transported to the intermembrane space

ELECTRON TRANSPORT INHIBITORS

Inhibitors have been used to elucidate the order of the ETC components.

ANTIMYCIN A inhibits cyt b (Complex III).

ROTENONE and **AMYTAL** inhibit NADH dehydrogenase (Complex I).

CARBON MONOXIDE, **AZIDE**, and **CYANIDE** (CO, N^{3-}, CN^{-}) inhibit cytochrome oxidase (Complex IV).

10.2 OXIDATIVE PHOSPHORYLATION = USING ETC ENERGY TO MAKE ATP

THE CHEMIOSMOTIC THEORY

H^+ ions are transported from the matrix to the intermembrane space during ETC electron transfer. This creates a proton gradient (ΔpH) that is also an electrical potential, the **PROTONMOTIVE FORCE** (Δp), across the inner membrane.

The H^+ ions in the intermembrane space can only flow back to the matrix (down the proton gradient) through special channels. As they pass through the channels, ATP synthesis occurs.

Evidence for the chemiosmotic theory:

1. Working mitochondria take in O_2 and expel H^+.
2. If the mitochondrial membrane is disrupted, the ETC continues but ATP synthesis stops.
3. **UNCOUPLERS** and **IONOPHORES** allow H^+ to leak across the membrane. Uncouplers carry H^+ across, and ionophores form channels through the membrane. Examples: Dinitrophenol is an uncoupler; gramicidin A is an ionophore.

ATP SYNTHESIS

ATP SYNTHASE is a molecular motor that resembles a lollipop. Keep in mind:

ATP SYNTHASE MAY *LOOK* LIKE A LOLLIPOP, BUT IT *ACTS* LIKE A MOTOR.

Think of ATP synthase primarily as a motor. Knowing that it looks like a lollipop can be confusing when studying its function. Twirl a lollipop, and all of it spins. Turn the shaft, or rotor (the lollipop stick), in ATP synthase and the stout, roundish stator stays still. **THE "LOLLIPOP" CANDY PART IS HELD IN PLACE AND DOES NOT ROTATE.**

As a motor, it converts the **PROTONMOTIVE FORCE** (the energy from the H^+ gradient across the inner membrane) into a rotational force. This rotational force causes conformational changes that result in ATP synthesis.

STUDY HINT: Spend some time comparing Figures 10.14 and 10.15 on pp. 361 and 362, respectively of your text.[1] Think of ATP synthase as a MOTOR and

[1] Figure 10.14, *ATP Synthase from* Escherichia coli, shows the location of each subunit, with an arrow that identifies the 12-part rotor. Figure 10.15, *ATP Synthesis Model,* shows only the catalytic sites on the β subunits of the stator. Figure 10.14 labels each subunit, with the α,β-hexamer on the matrix side. Figure 10.15, a close-up of the catalytic action at the three β subunits, looking down at ATP synthase from the matrix side. The α-subunits and the central shaft (the γ and ε subunits) have been omitted for clarity.

refer to these figures to connect its structure (Figure 10.14) with its production of ATP (Figure 10.15). Again, *the α,β-hexamer* (located on the matrix side of the inner membrane) *is stationary—it does not rotate.*

For example, in Figure 10.15 the section that faces the top of the page changes its conformation from L to T in the second step—it did not turn with the shaft.

Draw your own version of these figures, with the subunits labeled. Use colors to identify the rotor vs. the stator subunits.

ATP SYNTHASE COMPONENTS AND THEIR FUNCTIONS

F_0 unit	transmembrane H^+ channel contains subunits a, b, and c in the ratio $a{:}b_2{:}c_{12}$ c ring = 12 c subunits
F_1 unit	active ATP synthase with three nucleotide-binding sites located on the matrix side of the inner mitochondrial membrane contains 5 subunits: 3 αβ dimers, γ, δ, and ε

Note that F_1 subunits are named with Greek letters, and the F_0 subunits are not. Also, the units F_0 and F_1 identify the structure of ATP synthase, but not the function. Both the rotor and the stator include subunits of both F_0 and F_1.

ROTOR	ε, γ, c_{12}	The c ring and the shaft (ε, γ) rotate 120° for each H^+ that passes through.
STATOR	a, b_2, δ, α_3, β_3	a contains the entry and exit channels for H^+ b and δ prevent the three αβ-hexamers from rotating β subunit—site of catalysis; has three possible conformations: L, T, and O

Rotation changes the protein conformations, which changes the binding affinity.

Conformation L:	Loose:	binds ADP and P_i weakly
Conformation T:	Tight:	brings substrates closer, facilitates ATP formation
Conformation O:	Open:	nonbinding, expels ATP ONLY AFTER another ADP and P_i bind to the adjacent β subunit

MECHANISM:

1. ADP and P_i bind to the β subunit in the L conformation.
2. H^+ enters through the channel in the a subunit and is passed from an Arg to an Asp residue in c, causing a conformational change and motion that results in the entire c ring to rotate by 120°. The torque causes the shaft to rotate, which causes a conformational change in β from L to T, and ATP synthesis occurs.
3. The shaft continues to rotate, changing the *adjacent* β from T to O, and an ATP is released. Meanwhile, the ATP made in step 2 is still in its β-subunit in the T conformation. After another ADP and P_i bind, this ATP will be released.

After three H^+ ions pass through, a full turn is made and an ATP is released.

Control of Oxidative Phosphorylation

P/O Ratio: moles P_i used (to make ATP) per O_2 reduced to H_2O. This ratio correlates the ETC with ATP synthesis. The max P/O ratio measured for NADH is 2.5 and for $FADH_2$ is 1.5.

Respiratory control: control of aerobic respiration by ADP (that is, there has to be enough ADP and P_i available to be able to make ATP)

ATP synthase: inhibited by ATP, activated by ADP and P_i.

Ratio of ATP and ADP controlled by:

ADP-ATP translocator: ships ATP out (of the matrix) and imports ADP (into the matrix from the intermembrane space). Remember that the intermembrane space is more positive (because of all of those H^+ ions) and the matrix is more negative. So, shipping out ATP (with its extra negative charge) is favored.

Phosphate translocase: imports $H_2PO_4^-$ (into the matrix) with an H^+ from the intermembrane space. (This is driven by the proton gradient.)

For each ATP synthesized, four H^+ total must pass through the membrane: three to turn the ATP synthase rotor plus one to provide the phosphate.

For the Complete Oxidation of Glucose:

The electrons captured by NADH (from glycolysis in the cytoplasm) has to be transported into the mitochondrion, since NADH cannot cross the inner mitochondrial membrane.

Glycerol phosphate shuttle

Converts cytoplasmic NADH (from glycolysis) to inner membrane $FADH_2$

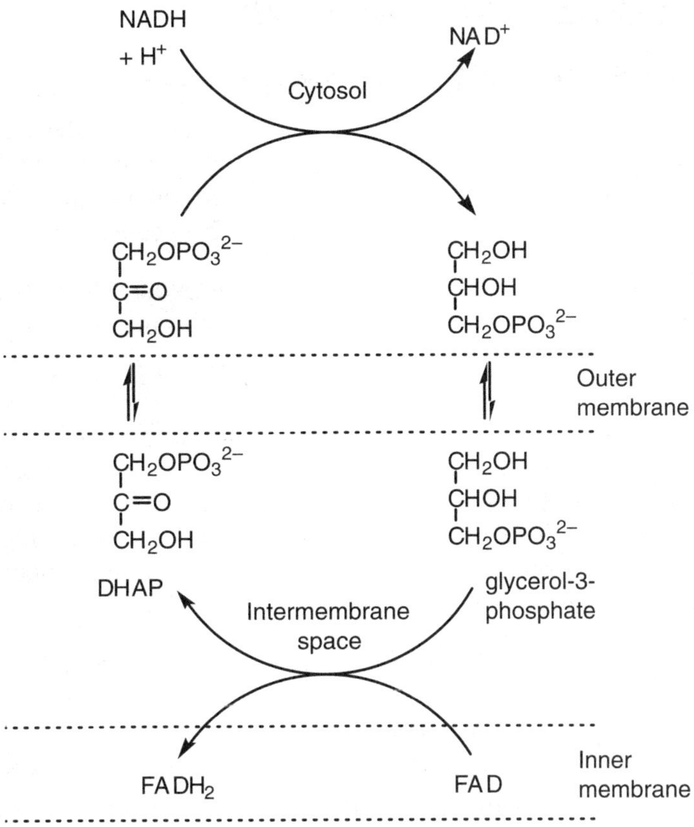

Malate-Aspartate Shuttle

Converts NADH and an H^+ from the intermembrane space into a matrix NADH and H^+, so it can enter the ETC. See Figure 10.17b, page 366, of your text. Because of that H^+ import, the number of ATP produced per NADH is less.

One More Shuttle

Transport of two citric-acid-cycle-produced ATPs from the matrix to the cytoplasm (so they can be used) costs $2H^+$ (total), so that also reduces the amount of ATP that can be produced from glucose oxidation.

The complete oxidation of glucose:

$$C_6H_{12}O_6 + 6O_2 + 30ADP + 30P_i \rightarrow 6CO_2 + 6H_2O + 30ATP$$

Uncoupled Electron Transport and Heat Generation

Sometimes a little leak is a good thing. When uncouplers and ionophores allow H^+ ions to cross the membrane, the energy that would have been used to make ATP is released as heat. That is **NONSHIVERING THERMOGENESIS**: "generating heat without shivering." Brown fat cells have many mitochondria that literally are little furnaces. In their inner membranes, they contain **THERMOGENIN** (**UNCOUPLING PROTEIN 1** or **UCP1**).

Norepinephrine is a neurotransmitter (released from neurons that end in brown adipose tissue) that ultimately causes fat molecules to hydrolyze. The fatty acids produced activate UCP1. Other uncoupling proteins are UCP2 (in most mammals, linked to body weight regulation) and UCP3 (linked to thermogenic effect of thyroid hormone; also detected in muscle).

10.3 Oxygen, Cell Function, and Oxidative Stress

Reactive Oxygen Species (ROS) Are Formed from O_2

$O_2^{\bullet-}$	**SUPEROXIDE RADICAL**	$^1O_2$	**SINGLET OXYGEN** (an unpaired electron absorbs energy and shifts to a higher orbital, i.e., an excited state); formed from superoxide or peroxide
$\bullet OH$	**HYDROXYL RADICAL**—very reactive, can initiate a chain reaction; can be formed from peroxide	H_2O_2	**HYDROGEN PEROXIDE**, H-O–O-H (can cross membranes) $Fe^{2+} + H_2O_2 \rightarrow Fe^{3+} + \bullet OH + OH^-$

These radicals may be new to you (with the possible exception of peroxide). They're so reactive they cannot be bought—to use them in a lab you would have to make them just prior to reacting them. Hydrogen peroxide is not a radical yet, but its O–O single bond can react to form radicals. (By the way, that is why drugstore peroxide is in a brown bottle because light energy would degrade it.) ROS formation is linked to aging, radiation exposure, smoking, and certain diseases.

Oxygen, or dioxygen (O_2), is itself a diradical and accepts electrons one at a time during reduction. So, ROS formation is also a normal byproduct of metabolism.

ROS react as soon as they are formed, oxidizing and damaging whatever they touch: inactivating enzymes, depolymerizing polysaccharides, breaking DNA, destroying membranes, etc. Macrophages and neutrophils seek out microorganisms and damaged cells and destroy them with a **RESPIRATORY BURST.** These cells create ROS specifically for this purpose.

REACTIVE NITROGEN SPECIES (RNS) are often classified as ROS because the synthesis of these species is often linked. Nitric oxide, •NO, can damage proteins with sulfhydryl groups or iron-sulfur clusters. Other examples of RNS are nitrogen dioxide ($\bullet NO_2$) and peroxynitrite ($ONOO^-$).

Antioxidant Enzyme Systems Destroy ROS

The major enzymatic defenses against oxidative stress are **SUPEROXIDE DISMUTASE (SOD)**, **GLUTATHIONE PEROXIDASE** [part of the **GLUTATHIONE-CENTERED SYSTEM (GSH/GSSG)]** **PEROXIREDOXIN (PRX)**, and **CATALASE.**

Superoxide dismutase (SOD): $2O_2^{\bullet-} + 2H^+ \rightarrow H_2O_2 + O_2$

Two types in humans: Cu-Zn isozyme in the cytoplasm, Mn isozyme in the mitochondrial matrix. One form of inherited Lou Gehrig's disease is caused by a mutation in the gene that codes for the cytosolic Cu-Zn isozyme of SOD.

Glutathione peroxidase: 2GSH → GS–SG

Controls cellular peroxide levels; contains selenium

$2GSH + R\text{–}O\text{–}O\text{–}H \rightarrow G\text{–}S\text{–}S\text{–}G + R\text{–}OH + H_2O$

(If R = H, then ROOH is simply hydrogen peroxide, which forms water.)

Glutathione reductase regenerates GSH:

$G\text{–}S\text{–}S\text{–}G + NADPH + H^+ \rightarrow 2GSH + NADP^+$

NADPH is produced by the pentose phosphate pathway, isocitrate dehydrogenase, and malic enzyme.

PEROXIREDOXINS (PRX) DETOXIFY PEROXIDES

by reacting peroxides with the side chain of a cysteine residue (–SH). Peroxide is reduced, and the thiol is oxidized to sulfenate. Cys –SH is regenerated by another –SH-containing protein like thioredoxin (TRX). The TRX-centered system includes TRX (thioredoxin), TR (thioredoxin reductase), and PRX.

CATALASE (CONTAINS HEME) CATALYZES $2H_2O_2 \rightarrow O_2 + 2H_2O$

Step 1. H_2O_2 + Fe(III)–enzyme $\rightarrow$ H_2O + O=Fe(IV)–enzyme

Step 2. H_2O_2 + O=Fe(IV)–enzyme $\rightarrow$ O_2 + H_2O + Fe(III)–enzyme

ANTIOXIDANT MOLECULES

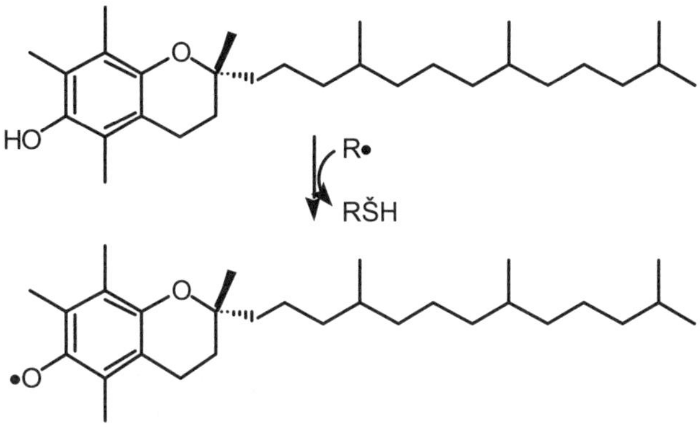

Antioxidant molecules can accept an electron from ROS to form a radical that is stabilized by resonance and are not as reactive. Great examples of the stabilized radicals are FMNH• and UQH• (see these radical structures in an earlier section in this chapter). **β-Carotene,** vitamin A precursor in animals, is an important antioxidant in membranes. Vitamin C (ascorbic acid) and GSH are also important antioxidants. Vitamin E is α-tocopherol, shown above.

ABBREVIATIONS AND ACRONYMS

ATP	adenosine triphosphate	NAD	nicotinamide adenine dinucleotide
cyt	cytochrome	RNS	reactive nitrogen species
DNP	dinitrophenol (an uncoupler)	ROS	reactive oxygen species
ETC	electron transport chain	SMP	submitochondrial particles
FAD	flavin adenine dinucleotide	SOD	superoxide dismutase
FMN	flavin mononucleotide	UQ	ubiquinone, or coenzyme Q

AFTER STUDYING THIS CHAPTER, YOU SHOULD BE ABLE TO:

- Explain how certain electron carriers such as UQ can transfer electrons one at a time.
- Describe the prominent features of each complex in the ETC, including the enzymes, cofactors, and coenzymes present.
- Trace electrons through the each complex of the ETC.
- Identify which metabolites would accumulate should a given ETC enzyme be inhibited.
- Describe how the proton gradient is formed and its function in metabolism.

- Describe the chemiosmotic coupling theory.
- Outline the mechanism of ATP synthesis and correlate each step with specific structural features of ATP synthase.
- Describe the mode of action of uncoupling agents and predict their effect on the metabolic pathways affected as well as on the organism.
- Describe ROS. Include how they are formed, their mode of action, and how they are destroyed.

Use this space to note any additional objectives provided by your instructor.

CHAPTER 10: SOLUTIONS TO REVIEW QUESTIONS

10.1 a. electron transport – the transfer of electrons from one electron carrier to another in a series of oxidation reduction reactions

b. ETC – electron transport chain – a series of electron carrier proteins that bind reversibly to electrons at different energy levels

c. Q cycle – the movement of electrons from reduced coenzyme Q (UQH_2) to cytochrome c during electron transport

d. protonmotive force – the force arising from a gradient of protons and a membrane potential

e. UQH_2 – reduced coenzyme Q

10.2 a. Complex I – NADH dehydrogenase complex – in the mitochondrial ETC catalyzes the transfer of electrons from NADH to UQ

b. Complex II – succinate dehydrogenase complex – in the mitochondrial ETC mediates the transfer of electrons from succinate via $FADH_2$ to UQ

c. Complex III – cytochrome bc_1 complex – in the mitochondrial ETC mediates the transfer of electrons from UQH_2 to cytochrome c

d. Complex IV – cytochrome oxidase – catalyzes the four electron reduction of oxygen to water

e. respirasome – functional aerobic respiration unit in the inner mitochondrial membrane; the I , III_2, $IV_{1\text{-}2}$ supercomplex has been identified in animals, plants and fungi

10.4 a. stator – a stationary component of a motor

b. rotor – a rotary motor; a device that revolves around a central axis

c. torque – twisting force that causes rotation

d. ATP synthase – the enzyme complex that synthesizes ATP; in mitochondria; ATP synthase consists of F_1, the ATPase activity and F_0, a transmembrane channel for protons

e. respiratory control – the control of aerobic respiration by ADP concentration

10.5 a. cytochrome – one of a family of electron transport proteins that possess a heme prosthetic group; the iron in the heme is alternately oxidized and reduced

b. glycerol phosphate shuttle – a metabolic process that uses glycerol-3-phosphate to transfer electrons from NADH in the cytosol to mitochondrial FAD

c. malate-aspartate shuttle – a metabolic process in which the electrons from NADH in the cytosol are transferred to mitochondrial NAD^+

d. uncoupling protein – a molecule that dissipates the proton gradient across the inner mitochondrial membrane by translocating protons

e. oligomycin – an antibiotic molecule that blocks the proton channel of ATP synthase

10.7 a. superoxide dismutase – one of a class of enzymes that catalyze the formation of hydrogen peroxide and oxygen from the superoxide radical

b. peroxiredoxin – one of a class of enzymes that detoxify peroxides

c. thioredoxin – a protein that acts as an antioxidant; contains a dithiol-disulfide active site that facilitates the reduction of other proteins

d. β-carotene – an antioxidant found in yellow-orange and dark green fruits and vegetables

e. ascorbate – an antioxidant that scavenges ROS in the aqueous compartments of cells and extracellular fluids

10.8 a. ischemia – inadequate blood flow

b. reperfusion injury – damage to cells in an ischemic tissue after it is reoxygenated

c. MPTP – mitochondrial permeability transition pore – a channel in mitochondria that opens when ROS form in reperfusion injury; MPTP opening leads to membrane potential collapse

d. α-tocopherol – a radical scavenger belonging to a class of compounds called phenolic antioxidants

e. glutathione – an intracellular tripeptide reducing agent

10.10 The chemiosmotic coupling theory has the following principal features: As electrons pass through the electron transport chain (ETC), protons are transported from the matrix and released into the inner membrane space. As a result, an electrical potential (Ψ) and a proton gradient (ΔpH) are created across the inner membrane. The electrochemical proton gradient is sometimes referred to as the protonmotive force (Δp). Protons, which are present in the intermembrane space in great excess, can pass through the inner membrane and back into the matrix down their concentration gradient only through the proton-translocating ATP synthase.

10.11 According to the chemiosmotic theory, an intact inner mitochondrial membrane is required to maintain the proton gradient required for ATP synthesis.

10.13 In addition to ATP synthesis, the mitochondrial electron transport-driven proton gradient also drives transport of substances across the inner membrane (e.g., phosphate ADP and ATP) and when it diminished it is involved in heat generation (nonshivering thermogenesis).

10.14 ATP synthesis requires the presence of a proton gradient across the mitochondrial membrane. Dinitrophenol, which has an ionizable hydrogen, can diffuse across the mitochondrial membrane. As it diffuses across the membrane, protons are transported from one side of the mitochondrial membrane to the other, thereby disrupting the proton gradient and interfering with ATP synthesis.

10.16 Oxygen is widely used as an energy source because it makes possible greater energy yields from food molecules and it is readily available.

10.17 In glycolysis, the net production of energy from one mole of glucose is two moles of ATP. If the two molecules of pyruvate are further oxidized by oxidative phosphorylation mechanism the total number of ATPs generated from the mole of glucose will be 29.5 to 31, depending on the shuttle mechanism used to transfer the electrons from the two cytoplasmic NADHs to the mitochondrial ETC.

10.19 Examples c, d, and f are all reactive oxygen species. Each of these species can act as a free radical, that is, it can attack various cell components producing such effects as enzyme inactivation, polysaccharide depolymerization, DNA breakage, and membrane destruction.

10.20 ROS damage cells by inactivating enzymes, depolymerizing polysaccharides, breaking DNA, and destroying membranes.

10.22 A genetic defect with survival value is G-6-PD deficiency. The lower NADPH and GSH concentrations in the red blood cells of G-6-PD-deficient individuals inhibit the growth of *Plasmodium* (the malarial parasite), a major killer of humans.

10.23 Although the end products of the following substances in an electron transport system are not named specifically in the text, we can speculate based upon the most reduced forms of the atom with the highest oxidation state. The nitrogen atom in nitrate (NO_3^-) has an oxidation state of +5; possible reduction products include: NO_2^-, NO, N_2, with NH_3 being the most reduced form. The ferric ion (Fe^{3+}) may reduce to ferrous ion (Fe^{2+}) or metallic iron. From carbon dioxide, CH_4 is the most reduced form of carbon. The most reduced form of sulfur, sulfide, S^{2-}, would likely be the final product of the reduction of both sulfate (SO_4^{2-}) and elemental sulfur.

10.25 Reaction equations that illustrate the production of ROS from electrons leaking from the electron transport system include:

$$O_2 + e^- \rightarrow O_2^{\bullet -} \text{ (superoxide)}$$

$$2H^+ + 2\,O_2^{\bullet -} \rightarrow O_2 + H_2O_2 \text{ (peroxide)}$$

$$Fe^{2+} + H_2O_2 \rightarrow Fe^{3+} + OH^- + \cdot OH \text{ (hydroxyl radical)}$$

$$2\,O_2^{\bullet -} + 2H^+ \rightarrow H_2O_2 + {}^1O_2 \text{ (singlet oxygen)}$$

$$2\text{ ROOH} \rightarrow 2\text{ROH} + {}^1O_2$$

10.26 The many causes of reperfusion-induced cardiac cell damage are primarily the result of ROS production and the opening of MPTP. ROS are generated by a reenergized mitochondrial ETC, the neutrophil enzymes NADPH oxidase and xanthine oxidase, and iron-induced hydroxyl ion production. Acidosis is caused by lactate accumulation, which is triggered by increased oxygen unloaded by hemoglobin. Reperfusion also promotes increased nitric oxide synthesis. MPTP opening causes the mitochondrial membrane potential to collapse.

10.28 The passage of K^+ across a membrane through an ionophore such as valinomycin lowers the electrical potential gradient. As a result, some of the energy that would have been used to synthesize ATP is dissipated as heat. This explains the rise in body temperature and sweating upon treatment with valinomycin.

10.29 Uncoupling agents disrupt phosphorylation by reducing (or collapsing) the proton gradient across the inner mitochondrial membrane. Uncoupling proteins such as UCP1 form a proton channel across this membrane, while other uncoupling agents such as dinitrophenol carry protons across the membrane. Both types of uncouplers provide a mechanism for protons to move across the membrane in a way that bypasses ATPase. The energy captured by the electron transport system and used to build the proton gradient is dissipated as heat instead of being used to do work, i.e., to synthesize ATP. UCP1 is activated by fatty acids that are produced by the hydrolysis of fats in brown adipose tissue. Fat hydrolysis is initiated by a cascade mechanism, which is triggered by the neurotransmitter norepinephrine.

10.31 When added to actively respiring mitochondria, azide inhibits cytochrome oxidase in complex IV, resulting in the accumulation of oxygen and H^+ on the matrix side, and reduced cytochrome c in the intermembrane space.

10.32 The minimum voltage drop for individual electron transfer events in the mitochondrial electron transport system that is necessary for ATP synthesis can be ascertained by referring to Figure 10.8. Note that of the three steps in the oxidation of NADH that account for the synthesis of an ATP molecule, the minimum voltage drop (+0.18V) occurs between UQH_2 and cyt c.

10.34 In nonshivering thermogenesis the energy generated by the oxidation of fatty acids is released as heat. UCP1, a protein dimer with a proton channel, dissipates the proton gradient across the inner membrane, thereby causing reduced ATP synthesis and the release of heat.

CHAPTER 10: SOLUTIONS TO FILL-IN-THE-BLANK QUESTIONS

10.35 FADH2

10.37 Proton motive

10.38 Ionophores

10.40 Respiratory control

10.41 Oxidative stress

10.43 Radical

10.44 Catalase

CHAPTER 10: SOLUTIONS TO SHORT-ANSWER QUESTIONS

10.46 Nonshivering thermogenesis is a process regulated by norepinephrine that occurs most commonly in brown fat. UCP1 (uncoupler protein 1) or thermogenin is a mitochondrial inner-membrane protein that dissipates the proton gradient so that fatty acid oxidation releases energy in the form of heat instead of capturing it in ATP molecules.

10.47 Nystatin is an ionophore because of its capacity to dissipate an ion gradient.

10.49 The consequences of rising cytoplasmic calcium levels include the activation of calcium-dependent enzymes such as proteases and phospholipases that can damage cell structures. The increase in osmotic pressure caused by higher cytoplasmic calcium levels cause cell swelling and possible leakage of the cell's contents.

CHAPTER 10: SOLUTIONS TO THOUGHT QUESTIONS

10.50 Once a labeled acetate molecule is converted to acetyl-CoA it is processed through the citric acid cycle (Figure 9.8 on p. 327). Because of the symmetrical structure of the intermediate succinate, $^{14}CO_2$ is not released until two or more turns of the cycle. The number of moles of ATP produced from 1 mol of acetate is calculated as follows: Each turn of the cycle produces 10 ATP. Assuming that the release of $^{14}CO_2$ requires two turns of the cycle, a total of 20 ATP are generated. Because 2 ATP are required to convert acetate to acetyl-CoA, the net production of ATP is 18 mol.

10.52 The oxidation of one mole of ethanol to acetyl-CoA produces two moles of NADH. The conversion of acetate to carbon dioxide and water through the citric acid cycle produces 3 NADH, 1 $FADH_2$, and 1 GTP. Assuming that the aspartate-malate shuttle is in operation, each cytoplasmic NADH yields 2.25 ATP for a total of 4.5 ATP. Each mitochondrial NADH yields 2.5 ATP for a total of 7.5 ATP. Each molecule of $FADH_2$ yields 1.5 ATP for a total of 1.5 ATP. Each GTP yields 0.75 ATP for a total of 0.75 ATP. The total ATP produced by the oxidation of ethanol is therefore 14.25 ATP.

10.53 Glutamine is readily converted to glutamate which then enters the citric acid cycle as α-ketoglutarate. The first turn of the cycle liberates one CO_2 and produces 2 NADH, 1 $FADH_2$, and 1 ATP. The four-carbon product is consumed in two further turns of the cycle, each of which generates 3 NADH, 1 $FADH_2$, and 1 ATP. The total yield is 8 NADH, 3 $FADH_2$ and 3 ATP. Assuming that each NADH yields 2.5 ATP and each $FADH_2$ yields 1.5, a grand total of 27.5 mol ATP are generated.

10.55 An organic peroxide (ROOH) is reduced by a glutaridoxin to an alcohol (ROH)

and water. GSH, the source of the electrons, is oxidized in the reaction to yield GSSG. Regeneration of the reduced GSH from GSSG requires an electron source, in this case NADPH.

10.56 Other ions such as sodium or potassium moving across the membrane would disrupt the electrical potential and reduce or eliminate oxidative phosphorylation.

10.58 In order for the system to operate efficiently energy should be released in stages. Each relatively large decrease in potential corresponds to thc energy required to generate an ATP molecule. If the energy is released all at once much of the energy would be released as heat and little ATP would be produced. The gradual change in voltage of the cytochromes makes this gradual release of energy possible.

10.59 Complex I transfers protons from NADH to UQ. Inhibition of this process will prevent the oxidation of NADH to NAD^+. However, NADH will accumulate and the NADH/NAD ratio will increase. UQH_2 will be oxidized by the remainder of the cytochrome system thus generating more UQ. As a result the UQ/UQH_2 ratio will also increase.

10.61 Rotenone inhibits complex I which oxidizes NADH. NADH is provided by the citric acid cycle with acetyl-CoA, a derivative of pyruvate, as the substrate. Succinate, the substrate for complex II, enters the series below the point where rotenone inhibits.

10.62 The reaction pathway is as follows:

Cys-SH + ROOH → Cys-SOH + ROH

Cys-SOH + Cys-SH → Cys-S-S-Cys + H_2O

where Cys-SOH = a sulfenic acid

10.64 Dehydroascorbic acid is a diketo acid. Glutathione reduces the carbonyls to regenerate ascorbic acid.

10.65 Oxygen acts as a radical and removes hydrogen from the alpha position of the amino acid. This produces a radical at the alpha position of the amino acid. This radical then reacts with another oxygen molecule to form an alkylperoxyl radical, $RCOO^{\bullet}$.

11 Lipids and Membranes

Brief Outline of Key Terms and Concepts

11.1 LIPID CLASSES

FATTY ACIDS
Long-chain carboxylic acids; components of triacylglycerols, membrane-bound lipids, and acylated membrane proteins.
***CIS*-ISOMERS VS. *TRANS*-ISOMERS**
ESSENTIAL VS. NONESSENTIAL FATTY ACIDS
OMEGA-3 AND OMEGA-6 FATTY ACIDS

THE EICOSANOIDS
AUTOCRINE REGULATORS derived from arachidonic acid or EPA
PROSTAGLANDIN **THROMBOXANE**
LEUKOTRIENE

TRIACYLGLYCEROLS (**NEUTRAL FATS**) consist of glycerol esterified to three fatty acids.
MONOUNSATURATED; POLYUNSATURATED
In both animals and plants they are a rich energy source.

WAX ESTERS

PHOSPHOLIPIDS are amphipathic molecules with a POLAR HEAD GROUP and long, nonpolar hydrocarbon chains. Functions include membrane components, emulsifying agents, and surface active agents.
Phospholipid types: **PHOSPHOGLYCERIDES** and **SPHINGOMYELINS.**
PHOSPHOLIPASES hydrolyze bonds in phosphoglycerides

SPHINGOLIPIDS are important membrane components of animals and plants and contain a complex long-chain amino alcohol (either **SPHINGOSINE** or **PHYTOSPHINGOSINE**). The core of each sphingolipid is **CERAMIDE**, a fatty acid amide derivative of the alcohol molecule.
GLYCOLIPIDS are derivatives of ceramide that possess a carbohydrate component.
Example: **GPI ANCHORS**

SPHINGOLIPID STORAGE DISEASES

ISOPRENOIDS have repeating units derived from isopentenyl pyrophosphate.
- **TERPENES**
- **CAROTENOIDS**
- **STEROIDS**
- **MIXED TERPENOIDS**
- **PRENYLATION**

LIPOPROTEINS
Plasma lipoproteins transport lipids through the bloodstream. Lipoproteins are classified by density (very low density, intermediate density, low density, and high density):
CHYLOMICRONS, VLDL, IDL, LDL, AND HDL.

11.2 MEMBRANES

MEMBRANE STRUCTURE
LIPID BILAYER, FLUID MOSAIC MODEL, LIPID RAFTS
Lipid bilayers consist of **PHOSPHOLIPIDS** and other **AMPHIPATHIC** lipid molecules.
PERIPHERAL PROTEIN; INTEGRAL PROTEIN

MEMBRANE FUNCTION
Membrane transport mechanisms:
PASSIVE TRANSPORT: does not require energy; solutes cross the membrane down (or with) their concentration gradient; Types of passive transport: **SIMPLE DIFFUSION**, **FACILITATED DIFFUSION**
ACTIVE TRANSPORT: requires energy (directly or indirectly from ATP hydrolysis or another energy source); solutes cross the membrane against their concentration gradient.
CYSTIC FIBROSIS; CYSTIC FIBROSIS TRANSMEMBRANE CONDUCTANCE REGULATOR (CTFR)

ADDITIONAL TERMS
- **HETEROKARYON**
- **AQUAPORINS**
- **NEPHROGENIC DIABETES INSIPIDIS**

11.1 LIPID CLASSES

FATTY ACIDS

Fatty acids have a long hydrocarbon chain (12-20 carbons) and one carboxylic acid. They are components of lipid molecules, primarily triacylglycerols and membrane-bound lipids. Fatty acids are **AMPHIPATHIC**, meaning they contain both hydrophobic and hydrophilic ends.

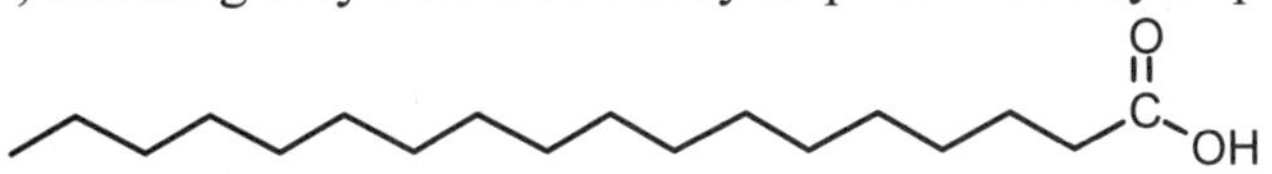

SATURATED VS. UNSATURATED

If the hydrocarbon chain contains at least one alkene, the fatty acid is **UNSATURATED** (not completely filled with hydrogens). If the hydrocarbon chain contains no double bonds, it is **SATURATED** (completely filled with hydrogens). **MONOUNSATURATED** fatty acids have one double bond, and **POLYUNSATURATED** fatty acids have two or more double bonds. Double bonds can be oxidized.

MELTING POINTS AND FLUIDITY OF A FATTY ACID; *CIS*- VS. *TRANS*-ISOMERS

The lower the melting point, the more fluid a fatty acid (or lipid) is likely to be at a given temperature. What determines whether the fatty acid is a solid or liquid at a given temperature? (1) Number of carbons and (2) Number and type of double bonds.

The longer the hydrocarbon chain, the higher the melting point (that is, it takes more energy to put those long chains in motion).

The greater the number of double bonds, the lower the melting point. A *cis*-alkene causes a "kink" in the chain that prevents *cis*-unsaturated fatty acids from packing as closely as saturated fatty acids do. So, it takes less energy to melt *cis*-unsaturated fatty acids (than a saturated fatty acid with the same number of carbons), and they have lower melt points. (*Trans*-unsaturated fatty acids are similar in conformation to saturated fatty acids.)

Example: Compare the melting points of these fatty acids. Why does stearic acid have the highest melting point and linoleic acid have the lowest?

myristic acid (14 carbons)	$CH_3(CH_2)_{12}COOH$	54°C
palmitic acid (16 carbons)	$CH_3(CH_2)_{14}COOH$	63°C
stearic acid (18 carbons)	$CH_3(CH_2)_{16}COOH$	70°C
oleic acid	$CH_3(CH_2)_7CH{=}CH(CH_2)_7COOH$	4°C
linoleic acid	$CH_3(CH_2)_4CH{=}CHCH_2CH{=}CH(CH_2)_7COOH$	–12°C

Naturally occurring fatty acids are typically:

unbranched and have an even number of carbon atoms; if unsaturated, they have a *cis*-isomeric form

DESIGNATION:

(total number of carbons) **:** (number of double bonds)$^{\Delta(\text{location of double bond})}$

Location of double bond:

Counting from the carboxyl carbon, the first carbon that begins the double bond

Alternatively: The carbon furthest from the carboxyl carbon is the omega (**ω**) carbon. When one of the double bonds begins 3 or 6 carbons from the end, its designation is **ω-3** or **ω-6. OMEGA-3 FATTY ACIDS** and **OMEGA-6 FATTY ACIDS** are now believed to promote cardiovascular health.

Example: Linoleic acid, $18{:}2^{\Delta 9,12}$, has 18 total carbons and two double bonds, one located between carbons 9 and 10 and the second located between carbons 12 and 13. The second double bond is located at the sixth carbon from the last so it may be designated 18:2ω-6.

NONESSENTIAL VS. ESSENTIAL FATTY ACIDS

NONESSENTIAL FATTY ACIDS can be synthesized by the body.

ESSENTIAL FATTY ACIDS must be obtained from the diet (sources include some vegetable oils, nuts, and seeds). Examples include linoleic and linolenic acids which serve as metabolic precursors. Symptoms of dietary deficiency include dermatitis, poor wound healing, reduced resistance to infection, alopecia, and thrombocytopenia.

CHEMICAL REACTIONS OF FATTY ACIDS

- esterification: carboxylic acid + alcohol = ester (reversible)
- hydrogenation of unsaturated fatty acids = saturated fatty acids
- oxidation of unsaturated fatty acids (Chapter 12)
- protein acylation—fatty acid groups (ACYL GROUPS) covalently attach to proteins to create ACYLATED PROTEINS

Example: Fatty acids undergo reactions in the body in order to be transported. Esterification to serum proteins allows transport from fat cells to body cells. Acyl transfer reactions allow fatty acids to enter cells.

EICOSANOIDS

FUNCTION: AUTOCRINE REGULATORS (rather than hormones, since they tend to be active in the cell in which they are produced)

NAMING: type of eicosinoid + type of modification + number of double bonds
Type of eicosinoid: PG = PROSTAGLANDIN TX = THROMBOXANE LT = LEUKOTRIENE
Type of modification: A = –OH, ether ring B = two –OH groups

PROSTAGLANDINS

STRUCTURE: contain a cyclopentane ring, –OH groups at C-11 and C-15.
E-series: have a carbonyl at C-9
F-series: have an –OH at C-9
2-series: derived from arachidonic acid; most important group for humans
3-series: derived from EPA (eicosapentaenoic acid)

FUNCTION: involved in wide range of regulatory functions, including involvement in inflammation, reproduction, and digestion.

THROMBOXANES contain a cyclic ether and are derivatives of arachidonic acid (TXA_2, TXB_2 in platelets) or EPA (TXA_3, TXB_3 in polymorphonuclear leukocytes).

LEUKOTRIENES

STRUCTURE: linear (noncyclic); synthesis begins with peroxidation reaction (lipoxygenase); contains a thioether near the site of this peroxidation.

EXAMPLES of leukotrienes: LTC_4, LTD_4, and LTE_4 are components of SRS-A—SLOW-REACTING SUBSTANCE OF ANAPHYLAXIS (an unusually severe allergic reaction). LTB_4 is a potent **CHEMOTACTIC AGENT**, or **CHEMOATTRACTANT**; it attracts white blood cells to damaged tissue.

TRIACYLGLYCEROLS—ESTERS OF GLYCEROL WITH THREE FATTY ACIDS

– also called NEUTRAL FATS (they are not charged)

STRUCTURE: All three hydroxyl groups of glycerol have a fatty acid attached via an ester bond. For each ester bond that is formed, a water molecule is removed.

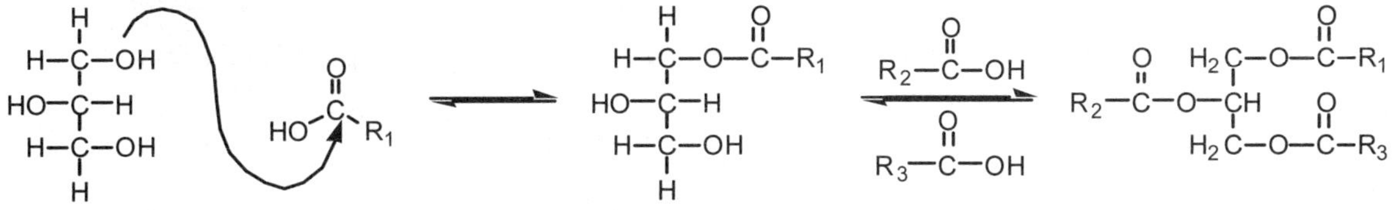

glycerol fatty acid Remember: To be fatty acids, the R groups ***must*** be ***long*** hydrocarbon chains. triacylglycerol

FATS VS. OILS: AT ROOM TEMPERATURE, FATS ARE SOLID AND OILS ARE LIQUID. WHY?

FATS have more saturated fatty acids, and OILS have more *cis*-unsaturated fatty acids.
PARTIAL HYDROGENATION: commercial hydrogenation of double bonds in oils converts them to fats.

FUNCTIONS OF TRIACYLGLYCEROLS:

- To store and transport fatty acids

  Triacylglycerols store energy more efficiently than glycogen because:

  1. they are hydrophobic; they coalesce into compact anhydrous droplets in adipocytes; and they take up 1/8 the volume of glycogen (because glycogen binds a substantial amount of water)
  2. when degraded (oxidized), triacylglycerols release more energy per gram than glycogen
- Insulation; poor conductor of heat, prevents heat loss
- Water repellent—secreted by specialized glands to make fur or feathers water repellent
- Plants—important energy reserve in fruits (avocados and olives) and seeds (peanut, corn, palm, safflower, soybean)
- Soapmaking (SAPONIFICATION): SOAP = sodium or potassium salts of fatty acid anions; EMULSIFYING AGENTS

WAX ESTERS

Waxes are complex mixtures that consist primarily of wax esters made from long-chain fatty acids and long-chain alcohols and may also contain hydrocarbons, alcohols, fatty acids, aldehydes, and sterols.

FUNCTION: Protective coatings on leaves, stems, fruits, and animal skin and fur.

Examples: carnauba wax (melissyl cerotate, a 26-carbon carboxylic acid esterified to a 30-carbon alcohol) and beeswax (triacontyl hexadecanoate)

PHOSPHOLIPIDS: **PHOSPHOGLYCERIDES AND SPHINGOMYELINS**

STRUCTURE: **AMPHIPATHIC:** hydrophilic (polar head group) and hydrophobic (long hydrocarbon chains); polar head group contains a phosphoryl group

In water, phospholipids spontaneously rearrange into micelles and/or bilayers.

Other types of polar or charged molecules can be attached to the phosphate group via a phosphodiester bond linkage.

PHOSPHATIDIC ACID is the simplest phospholipid and consists of the following:
Glycerol-3-phosphate esterified to two fatty acids, typically with 16-20 carbons each and:

- R1 = saturated fatty acid esterified to C-1
- R2 = unsaturated fatty acid esterified to C-2

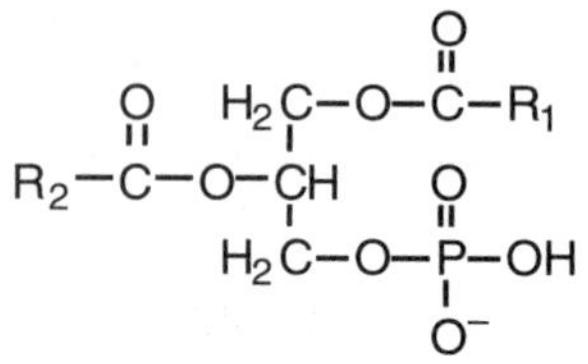

Phosphatidic acid

FUNCTIONS: Structural components of all cell membranes
EMULSIFYING AGENTS
SURFACE ACTIVE AGENTS or **SURFACTANTS** (lower the surface tension of a liquid so it can spread out over a surface)
intracellular signal transduction; Example: phosphatidylinositol-4,5-bisphosphate (PIP_2) phosphatidylinositol cycle
component of **GPI** (glycosyl phosphatidylinositol) **ANCHORS**—attach proteins to the external surface of the plasma membrane by embedding the phosphatidylinositol fatty acid chains in the membrane and connecting to the carboxy-terminal end of a protein through a linker component

TYPES: **PHOSPHOGLYCERIDES:** glycerol, fatty acids, phosphate, and an alcohol
SPHINGOMYELINS: sphingosine, fatty acids, phosphate, and an alcohol

PHOSPHOGLYCERIDES:

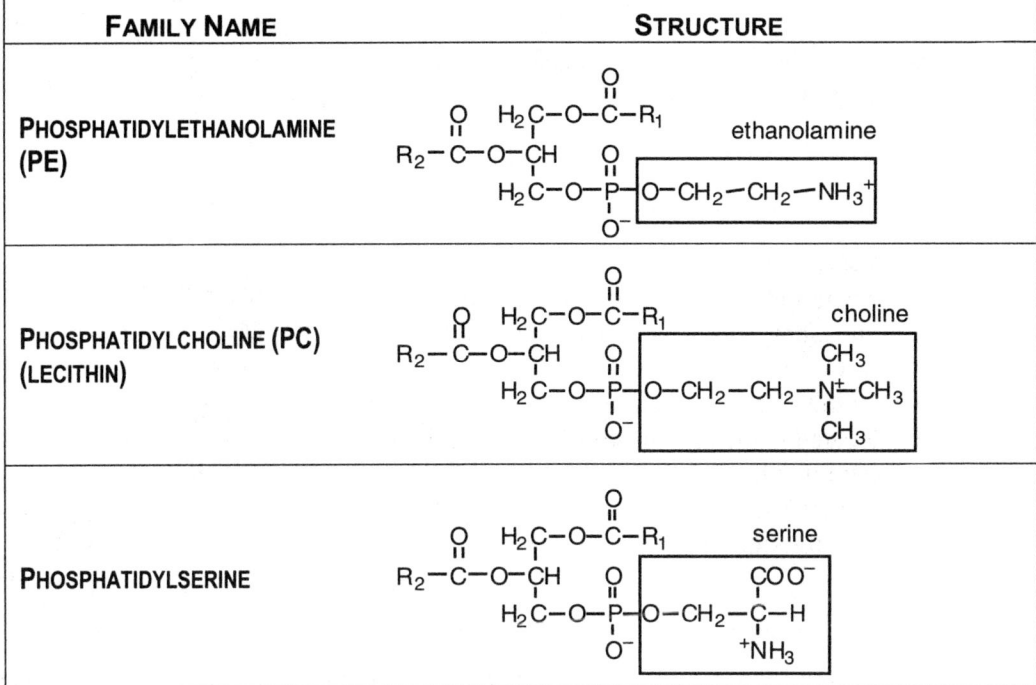

FAMILY NAME	STRUCTURE
PHOSPHATIDYLETHANOLAMINE (PE)	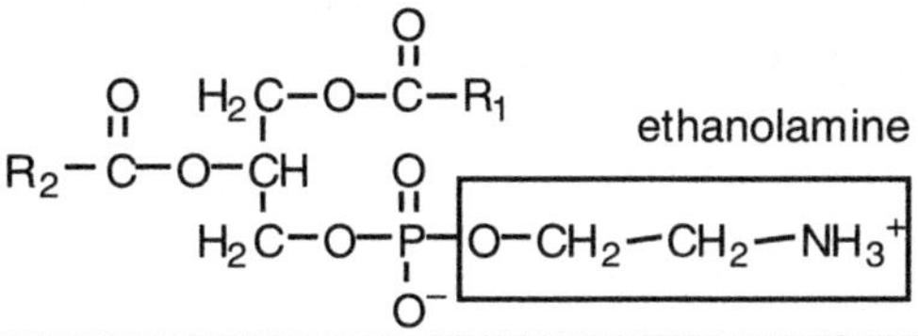
PHOSPHATIDYLCHOLINE (PC) (LECITHIN)	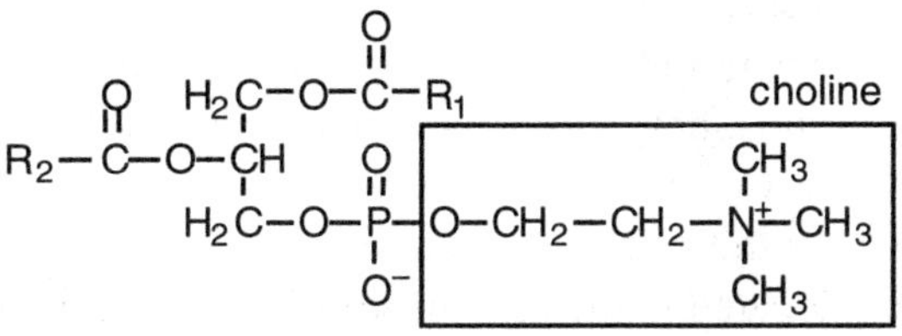
PHOSPHATIDYLSERINE	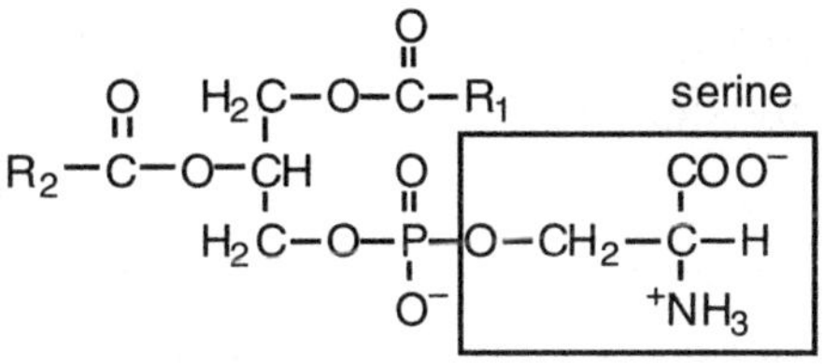

PHOSPHOGLYCERIDES, *CONTINUED*

PHOSPHATIDYLGLYCEROL	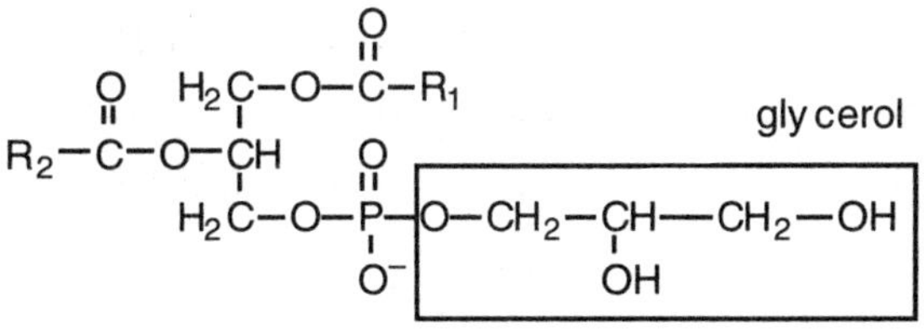
PHOSPHATIDYLINOSITOL a prominent component of GPI anchors	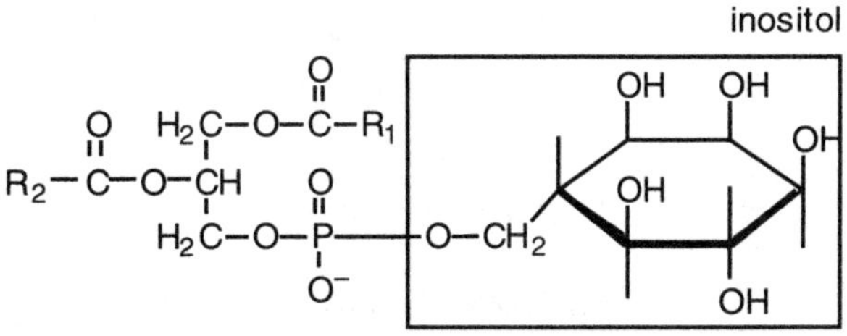

SPHINGOLIPIDS: ANOTHER LIPID COMPONENT OF BIOLOGICAL MEMBRANES

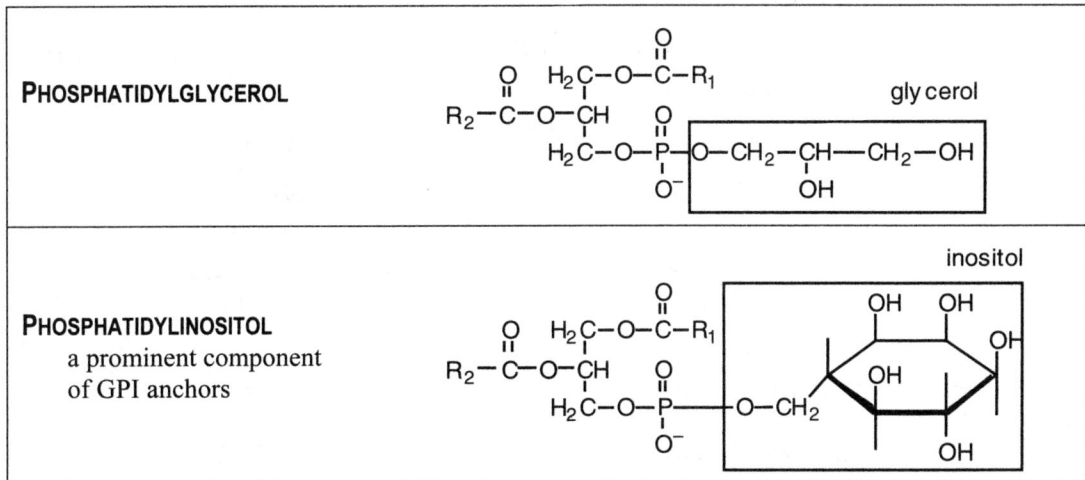

SPHINGOSINE is the "backbone" of sphingolipids (in animals; phytosphingosine is in plants).	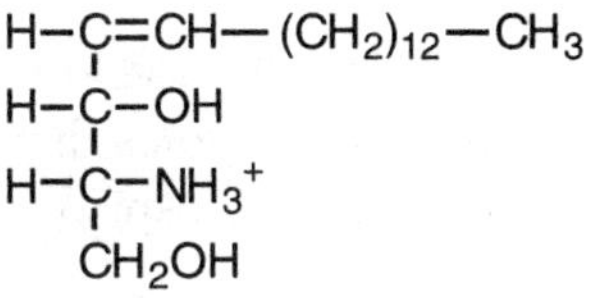	Sphingosine

(continued on next page)

CERAMIDES have a fatty acid attached to the amino group of sphingosine.

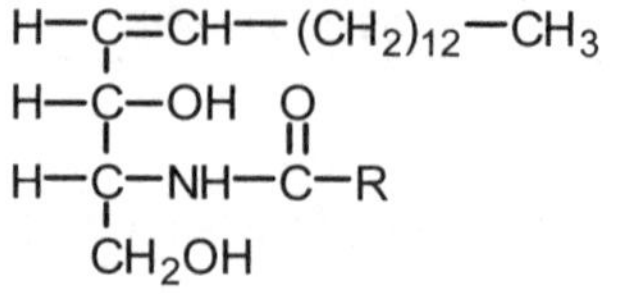

Ceramide

SPHINGOMYELIN

- a ceramide with a phosphorylcholine or phosphorylethanolamine attached to the last –OH group
- an important brain lipid
- located in the myelin sheath of nerve cells (and in other cell membranes)
- insulates nerve cells to allow for rapid transmission of nerve impulses

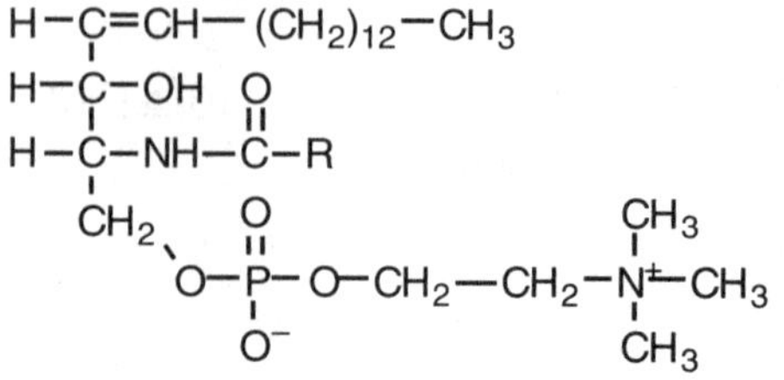

Sphingomyelin

Sphingomyelin

GLYCOLIPIDS (sugar + lipid) (also **GLYCOSPHINGOLIPIDS**)—monosaccharide, disaccharide, or oligosaccharide *O*-glycosidic linkage to ceramide; contain no phosphoryl group

CEREBROSIDES—polar head group is a monosaccharide (nonionic)

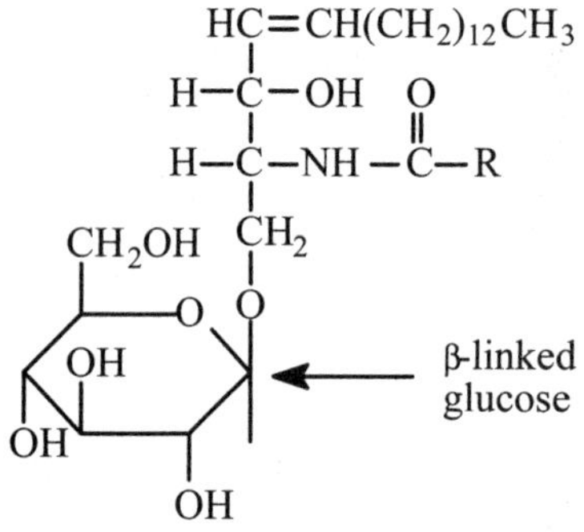

Glucocerebroside

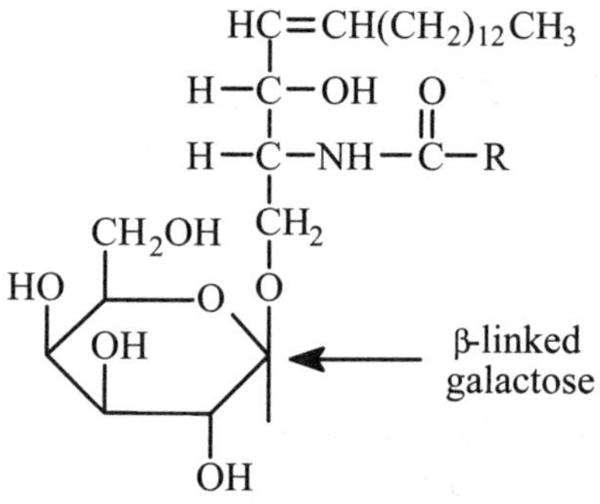

Galactocerebroside

GALACTOCEREBROSIDES—found in cell membranes of the brain

SULFATIDES—sulfated cerebroside, negatively charged at pH 7

GANGLIOSIDES—sphingolipids with oligosaccharide groups with at least one sialic acid residue; names: subscript letter (M, D, or T = 1, 2, or 3 sialic acid residues) and numbers (the sequence of sugars that are attached to ceramide)

SPHINGOLIPID STORAGE DISEASES

Sphingolipidoses (enzymes that metabolize sphingolipids): an enzyme deficiency results in the accumulation of a specific sphingolipid; most of these diseases are fatal.

Examples: Tay-Sachs disease, Gaucher's disease, Krabbe's disease, Niemann-Pick disease. Tay-Sachs disease is caused by a deficiency of β-hexosaminidase A, which degrades G_{M2}, a ganglioside.

ISOPRENOIDS: TERPENES AND STEROIDS

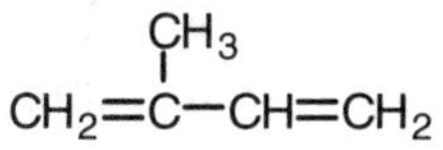

isoprene

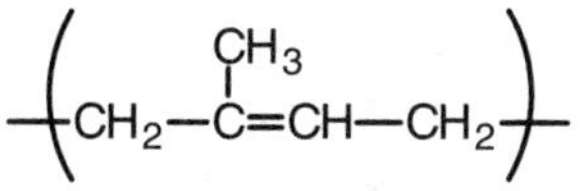

isoprene unit or prenyl group

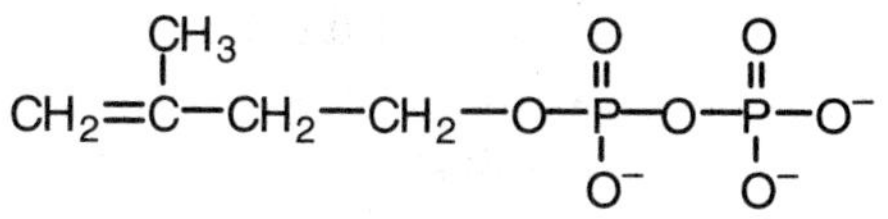

isopentenyl pyrophosphate

ISOPRENOIDS contain repeating **ISOPRENE** units and are synthesized from acetyl-CoA (not isoprene), beginning with the formation of isopentenyl pyrophosphate.

TERPENES—classified according to number of isoprene residues; found in essential oils of plants.

NO. OF ISOPRENE UNITS	TYPE
2	monoterpene
3	sesquiterpene
4	diterpene
6	triterpene
8	tetraterpene

CAROTENOIDS—orange pigments in most plants; the only tetraterpenes (examples: CAROTENES, xanthophylls)

MIXED TERPENOIDS—nonterpene components attached to isoprenoid groups (PRENYL or ISOPRENYL groups) *Examples*: vitamins E and K, ubiquinone, some cytokinins

PROTEIN PRENYLATION—prenyl groups covalently attached to proteins; function is not clear; typically farnesyl and geranylgeranyl groups

STEROIDS—derivatives of triterpenes, found in all eukaryotes, few bacteria

STEROID STRUCTURE:
Four fused rings; may contain double bonds and various substituents (OH, C=O, alkyl groups); sterols contain an OH

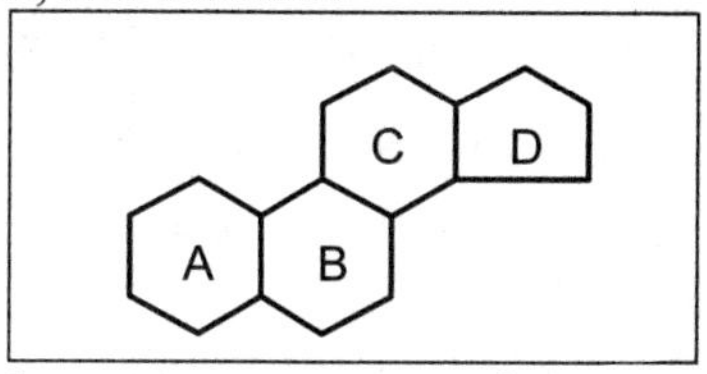

Examples: **CHOLESTEROL**: functions as an animal cell membrane component and as a precursor to steroid hormones, vitamin D, and bile salts (emulsifying agents that aid in the digestion of fats in the small intestine); cholesterol usually stored within cells as a fatty acid ester

Many hormones and vitamins are steroids. Testosterone, progesterone, and cortisol are examples of steroid hormones.

Example of a steroid derivative:
Cardiac glycosides increase the force of cardiac muscle contraction; digitoxin is a glycoside in digitalis, an extract from the foxglove plant; used to treat congestive heart failure; toxic in higher than therapeutic doses; inhibits Na^+-K^+ ATPase.

LIPOPROTEINS TRANSPORT LIPID MOLECULES THROUGH THE BLOODSTREAM

Molecular complexes found in mammalian blood plasma; also contains lipid-soluble antioxidants (α-tocopherol, carotenoids)

APOLIPOPROTEINS (or **APOPROTEINS**) are the protein components of lipoproteins (that is, the lipid portion is *A*bsent). Five classes of apolipoproteins: A, B, C, D, and E

CLASSIFIED ACCORDING TO DENSITY

Higher lipid content = lower density; higher protein content = higher density
As the lower-density lipids are removed (and the proteins are not), what remains has a higher percentage of protein that has a higher density.

LIPOPROTEIN	FUNCTION
CHYLOMICRONS (large, extremely low density)	Transports dietary triacylglycerols and cholesteryl esters from intestine to muscle and adipose tissues
VERY LOW DENSITY LIPOPROTEINS (VLDL) (produced in the liver)	Transports lipids to tissues
INTERMEDIATE-DENSITY LIPOPROTEINS (IDL) (VLDLs after some depletion of lipid components)	Continue to transport lipids to tissues; may be removed by the liver
LOW-DENSITY LIPOPROTEINS (LDL)[1] (converted from lipid-depleted VLDLs)	Transports cholesterol to tissues; engulfed by cells after binding to LDL receptors
HIGH-DENSITY LIPOPROTEIN (HDL) (produced in the liver)	Scavenges excess cholesterol from cell membranes

A plasma enzyme transfers a fatty acid residue from lecithin to cholesterol, forming a cholesteryl ester that is transported by HDL to the liver, which converts most of it to bile acids.

11.2 MEMBRANES

FLUID MOSAIC MODEL—proteins float within a **LIPID BILAYER**; the types and proportion of proteins and lipids depend on the type of cell (or organelle).

MEMBRANE STRUCTURE

MEMBRANE LIPIDS

1. **MEMBRANE FLUIDITY** is determined by percentage of unsaturated fatty acids in phospholipids; more unsaturated chains = greater fluidity.
   Cholesterol moderates fluidity because it has both rigid and flexible components.
   Lateral diffusion is rapid; flipping across the membrane is rare.
   HETEROKARYON—formed by fusion of cells from two different species; evidence that proteins move freely in the lipid bilayer
2. **SELECTIVE PERMEABILITY**: A bilayer is a good barrier to charged molecules like Na^+, K^+, Cl^-, and most polar molecules. Nonpolar substances diffuse through the lipid bilayer down their concentration gradients. Transport across the membrane is controlled by carrier proteins or protein channels.
3. **SELF-SEALING CAPABILITY**: Resealing small breaks is spontaneous and immediate; repairing larger tears is an energy-requiring, Ca^{2+} dependent process.
4. **ASYMMETRY**: different components face the interior vs. the exterior of the cell

[1] There are two LDL classes: small dense LDL (sdLDL) and large buoyant LDL. SdLDL are more atherogenic than the large buoyant LDL.

MEMBRANE PROTEINS

CLASSIFICATION

- by function: structural, enzymes, hormone receptors, transport, signal transduction, immunological identity
- by structural relationship to membrane: INTEGRAL vs. PERIPHERAL

INTEGRAL PROTEINS in red blood cell membrane:

Glycophorin—60% carbohydrate (by weight) includes the ABO and MN blood group antigens, function is unknown

Anion channel protein (band 3)—CO_2 transport in blood; HCO_3^- exchanged with Cl^- (chloride shift)

PERIPHERAL PROTEINS in red blood cell membrane:

spectrin, ankyrin, band 4.1

Preserves biconcave cell shape—allows rapid diffusion of O_2

Links cytoskeleton and membrane

Anion channel protein linked to ankyrin linked to spectrin linked to band 4.1 linked to actin filaments (a cytoskeleton component); band 4.1 also binds to glycophorin

MEMBRANE MICRODOMAINS

– nonuniform distribution of membrane components; example: lipid rafts

LIPID RAFTS:

- specialized microdomains in the external leaflet of eukaryotic plasma membranes
- have a higher degree of order and less fluidity than surrounding membrane regions
- components are primarily cholesterol, sphingolipids, and certain membrane proteins (GPI-anchored proteins, doubly acylated tyrosine kinases, etc.)

MEMBRANE FUNCTION

MEMBRANE TRANSPORT OF NUTRIENTS, WASTES, IONS, ETC.

PASSIVE TRANSPORT—transport down a concentration gradient with no energy input needed

SIMPLE DIFFUSION—spontaneous transport down a concentration gradient; the greater the gradient, the faster the rate; small nonpolar molecules (example: diffusion of gases such as O_2 and CO_2)

FACILITATED DIFFUSION—special **CHANNELS** or **CARRIERS** increase the rate at which certain solutes (large or charged molecules) move down their concentration gradients

CHANNELS—tunnel-like transmembrane proteins, designed for a specific solute, often chemically or voltage-regulated or "gated"

CHEMICALLY REGULATED CHANNELS—open or close in response to a specific chemical signal (example: chemically gated Na^+ channel)

VOLTAGE-REGULATED CHANNELS

Membrane potential—electrical gradient across a membrane: decrease = **DEPOLARIZATION**; reestablishment = **REPOLARIZATION** (example: voltage-gated Na^+ or K^+ channel)

CARRIERS OR PASSIVE TRANSPORTERS—solute binds to carrier on one side, causing a conformational change in the carrier that results in translocation across the membrane; carrier releases solute (example: red blood cell glucose transporter)

ACTIVE TRANSPORT needs energy to transport against a concentration gradient

PRIMARY ACTIVE TRANSPORT—ATP hydrolysis provides the energy; example: Na^+-K^+ pump (or Na^+-K^+ ATPase pump)

SECONDARY ACTIVE TRANSPORT—primary active transport generates a concentration gradient that is used to move substances across membranes

IMPAIRED MEMBRANE TRANSPORT MECHANISMS

Example: **CYSTIC FIBROSIS** (CF) is a fatal autosomal recessive disease caused by a missing or defective plasma membrane glycoprotein, CFTR.

CF—most common fatal genetic disorder in Caucasians; affects lungs and other organs; marked by lung disease and pancreatic insufficiency

CFTR—CYSTIC FIBROSIS TRANSMEMBRANE CONDUCTANCE REGULATOR

- Cl^- channel in epithelial cells
- ABC transporter (contains a polypeptide segment called an **A**TP-**b**inding **c**assette
- failure of CFTR results in retention of Cl^- within cells; osmotic pressure causes excessive water uptake, resulting in a thick mucus layer or other secretion.

TRANSPORT MECHANISMS:

TYPE:	TRANSPORT:
UNIPORTERS	one solute
SYMPORTERS	two different solutes simultaneously in the same direction
ANTIPORTERS	two different solutes simultaneously in opposite directions

MEMBRANE RECEPTORS

- Ligand-receptor binding results in a conformational change that causes a programmed response
- Functions: allow cells to monitor and respond to changes in the environment; intracellular communication; cell-cell recognition or adhesion
- **RECEPTOR-MEDIATED TRANSPORT**—receptor and ligand in coated pits are engulfed by endocytosis. *Example*: LDL receptor-mediated endocytosis

AFTER STUDYING THIS CHAPTER, YOU SHOULD BE ABLE TO:

- Predict relative melt points of fatty acids, given their names or structures.
- Draw a fatty acid given its designation. Example: $18{:}3^{\Delta 9,\,12,\,15}$
- Draw a triacylglycerol given the fatty acids' names, structures, or designations.
- Draw a phospholipid or sphingomyelin given the fatty acids and the alcohol.
- Determine the terpene class; locate isoprene units in terpenes.
- Given the structure of a lipid, determine its class.
- Describe membrane structure and membrane fluidity.
- Differentiate between types of transport across a membrane. Identify the type of transport given the molecule to be transported.
- Differentiate between types and functions of lipoproteins.

Use this space to note any additional objectives provided by your instructor.

CHAPTER 11: SOLUTIONS TO REVIEW QUESTIONS

11.1 a. fatty acid – a monocarboxylic acid; RCOOH where R is an alkyl group

b. monounsaturated fatty acid – fatty acid with one carbon-carbon double bond

c. polyunsaturated fatty acid – a fatty acid with two or more carbon–carbon double bonds, usually separated by a methylene group

d. saturated fatty acid – a fatty acid that contains no carbon-carbon double bonds

e. acyl group – any molecular group derived from a carboxylic acid by the removal of a hydroxyl group

11.2 a. essential fatty acid – a fatty acid that must be supplied in the diet because it cannot be synthesized by the body; linoleic and linolenic acids in humans

b. nonessential fatty acid – a fatty acid that can be synthesized by the human body

c. omega-3-fatty acid – α-linolenic acid and its derivatives such as eicosapentaenoic acid and docosahexaenoic acid

d. omega-6-fatty acid – linoleic acid and its derivatives

e. eicosanoids – a hormonelike molecule that contains 20 carbons; most are derived from arachidonic acid; examples include prostaglandins, thromboxanes, and leukotrienes

11.4 a. triacylglycerol – esters of glycerol with three fatty acids

b. neutral fat – a term used to describe triacylglycerols, which do not bear any charge

c. partial hydrogenation – a commercial process in which enough double bonds in vegetable oil are hydrogenated that a liquid oil is converted to semisolid oleomargarine

d. autoimmune disease – a condition in which an immune response is directed against an individual patient's own tissues

e. soap – sodium or potassium salts of fatty acids; the product of a saponification reaction involving TGs

11.5 a. saponification – a reaction in which an ester reacts with a base (e.g., NaOH) to yield a salt of a carboxylic acid; the reverse of an esterification

b. wax ester – an ester synthesized from a long chain fatty acid and a long chain alcohol

c. wax – a complex mixture of nonpolar lipids including wax esters

d. phospholipid – an amphipathic molecule that has a hydrophobic domain (hydrocarbon chains of fatty acid residues) and a hydrophilic (a polar head group) domain

e. polar head group – a molecular group that contains phosphate or other charged or polar groups

11.7 a. membrane remodeling – changes in a membrane's lipid composition that alters its fluidity or replaces damaged molecules

b. phospholipase – an enzyme that hydrolyzes phospholipids into fatty acids and other component molecules

c. cerebroside – a sphingolipid in which the head group is a monosaccharide

d. ganglioside – a sphingolipid that possesses an oligosaccharide group with one or more sialic acid residues

e. sphingolipidoses – a class of inherited disorders in which an enzyme required to degrade a specific sphingolipid is missing; a type of lysosomal storage disease

11.8 a. isoprenoid – one of a class of molecules that contain repeating five-carbon structural units known as isoprene units; examples include terpenes and steroids

b. isoprene unit – a five-carbon structural unit used to synthesize terpenes and steroids

c. terpene – one of a class of isoprenoids classified according to the number of isoprene residues it contains; examples include carotenoids, rubber, and many of the molecules found in essential oils of plants.

d. carotenoid – orange pigment found in many plants; classified as a tetraterpene

e. mixed terpenoid – a biomolecule composed of a nonterpene component attached to isoprene groups; examples include α-tocopherol and plastoquinone

11.10 a. chylomicron – a large lipoprotein of extremely low density; transports dietary TGs and cholesteryl esters from the intestine to muscle and adipose tissue

b. VLDL – very low density lipoprotein – a type of lipoprotein with a very high relative concentration of lipids; transports lipids from the liver to tissues

c. IDL – intermediate density lipoprotein – a lipoprotein formed when a very low density lipoprotein shrinks in size and becomes more dense as a result of depletion of TGs, apolipoprotein and phospholipids

d. LDL – low density lipoprotein – a type of lipoprotein that contains cholesterol, TGs and phospholipids; transports cholesterol to peripheral tissues

e. HDL – high density lipoprotein – a type of lipoprotein with a high protein content that is believed to scavenge excess cholesterol from cell membranes and transports it from peripheral tissues to the liver

11.11 a. lipid bilayer – a biomolecular lipid layer that constitutes the structural framework of cell membranes

b. membrane fluidity – the viscosity of the lipid bilayer; the degree of resistance of membrane components to movement

c. flippase – a protein that transfers phospholipids from the outer to the inner leaflet of a membrane

d. floppase – a protein that transfers phospholipids from the inner to the outer leaflet of a membrane

e. AE1 – the band 3 anion exchanger protein – an integral membrane protein found in red blood cells; a channel protein that allows the transport of chloride in exchange for bicarbonate ion

11.13 a. CFTR – cystic fibrosis transmembrane conductance regulator – the plasma membrane glycoprotein that functions as a chloride channel in epithelial cells

b. simple diffusion – a process in which each type of solute, propelled by random molecular motion, moves down a concentration gradient

c. facilitated diffusion – diffusion of a substance across a membrane that is aided by a carrier molecule

d. Na^+-K^+ pump – a membrane protein complex that uses the energy derived from ATP hydrolysis to drive the transport of sodium and potassium ions against their concentration gradients; also referred to as the Na^+-K^+ ATPase

e. aquaporin – one of a class of water channels in cell membranes

11.14 a. nephrogenic diabetes insipidis – a disease in which the kidneys cannot form a concentrated urine; in one form of the disease the mutated AQP2 water channel does not respond to the antidiuretic hormone vasopressin

b. LDL receptor – a plasma membrane protein that binds to low density lipoproteins; responsible for the uptake, via endocytosis, of LDLs into cells

c. coated pits – in endocytosis, concave regions of membrane surrounded by a clathrin cage

d. clathrin – the protein that forms coated pits in endocytosis

e. familial hypercholesterolemia (FH) – an inherited disease in which there are missing or defective LDL receptors

11.16 a. Phospholipids perform major roles as structural components of membranes, emulsifiers, and surface active agents.

b. Plant and animal membranes contain large amounts of sphingolipids.

c. Oils serve as an important energy reserve of fruits and seeds.

d. Waxes serve as protective coatings on the surface of leaves and stems, on the fur of animals, and on the shells of insects.

e. Steroids play an important structural role in animal membranes. Certain steroids act as hormones.

f. By acting as light-trapping pigments, carotenoids play an important role in photosynthesis.

11.17 Plasma lipoproteins improve the solubility of hydrophobic lipid molecules as they are transported in the bloodstream to the organs.

11.19 Unsaturated fatty acid content increases fluidity; cholesterol decreases fluidity.

11.20 Both a and c are true. (Ionophores are discussed on page 348.)

11.22 Diets in which there are insufficient fats and oils can have serious health consequences. Among these are deficiencies in fat soluble vitamins (A,D,E, and K) and the essential fatty acids linoleic and linolenic acids, which can result in dry skin, brittle hair, fatigue, high blood pressure, atherosclerosis, depressed immunity, poor wound healing and depression. In children essential fatty acid deficiencies have been linked to impaired brain development.

11.23 ACAT is an abbreviation for Acyl-CoA: cholesterol acyltransferase, the enzyme responsible for converting cholesterol to its acyl ester which is stored within cells. LCAT is an abbreviation for the plasma enzyme lecithin: cholesterol acyltransferase, which transfers a fatty acid residue from lecithin to cholesterol. The cholesterol ester product is then sequestered within a plasma lipoprotein.

11.25 When acetylcholine binds to the acetylcholine receptor complex in the nerve cell membrane, sodium ions flow into the nerve cell and a smaller number of potassium ions flow out. During the repolarization phase, potassium ions flow out of the cell through voltage-regulated potassium channels.

11.26 Many eicosanoids are derived from arachidonic acid. Medical conditions in which it is advantageous to suppress the synthesis of eicosanoids are anaphylaxis, allergies, pain, the inflammation caused by injury, and high fever.

11.28 An autocrine regulator is a hormone-like molecule that triggers a response within the same cell that produces it. A hormone is a molecule produced by one type of cell that elicits a response in target cells in another part of the body

11.29 Water molecules can penetrate through hydrophobic cell membrane by moving through water channel complexes called the aquaporins.

11.31 Triacylglycerol functions include energy storage, insulation, and shock absorption. Their highly reduced, long hydrocarbon chains are a very efficient storage form of energy; in addition, their hydrophobicity results in compact storage in adipocytes. The relatively low heat conductivity helps to prevent heat loss and serves organisms as insulation.

11.32 The polar head region containing the ester and amide linkages in a sphingomyelin are hydrophilic; the hydrocarbon tail region is hydrophobic.

11.34 With their smaller diameters sdLDL are more atherogenic than buoyant LDLs because they easily enter artery walls where they are susceptible to oxidation.

11.35 Lipid rafts are plasma membrane microdomains, which are enriched in cholesterol, sphingolipids and certain membrane proteins. Lipid rafts are more rigid than the surrounding membrane. Lipid rafts have been implicated in such cellular processes as exocytosis, endocytosis and signal transduction.

11.37 The sodium gradient created by the Na^+-K^+-ATPase in the plasma membrane of kidney tubule cells transports the glucose. This is secondary active transport.

11.38 In facilitated diffusion, polar, charged, or large molecules that normally cannot penetrate the cell membrane diffuses across the membrane through protein channels or carriers that "facilitate" this diffusion. Because facilitated diffusion occurs with (or down) a concentration gradient, this is a spontaneous process, so a coupled reaction that provides energy (such as the hydrolysis of ATP) is not needed. The transport of glucose across red blood cell membranes is an example of facilitated diffusion by a carrier protein, and anion channels in red blood cell membranes are examples of protein channels. (Note that the transport of glucose across the plasma membrane of kidney tubule cells is an example of secondary active transport, a distinctly different transport mechanism. See Review question 11.37.)

11.40 The outer layer of lipoproteins consist of a single layer of phospholipids and proteins. The hydrophobic hydrocarbon chains of the phospholipids face inward, towards the neutral lipids contained within. The hydrophilic group of the phospholipids face outward and are solvated by water molecules, allowing the lipoprotein to dissolve in the bloodstream.

11.41 Like saturated fatty acids, *trans* fatty acids, can have fully extended chains that pack together well, resulting in an increased melting point relative to unsaturated fatty acids that contain *cis*-alkenes.

11.43 To form bilayers, molecules need to be amphipathic. Although triacylglycerols have three polar ester bonds, their three hydrocarbon chains are sufficiently long to cause these molecules to be nonpolar overall. Their hydrophobic nature causes them to coalesce into droplets rather than forming bilayers.

CHAPTER 11: SOLUTIONS TO FILL-IN-THE-BLANK QUESTIONS

11.44 Steroid

11.46 Fats

11.47 Passive

11.49 Facilitated diffusion

11.50 Fluidity

11.52 Inflammation

11.53 Digitalis

CHAPTER 11: SOLUTIONS TO SHORT-ANSWER QUESTIONS

11.55 Terpenes are a class of isoprenoids that most commonly occur in plants. Farnesene, found in the oil of citronella, is used in soap and perfumes. Squalene, an intermediate in the synthesis of steroids, is found in shark liver oil and olive oil. The carotenoids, which include the carotenes and the xanthophylls, are plant pigments.

11.56 The sphingolipid storage diseases, also called the sphingolipidoses, are a class of lysosomal storage disease in which an enzyme required to degrade a specific sphingolipid is either missing or defective. Tay-Sachs, the best known example, is caused by a deficiency of the enzyme β-hexaminidase A, the enzyme that degrades the ganglioside GM2. Without the capacity to degrade ganglioside GM2, cells accumulate this molecule, swell, and eventually die.

11.58 Endorphins are endogenous opiate-like proteins that act on the brain to promote a sense of pleasure. Dietary fat elicits a sense of pleasure by stimulating endorphin release.

CHAPTER 11: SOLUTIONS TO THOUGHT QUESTIONS

11.59 The fluidity of the membrane allows for flexible movement. Any breaks that do occur expose the hydrophobic core of the membrane to an aqueous environment. Hydrophobic interactions spontaneously move the broken ends together and, in combination with certain other components of cell membrane resealing mechanisms (e.g., cytoskeleton and calcium ions), the membrane reseals.

11.61 The carbohydrate portion of the glycolipid can form hydrogen bonds with the water. This carbohydrate is the polar group, and it is analogous to the charged portion of the phospholipid.

11.62 For a phospholipid to move from one side of the bilayer to the other, the polar head must move through the hydrophobic portion of the phospholipid membrane. This process requires a significant amount of energy and is therefore relatively slow.

11.64 The ordered water molecules surrounding each phospholipid molecule are released from the polar heads. Order is lost and entropy increases.

11.65 Many transmembrane and peripheral proteins are attached to the cytoskeleton and therefore are not free to move in the phospholipid bilayer.

11.67 High temperatures destabilize the membranes of prokaryotes adapted to "normal" temperatures. The membranes of thermophilic prokaryotes are resistant to high temperatures because they contain longer, saturated fatty acyl groups that pack closely together and lipids such as sterols or similar molecules that act as stiffening agents. In addition, they also contain glycero-ether lipids, which are difficult to hydrolyze.

11.68 The steroids are lipid soluble molecules and would tend to dissolve in adipose tissue. During a diet the fat content of this tissue is reduced and the dissolved steroid molecules are released. This causes an overall increase in blood steroid levels.

11.70 The myelin sheath is primarily composed of hydrophobic molecules that are not good conductors of electricity. They serve to insulate electrically active nerve cells from the highly conductive aqueous environment.

12 Lipid Metabolism

Brief Outline of Key Terms and Concepts

12.1 FATTY ACIDS, TRIACYLGLYCEROLS, AND THE LIPOPROTEIN PATHWAYS

LIPOLYSIS VS. LIPOGENESIS

BILE SALTS; CHYLOMICRON REMNANTS; HORMONE-SENSITIVE LIPASE; FATTY ACID–BINDING PROTEIN

GLYCERONEOGENESIS AND THE TG CYCLE

The triacylglycerol cycle regulates the level of fatty acids that are available to the body

FATTY ACID DEGRADATION VIA β-OXIDATION

CARNITINE is needed to transport acyl-CoA into the mitochondrion, where β-oxidation oxidizes fatty acids in a series of four reactions. The fourth step is a THIOLYTIC CLEAVAGE (addition of CoASH with the breakage of the bond between the α- and β-carbons), giving an acetyl-CoA and an acyl-CoA that is two carbons shorter. β-Oxidation repeats until only acetyl-CoA is left.

THE COMPLETE OXIDATION OF A FATTY ACID

The number of ATP from each β-oxidation cycle (followed by ETC and oxidative phosphorylation) is approximately:

1.5 ATP/$FADH_2$ x number of β-oxidation cycles
2.5 ATP/NADH x number of β-oxidation cycles
10 ATP/acetyl CoA x total acetyl-CoA
= Total ATP produced during β-oxidation
–2 ATP equivalents to form the fatty acyl-CoA
= Total ATP produced from a fatty acid

β-OXIDATION IN PEROXISOMES shorten very-long-chain fatty acids.

THE KETONE BODIES (acetoacetate, β-hydroxybutyrate, acetone) are produced from excess acetyl-CoA from β-oxidation.

KETOGENESIS; KETOSIS

FATTY ACID OXIDATION: DOUBLE BONDS, AND ODD CHAINS

Additional reactions are needed to oxidize fatty acids that are unsaturated, have an odd number of carbons, and/or are branched (α-OXIDATION).

FATTY ACID BIOSYNTHESIS

In animals, fatty acids are synthesized in the cytoplasm from acetyl-CoA, beginning with the carboxylation of acetyl-CoA by **ACC** to form malonyl-CoA, which is converted to malonyl-ACP (ACYL CARRIER PROTEIN). ACC, a key enzyme in fatty acid metabolism, is regulated by allosteric modulators and phosphorylation. The remaining reactions of fatty acid synthesis take place on the fatty acid synthase multienzyme complex.

FATTY ACID ELONGATION AND DESATURATION are carried out by mitochondrial and ER enzymes.

EICOSANOID METABOLISM

REGULATION OF FATTY ACID METABOLISM

Short term: allosteric modulators (malonyl-CoA inhibits CAT-I), covalent modification (**AMPK**-catalyzed phosphorylation of ACC1 and glycerol-3-phosphate acyltransferase), and hormones (e.g., insulin, glucagon, and epinephrine). Long-term regulation: gene expression triggered by TRANSCRIPTION FACTORS (e.g., the **SREBPs** and the **PPARs)** and certain hormones.

Lipoprotein Metabolism: The Endogenous Pathway

Newly synthesized lipids are incorporated into very low density lipoproteins and then transported throughout the body.

12.2 MEMBRANE LIPID METABOLISM

PHOSPHOLIPID METABOLISM

After phospholipids have been synthesized at the interface of the SER and the cytoplasm, they are often “remodeled”; that is, their fatty acid composition is adjusted. TURNOVER (i.e., the degradation and replacement) of phospholipids is rapid and is mediated by the phospholipases.

SPHINGOLIPID METABOLISM

Synthesis begins with the production of ceramide. Ceramide reacts with phosphatidylcholine to form sphingomyelin, or UDP-glucose (or UDP-galactose) to form glucosylcerebroside (or galactocerebroside). Reaction of galactocerebrosides with the sulfate donor **PAPS (3′-phosphoadenosine-5′-phosphosulfate)** produces the sulfatides.

12.3 ISOPRENOID METABOLISM

CHOLESTEROL METABOLISM

Cholesterol is the precursor for all steroid hormones and the bile salts.

CHOLESTEROL SYNTHESIS

Three phases of cholesterol synthesis: acetyl-CoA to HMG-CoA, HMG-CoA to squalene, and squalene to cholesterol. The rate-limiting step, the reduction of HMG-CoA to form mevalonate, is catalyzed by **HMGR**.

CHOLESTEROL DEGRADATION: primarily by conversion to **BILE SALTS** that emulsify dietary fat.

CHOLESTEROL HOMEOSTASIS: via intricate mechanisms that regulate cholesterol biosynthesis, LDL receptor activity, and bile acid biosynthesis.

STEROID HORMONE SYNTHESIS

STEROL CARRIER PROTEIN; CONJUGATION REACTION GLUCOCORTICOID; MINERALOCORTICOID; PREGNENOLONE

BIOCHEMISTRY IN PERSPECTIVE:

ATHEROSCLEROSIS: ATHEROMA; CHEMOKINES; PLAQUE

BIOTRANSFORMATION: PHASE I AND PHASE II REACTIONS; CYTOCHROME P450 SYSTEM; MONOOXYGENASE; CONJUGATION REACTIONS; DETOXICATION; DETOXIFICATION

12.1 FATTY ACIDS AND TRIACYLGLYCEROLS

COMPARISON OF LIPOLYSIS AND LIPOGENESIS	
LIPOLYSIS (degradation/catabolism) triacylglycerol + $3H_2O$ → 3 fatty acids + glycerol	**LIPOGENESIS** (synthesis/anabolism) 3 fatty acids + glycerol → triacylglycerol + $3H_2O$
Energy reserves are low and energy (ATP) is needed: fasting, exercise, in response to stress	Energy reserves are high and ATP levels are fine
Mobilizes fat stores to fatty acids + glycerol	Stores dietary fat in adipocytes
PRODUCE energy—ATP—by oxidizing fatty acids	Store energy by synthesizing triacylglycerols from excess nutrients and metabolites

HOW IS DIETARY FAT (TRIACYLGLYCEROLS) TRANSPORTED TO THE CELLS TO BE METABOLIZED?

1. **BILE SALTS** emulsify (solubilize) fats in the lumen of the small intestine because they are amphipathic, with detergent-like properties. Bile salts are made by the liver, stored in the gall bladder, and secreted into the small intestine.

Sodium cholate, a bile salt

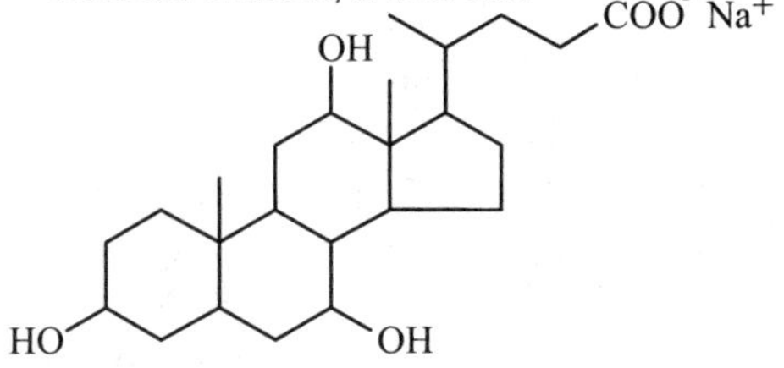

2. **PANCREATIC LIPASE** (and other intestinal lipases) hydrolyzes triacylglycerols to free two fatty acids, leaving a monoacylglycerol.

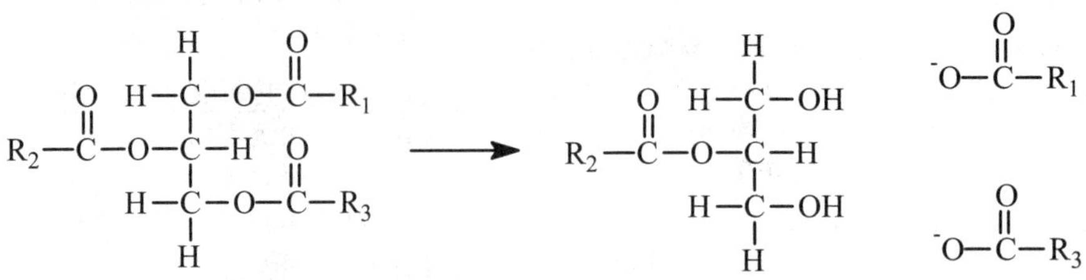

intestinal lumen →

← enterocytes

3. Bile salts form mixed micelles with these and with other fats. Micelles are then taken up into enterocytes (intestinal wall cells) and reconverted to triacylglycerols.
4. These triacylglycerols are combined with cholesterol, phospholipids, and protein to form **CHYLOMICRONS** which are transported out to the lymph and then into the bloodstream for distribution to cells. (Chylomicrons and other lipoproteins are described on pp. 401-404 of your text.)

   The use of fatty acids for energy depends on the tissue.
   Fatty acids are: —a major source of energy in muscle tissue (cardiac and skeletal)
   —used minimally in nervous tissue.
   During fasting, many tissues use fatty acids or ketone bodies for energy.

   Target tissue: primarily muscle cells (cardiac and skeletal) and adipocytes; also lactating mammary gland cells. All of these cells synthesize lipoprotein lipase, transferred to the endothelial surface of capillaries

5. Lipoprotein lipase converts triacylglycerols in chylomicrons (and VLDLs) to fatty acids and glycerol. Glycerol must be transported back to the liver to be metabolized.
6. In the liver:
   - Glycerol $\xrightarrow{\text{glycerol kinase}}$ glycerol-3-phosphate, which is used in the synthesis of triacylglycerols, phospholipids, or glucose.
   - Chylomicron remnants are removed from the blood via receptor-mediated endocytosis.
   - VLDLs are synthesized.

FATTY ACID DEGRADATION (IN THE MITOCHONDRIAL MATRIX)

1. **Activation** by acyl-CoA synthetase (in the outer mitochondrial membrane)
2. **Transport** (through the inner membrane into the matrix)
3. **β-oxidation**

ACTIVATION BY ACYL-COA SYNTHETASE

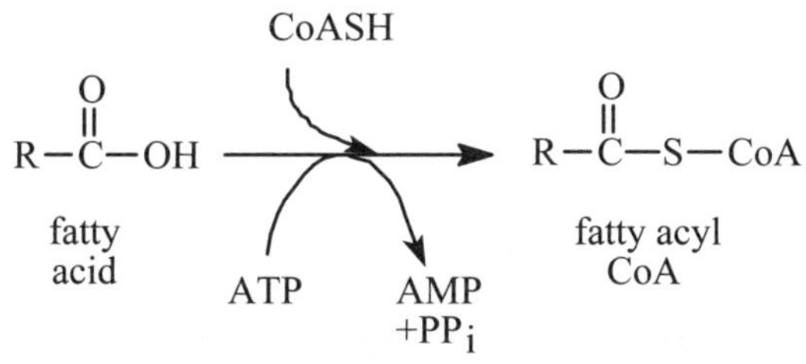

The ATP used to drive this reaction might be thought of as an "activation fee." The production of AMP means that two ATP equivalents were used. PP_i hydrolysis helps to drive the reaction to completion.

TRANSPORT: CARNITINE TRANSPORTS ACYL-COA INTO THE MATRIX

Why is this necessary?
Fatty acid catabolism occurs inside the mitochondrial matrix. Fatty acyl-CoA cannot cross the inner membrane, but carnitine can. Carnitine forms an ester bond with the fatty acyl group and carries it into the matrix (with the help of a carrier protein).

Consider the acyl-carnitine structure. Can you see why it needs a carrier protein to get through the membrane?

Carnitine

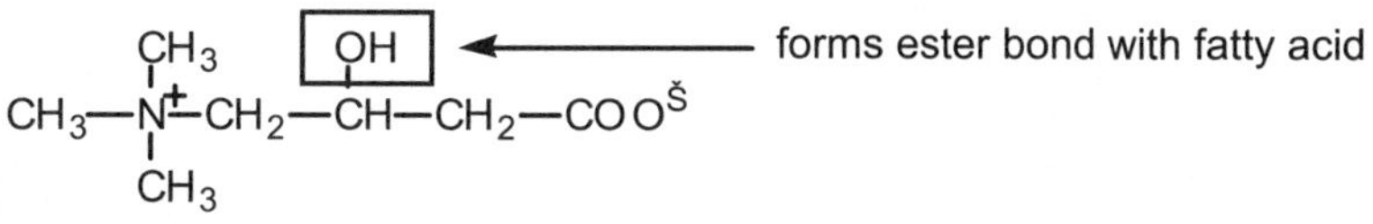

The following figure shows the transport of palmitoyl-CoA through the inner membrane. CoASH does not pass through the membrane, so the cell maintains separate cytoplasmic and mitochondrial pools of CoASH.

CARNITINE TRANSPORT OF ACYL-COA ACROSS THE MITOCHONDRIAL MEMBRANE

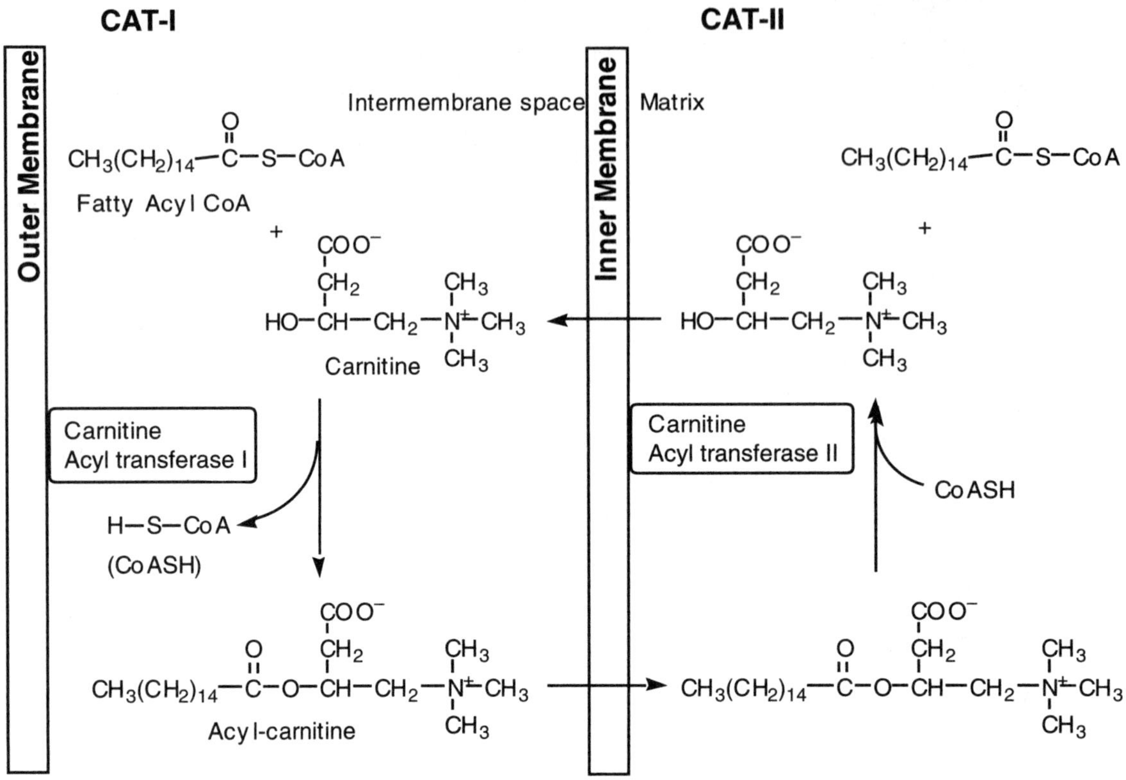

Author's note: I thought it worthwhile to see a specific example. It is so convenient to use "R" for the long hydrocarbon chain that it can be easy to forget just how long these molecules are.

THE RATE OF β-OXIDATION IN

TISSUE:	DEPENDS UPON:
MUSCLE	—current energy requirements. —substrate availability (fatty acid concentration in the blood).
LIVER	—the rate of fatty acid transport into mitochondria. CAT-I is strongly inhibited by malonyl-CoA (the product of the first committed step in fatty acid synthesis). —blood glucose levels. Increased blood glucose levels promote fatty acid synthesis. The product of the first committed step is malonyl-CoA. Malonyl-CoA inhibition of β-oxidation prevents fatty acid synthesis and oxidation from occurring simultaneously.

β-Oxidation

OF A SATURATED FATTY ACYL-COA WITH AN EVEN NUMBER OF CARBONS

Additional steps and enzymes are needed if the fatty acid is unsaturated, branched, and/or has an odd number of carbons.

The following four steps repeat until the entire fatty acid is converted to acetyl-CoA molecules.

Fatty acyl-CoA

1. **Acyl-CoA dehydrogenase**

   Oxidation (removes $2e^-$ and $2H^+$)
   Forms a *trans*-double bond between the α- and β-carbons.

   trans-α,β-enoyl-CoA

2. **Enoyl-CoA hydrase**

   Hydration (adds water)
   Forms –OH at β-carbon

   L-β-hydroxyacyl-CoA
   (3-hydroxyacyl-CoA)

3. **3-Hydroxyacyl-CoA dehydrogenase**

   Oxidation at the β-carbon to form C=O
   Now we officially have β-oxidation!
   (Recall that NAD^+ transfers $2e^-$ in the form of a hydride.)

   β-ketoacyl-CoA

4. **Thiolase (β-ketoacyl-CoA thiolase)**

   Thiolytic cleavage adds a CoASH and cleaves a bond to release acetyl-CoA

   Acyl-CoA

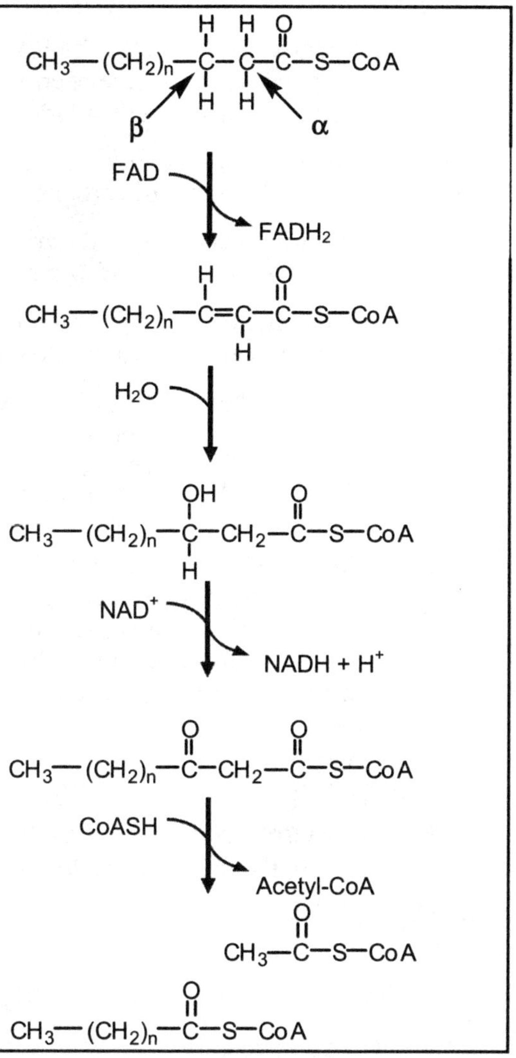

THE COMPLETE OXIDATION OF A FATTY ACID

TO CALCULATE THE TOTAL ATP PRODUCED PER FATTY ACID:

1. Determine the number of cycles of β-oxidation required. Each cycle produces: 1 $FADH_2$, 1 NADH, and 1 acetyl-CoA; the last cycle produces 2 acetyl-CoA. [The last cycle splits a 4-carbon acyl-CoA into 2 acetyl-CoA, so palmitate (16 carbons) produces 8 acetyl-CoA in 7 cycles (*not* 8).]
2. List the total number of $FADH_2$, NADH, and acetyl-CoA produced. [Palmitate: 7 $FADH_2$, 7 NADH, 8 acetyl-CoA]
3. Multiply each molecule above by the number of ATP it produces from the ETC: 1.5 ATP per $FADH_2$, 2.5 ATP per NADH, and 10 ATP per acetyl-CoA.

[The acetyl-CoA molecule can enter the citric acid cycle to give 3 NADH, 1 $FADH_2$, and 1 GTP. So, [(3 NADH × 2.5 ATP/NADH) + (1 $FADH_2$ × 1.5 ATP/$FADH_2$) + 1 ATP] = 10 ATP per acetyl-CoA.]

[Palmitate: (1.5 ×x 7 $FADH_2$) + (2.5 × 7 NADH) + (10 × 8 acetyl-CoA) = 108 ATP!]

4. Subtract 2 ATP from the total ATP to arrive at the final answer. Those 2 ATP were the "activation fee"—the energy cost to activate the fatty acid. It counts as *two* ATP because an AMP, not an ADP, was formed.
5. Make further adjustments if the fatty acid was unsaturated and/or branched and/or had an odd number of carbons. (Each of these is described later in the chapter.)

Palmitic acid (16 carbons) is described on pp. 436-437 of your text, and stearic acid (18 carbons) is Question 12.5 (on p. 437).

Try calculating how many NADH, $FADH_2$, and ATP molecules can be synthesized from one molecule of arachidic acid (20 carbons, saturated).

Solution:
1. Nine cycles of β-oxidation are required, producing 9 $FADH_2$, 9 NADH, and 10 acetyl-CoA.
2. (9 × 1.5) + (9 × 2.5) + (10 × 10) = 13.5 + 22.5 + 100 = 136 ATP
3. 136 – 2 = 134 ATP total from one arachidic acid.

β-OXIDATION IN PEROXISOMES *SHORTENS VERY LONG-CHAIN FATTY ACIDS*

Peroxisomal enzymes differ somewhat from the mitochondrial enzymes. $FADH_2$ produced in the first step donates its electrons directly to O_2 (instead of UQ) to produce H_2O_2, which is converted to H_2O by catalase. The last enzyme has a low affinity for medium-chain acyl-CoA molecules; these are transported to mitochondria to continue oxidation.

KETONE BODIES

KETOGENESIS: Excess acetyl-CoA molecules are converted into **KETONE BODIES** in liver cells, in the mitochondrial matrix.

KETONE BODIES:

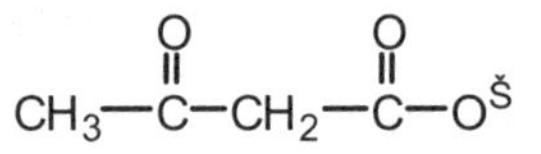
acetoacetate

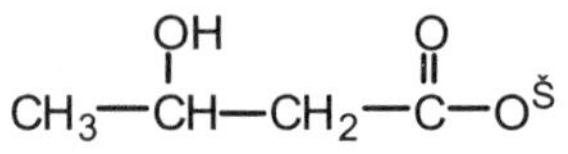
β-hydroxybutyrate

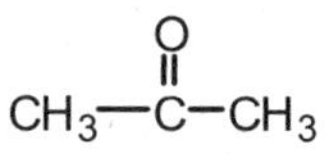
acetone

Ketone bodies are a result of excess acetyl-CoA from β-oxidation:

Intake: high high-lipid, low low-carb diet (and/or not enough oxaloacetate)

Starvation (body consumes fats)

Diabetes (problems with carbohydrate catabolism)

KETOGENESIS:

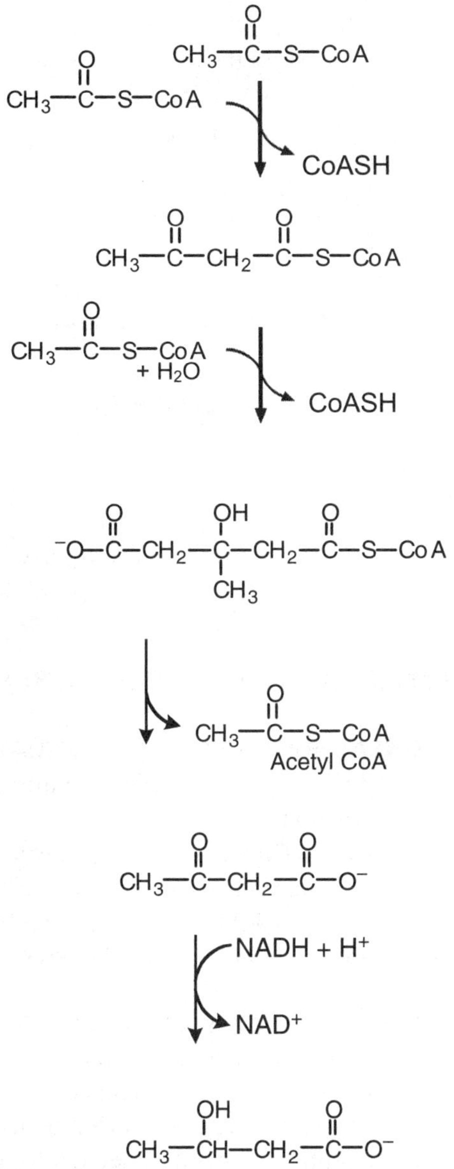

2 Acetyl-CoA

Acetoacetyl-CoA thiolase
Condensation of 2 acetyl-CoA
Reverse of the last step of β-oxidation

Acetoacetyl-CoA

HMG-CoA synthase
The acetyl-CoA added here will be regenerated in the next step.

HMG-CoA
(β-Hydroxy-β-methylglutaryl-CoA)

HMG-CoA lyase
Compare to the reaction with citrate synthase.

Acetoacetate

β-Hydroxybutyrate dehydrogenase

β-Hydroxybutyrate

Ketone bodies are used for energy by many tissues, most notably cardiac and skeletal muscle. The brain prefers glucose for its fuel, but can use ketone bodies when it is forced to (i.e., during starvation or no-carb diets). Ketone bodies are made in the liver, but the liver does not use them as fuel.

KETOSIS occurs when acetoacetate is made faster than it can be used. Under these conditions, acetoacetate spontaneously degrades (no enzyme is needed) to acetone, which is exhaled.

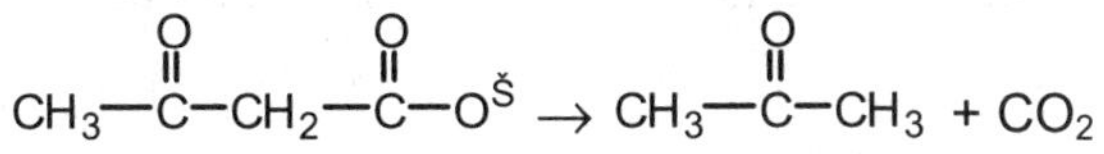

acetoacetate   acetone

In target tissues, acetoacetate is converted back to two molecules of acetyl-CoA:

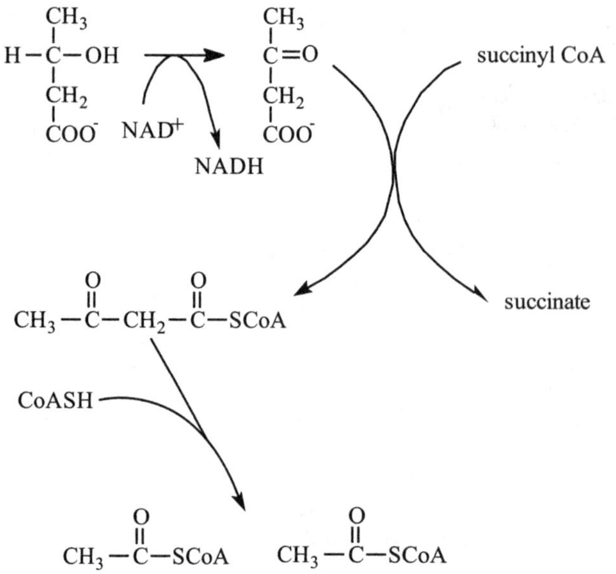

FATTY ACID OXIDATION: DOUBLE BONDS, ODD CHAINS, AND BRANCHES

UNSATURATED FATTY ACID OXIDATION

β-Oxidation proceeds normally until the alkene (usually *cis*-) is in the β,γ-β,γ-position.

Enoyl-CoA isomerase catalyzes a *cis*-β,γ-enoyl-CoA to a *trans*-α,β-enoyl-CoA.

β-Oxidation continues.

One less $FADH_2$ is produced because the double bond is already present. FAD is not needed to oxidize the acyl-CoA to produce a double bond.

ODD CHAIN FATTY ACID OXIDATION

β-Oxidation proceeds normally, but the products of the last cycle are acetyl-CoA and propionyl-CoA, which is carboxylated and isomerized to produce succinyl--CoA. The total energy produced depends upon the degradation pathway of succinyl-CoA. Additional enzymes needed for odd chain fatty acid oxidation are **propionyl-CoA carboxylase**, **methylmalonyl-CoA racemase**, and **methylmalonyl-CoA mutase.**

α-OXIDATION OF BRANCHED-CHAIN FATTY ACIDS

Fact: a β-methyl group blocks β-oxidation, but an α-methyl does not. α-Oxidation essentially oxidizes the α-carbon and removes the carboxyl group so that the product fatty acid has an α-methyl instead of a β-methyl (and one less carbon than the original). β-Oxidation can then proceed normally because the remaining branches occur in α-positions.

Phytanic acid is a common dietary branched-chain fatty acid.

Fatty Acid Biosynthesis Takes Place in the Cytoplasm

Fatty acid biosynthesis cannot be a simple reversal of β-oxidation because of thermodynamic considerations. (Both pathways must be spontaneous ($-\Delta G$) to go forward, but a simple reversal of any reaction also reverses the sign of ΔG.)

Fatty acids are typically made from glucose:

glucose → pyruvate $\xrightarrow{\text{transport into mitochondrion}}$

pyruvate → acetyl-CoA → citrate $\xrightarrow{\text{transport to the cytoplasm}}$

citrate → acetyl-CoA → fatty acids

Lipogenic tissues: liver, adipose tissue, lactating mammary gland.

1. Citrate transports acetyl-CoA from the mitochondrion to the cytoplasm.

   Problem: Acetyl-CoA cannot cross the mitochondrial membrane (where it is produced) so it must be transported to the cytoplasm to be used in fatty acid synthesis.

   Solution: Acetyl-CoA + oxaloacetate → citrate. Citrate crosses the membrane. In the cytoplasm: citrate → acetyl-CoA + oxaloacetate.

   Recall that this synthesis of citrate is the first reaction of the citric acid cycle. Why would citrate leave the citric acid cycle in the mitochondrion and venture out into the cytoplasm? Remember that the citric acid cycle exists to produce energy. If energy is not needed by the cell, citrate will accumulate.

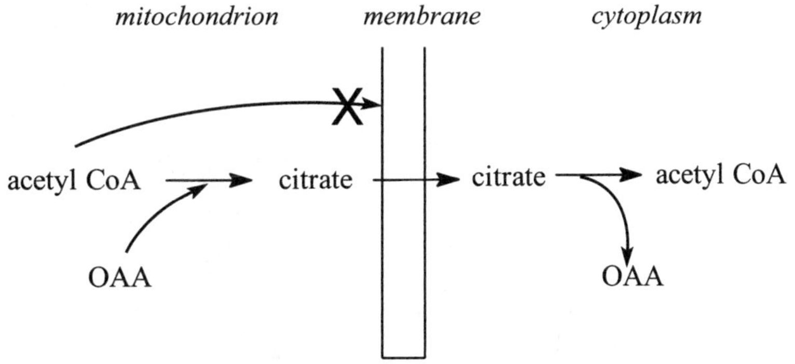

2. **Acetyl-CoA carboxylase (ACC)** carboxylates acetyl-CoA to form **malonyl-CoA**. ACC requires biotin, a CO_2 carrier. Eukaryotic ACC has three domains: **BCCP**, **BC**, and **CT** (**biotin carboxyl carrier protein, biotin carboxylase,** and **carboxyltransferase,** respectively.)
   Remember this step! Both ACC and malonyl-CoA are important regulators.
3. The malonyl group is transferred from CoA to **ACP (acyl carrier protein).** Malonyl-ACP serves as the carbon donor for fatty acid biosynthesis.
4. A second acetyl group is transferred from acetyl-CoA to synthase to form acetyl-S-synthase.
5. Synthase condenses malonyl-ACP with acetyl-S-synthase to form acetoacetyl--ACP. CO_2 is removed as well. This decarboxylation drives the process forward.
6. The reduction/dehydration/reduction reactions have the same intermediates as the β-oxidation of fatty acids, but use different enzymes.

7. The product butyryl-ACP transfers its acyl group to synthase, which condenses with another malonyl-ACP (again, with loss of CO_2). The process repeats until the fatty acid is synthesized (up to 16 carbons total).

SUMMARY OF FATTY ACID BIOSYNTHESIS:

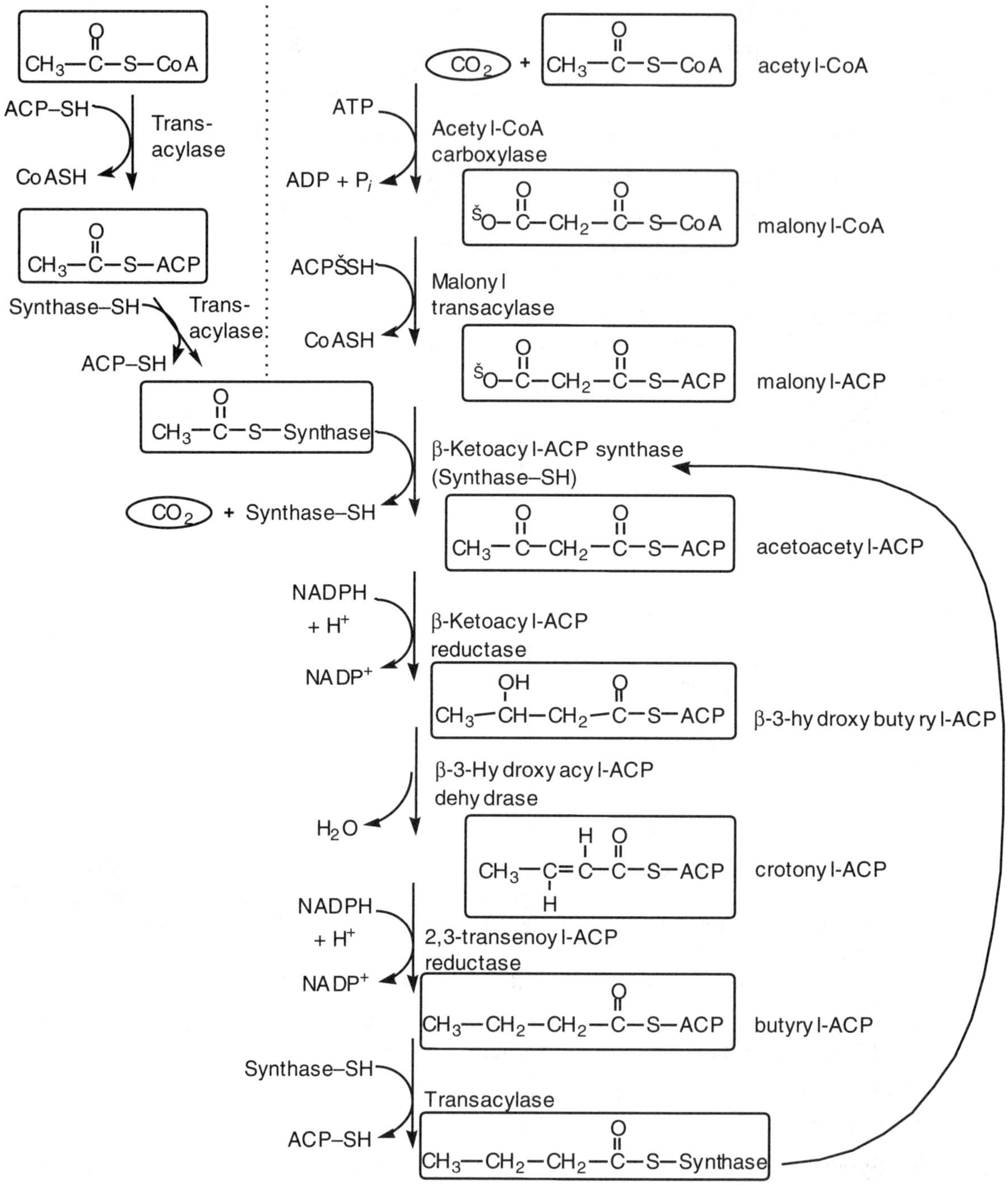

Note: Fatty acid synthase is a multienzyme complex and a dimer, so two fatty acid molecules are synthesized at the same time. (See Figure 12.19 on page 446 of your text.)

THE OVERALL REACTION TO SYNTHESIZE A FATTY ACID WITH 16 CARBONS:

$$8\ \text{acetyl-CoA} + 14\text{NADPH} + 14\text{H}^+ + 7\text{ATP} \rightarrow \text{palmitate} + 14\text{NADP}^+ + 7\text{ADP} + 7\text{P}_i + 8\text{CoASH} + 6\text{H}_2\text{O}$$

β-Oxidation of Fatty Acids	vs.	Fatty Acid Synthesis
• Location: mitochondria		• Location: cytoplasm (chloroplasts/plants)
• CoA is the acyl carrier.		• ACP is the acyl carrier.
• FAD and NAD^+ are electron acceptors.		• NADPH is the electron donor.
• $FADH_2$ and NADH are formed.		• NADPH is consumed; NAD^+ is formed.
• Acetyl-CoA is the two-carbon-unit product.		• Malonyl-CoA is the two-carbon-unit donor.
• Enzymes differ.		• Fatty acid synthase is a dimeric multienzyme complex. (See Figure 12.19, p. 446 of your text.)

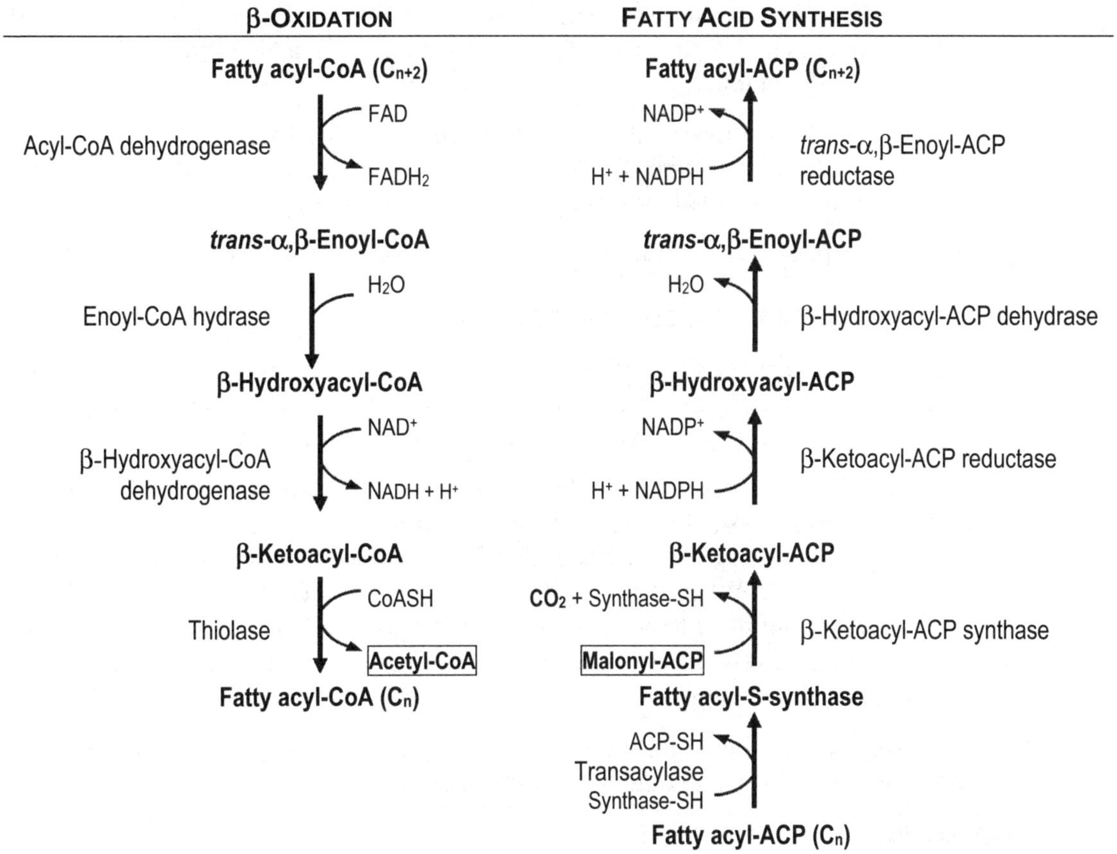

FATTY ACID ELONGATION AND DESATURATION—PRIMARILY BY ER ENZYMES

ER enzymes synthesize fatty acids longer than palmitate (16 carbons) and desaturates fatty acids (removes hydrogens to add double bonds) when these fatty acids are not obtained by the diet.

Why is this important? Cells need unsaturated (and perhaps longer) fatty acids to maintain proper membrane fluidity and as precursors for fatty acid derivatives (such as eicosanoids; cerebrosides, and sulfatides in myelin).

REGULATION OF FATTY ACID METABOLISM IN MAMMALS: SHORT-TERM VS. LONG-TERM

SHORT-TERM REGULATION OF FATTY ACID METABOLISM:

COVALENT MODIFICATION

ACC1[1] catalyzes the first step of fatty acid synthesis to store energy (and/or to lower glucose levels). ACC1 dimers aggregate to form polymers.

Inactive: ACC1 monomer; phosphorylated
Active: ACC1 polymer (NO phosphate)

Phosphorylation/dephosphorylation:

Phosphorylated-ACC1 = **inactive;** it will not polymerize.
Inactive ACC1 halts fatty acid synthesis (prevents energy storage).
Phosphorylation (depolymerization): AMPK (activated by upstream AMPK kinases and AMP), PKA (glucagon-stimulated), and palmitoyl-CoA accumulation.

Dephosphorylation of ACC1 = **active**; allows ACC1 to polymerize.
Dephosphorylation: PP2A—phosphoprotein phosphatase 2A—(activated by insulin, inhibited by glucagon and epinephrine)

Glycerol-3-phosphate acyltransferase phosphorylation is catalyzed by AMPK.

ALLOSTERIC REGULATION; SUBSTRATES AND PRODUCTS

Citrate stimulates **ACC**, the first step of fatty acid synthesis (its product is malonyl-CoA). (Excess citrate indicates that the citric acid cycle is not needed, energy level is good, and fatty acid synthesis can begin to store excess energy.)

Malonyl-CoA inhibits **CAT-I**, which is required to transport fatty acids into the mitochondrion for β-oxidation. (CAT-I is carnitine acyltransferase I).
Why malonyl-CoA? Remember that the first step of fatty acid synthesis makes malonyl-CoA. When that first step is inhibited, levels of malonyl-CoA drop. Since it is not there to act as inhibitor, the acyl-CoA can enter the mitochondrion to be degraded by β-oxidation. So, inhibiting the first step of fatty acid synthesis propels β-oxidation.

Palmitoyl-CoA inhibits fatty acid synthesis by causing the depolymerization of ACC. Palmitoyl-CoA also inhibits the pentose phosphate pathway.

HORMONES

INSULIN promotes lipogenesis and inhibits lipolysis.
GLUCAGON and **EPINEPHRINE** inhibit fatty acid synthesis.
Both increase hydrolysis of triacylglycerols so fatty acids enter the blood.
Both inhibit PP2A, and glucagon activates PKA, resulting in inactive, phosphorylated ACC1.

---

[1] Mammals have two forms of ACC: ACC1 is cytoplasmic, in lipogenic tissues. ACC2 is mitochondrial, in oxidative tissues such as cardiac and skeletal muscles, where the product, malonyl-CoA, inhibits CAT-I.

LONG-TERM REGULATION OF FATTY ACID METABOLISM:

—occurs in response to fluctuating nutrient availability and energy demand.
—involves changes in gene expression triggered by transcription factors (e.g., the **SREBPs** and the **PPARs**) and certain hormones.

For a summary, refer to Figure 12.22, Regulation of Intracellular Fatty Acid Metabolism, on p. 450, of your text.

12.2 MEMBRANE LIPID METABOLISM: PHOSPHOLIPIDS AND SPHINGOLIPIDS

PHOSPHOLIPID METABOLISM

Phospholipid turnover (degradation and replacement mediated by the phospholipases) is rapid.

PHOSPHOLIPID SYNTHESIS

Location: at the interface of the SER and the cytoplasm
Once choline (or ethanolamine) has entered a cell, it is phosphorylated and converted to a CDP derivative, which reacts with a diacylglycerol to form phosphatidylcholine (or phosphatidylethanolamine).

REMODELING: a process that allows a cell to adjust the fluidity of its membranes. Unsaturated fatty acids replace the original fatty acids that were incorporated during synthesis.

PHOSPHOLIPID DEGRADATION is catalyzed by several phospholipases.

SPHINGOLIPID METABOLISM

SPHINGOLIPID SYNTHESIS BEGINS WITH THE PRODUCTION OF CERAMIDE:

Palmitoyl-CoA combines with serine to form 3—ketosphinganine, which is reduced by NADPH to form sphinganine, which is converted to ceramide in a two-step process involving acyl-CoA and $FADH_2$. Ceramide can then be used to form sphingomyelin, glucosylceramide, or galactosylceramide.

SPHINGOLIPID DEGRADATION occurs in lysosomes by specific hydrolytic enzymes.

12.3 ISOPRENOID METABOLISM

Isoprenoids occur in all eukaryotes.

First phase of isoprenoid synthesis: the synthesis of **ISOPENTYL PYROPHOSPHATE**
Isopentyl pyrophosphate synthesis appears to be identical in all of the species in which this process has been investigated (despite the huge diversity of isoprenoids produced).

CHOLESTEROL METABOLISM

Location: within the ER membrane; also in the cytoplasm, at or near the ER.

CHOLESTEROL SYNTHESIS occurs primarily in the liver.
Three multi-step phases convert:

1. two acetyl-CoA to HMG-CoA (β-hydroxy-β-methylglutaryl-CoA)
2. HMG-CoA $\xrightarrow{\text{HMGR}}$ mevalonate → → → squalene
   HMGR is the rate-limiting enzyme for cholesterol synthesis.
3. squalene to cholesterol.

A sterol carrier protein binds squalene and its intermediate to lanosterol,[2] which binds to a second carrier protein.

CHOLESTEROL DEGRADATION

Cholesterol is degraded and eliminated primarily by conversion to **BILE SALTS**.
(Bile salts facilitate the emulsification and absorption of dietary fat.)

SYNTHESIS OF BILE ACIDS (OR BILE SALTS)

The rate-limiting reaction:

$$\text{cholesterol} \xrightarrow{\text{cholesterol-7-hydroxylase}} \text{7-}\alpha\text{-hydrocholesterol}$$

Subsequent reactions to form cholic acid and deoxycholic acid:
—rearrangement and reduction of double bond at C-5
—introduction of an additional –OH group
—reduction of the 3-keto group to a 3-α-hydroxyl group

ER enzymes catalyze **CONJUGATION REACTIONS** that convert cholic acid and deoxycholic acid to bile salts.

CHOLESTEROL HOMEOSTASIS

To maintain cholesterol homeostasis, regulate:
cholesterol synthesis, LDL receptor activity, and bile acid synthesis.

TO REGULATE CHOLESTEROL SYNTHESIS, REGULATE HMGR:

HMGR—the rate-limiting enzyme for cholesterol synthesis—catalyzes the formation of mevalonate.

Phosphorylated HMGR is **inactive** and is promoted by:
high AMP levels activate AMPK-catalyzed phosphorylation of HMGR
cAMP—activates PKA, which phosphorylates and activates phosphoprotein phosphatase inhibitor 1 (PPI-1), which inhibits phosphoprotein phosphatase, which prevents HMGR from becoming active by dephosphorylation. (Recall cAMP regulation by hormones.)

Endocytosis of LDL receptors release cholesterol that inhibits HMGR by negative feedback.

GENE EXPRESSION: mediated by sterols

SREBP2—ER membrane protein; stimulates expression of genes that code for: cholesterol biosynthesis enzymes, LDL receptor, three NADPH synthesis enzymes (G-6-PD, 6-phosphogluconate-dehydrogenase, malic enzyme).

Refer to Figure 12.33 (p. 466 in your text for an excellent summary figure of SREBP2 regulation, which involves:
SCAP—SREBP cleavage-activating protein; binds SREBP2
SSD—sterol-sensing domain
Insig—insulin-induced gene
NF-1—nuclear factor-1
CREB—cAMP response element binding protein
ACAT—acyl-CoA-cholesterol acyltransferase

STEROID HORMONE SYNTHESIS

Cholesterol is the precursor for all steroid hormones.

---

[2] The C-18 methyl shifts from C-8 to C-13 in the conversion of squalene-2,3-epoxide to lanosterol.

$$\text{Cholesterol} \xrightarrow{\text{Desmolase (mitochondria)}} \text{Pregnenolone} \xrightarrow{\text{(in the ER)}} \text{Progesterone}$$

Pregnenolone and progesterone are the precursors for all other steroid hormones.

BIOCHEMISTRY IN PERSPECTIVE

ATHEROSCLEROSIS

BIOTRANSFORMATION

AFTER STUDYING THIS CHAPTER, YOU SHOULD BE ABLE TO:

- State which pathway(s) (lipogenesis, lipolysis, β-oxidation, or fatty acid synthesis) will occur under given conditions (high- or low-energy needs, levels of ATP, AMP, cAMP, NADH, NAD^+, and blood glucose).
- Compare and contrast opposing pathways, including differences in intermediates, enzymes, location, transport mechanisms (if needed), and regulatory mechanisms.
- Write the β-oxidation pathway (including structures and enzymes) of a given fatty acid.
- Determine how many ATP equivalents are obtained from the complete oxidation of a given fatty acid.
- Describe how the oxidation of a fatty acid differs if it has a double bond, branches, or an odd number of carbons.
- Describe the regulation mechanisms of triacylglycerol, fatty acid, and/or cholesterol metabolism. Include how the location of these pathways impacts their regulation.
- Describe a specific transport mechanism required for a given pathway to proceed. For example, describe the role of carnitine in fatty acid metabolism.
- Explain the effects of insulin and glucagon on the pathways studied in this chapter.
- State the key enzymes in a given pathway that are either rate-limiting enzymes or highly regulated.
- Describe the roles of SREBPs and PPARs in lipid metabolism.

Use this space to note additional objectives provided by your instructor:

Chapter 12: Solutions to Review Questions

12.1 a. chylomicron remnants – chylomicrons after about 90% of the triacylglycerols have been removed by lipoprotein lipase

b. glyceroneogenesis – an abbreviated version of gluconeogenesis in which glycerol-3-phosphate (required for TG synthesis) is synthesized from substrates other than glucose or glycerol

c. exogenous lipoprotein pathway – the absorption of TGs and other lipid nutrients and their distribution to the body's tissues

d. apolipoprotein B-48 – the main lipoprotein component of nascent chylomicrons; synthesized from an mRNA that is a truncated version of apolipoprotein B-100 mRNA

e. enterocytes- small intestinal wall cells; absorb nutrients digested within the lumen of the small intestine

12.2 a. β-oxidation – fatty acid degradation by the removal of two-carbon fragments from the carboxyl end to yield acetyl-CoA

b. carnitine – a carrier molecule that transports fatty acids into the mitochondrial matrix across the mitochondrial inner membrane

c. ketogenesis – the condition in which excess acetyl-CoA molecules are converted to acetoacetate, β-hydroxybutyrate and acetone, which are known collectively as the ketone bodies

d. ketone bodies – three molecules (acetoacetate, β-hydroxybutyrate and acetone) that are produced in the liver from acetyl-CoA

e. ketosis – accumulation of ketone bodies in blood and tissues

12.4 a. sterol carrier protein – a cytoplasmic protein carrier for certain intermediates during cholesterol biosynthesis

b. fatty acid binding protein – an intracellular water-soluble protein whose function is to bind and transport hydrophobic fatty acids

c. biotransformation- a series of enzyme-catalyzed processes in which toxic and/or hydrophobic molecules are converted into more soluble metabolites

d. *de novo* – new synthesis of a specific molecule, e.g., *de novo* cholesterol synthesis

e. turnover – the rate at which a molecule is degraded and then replaced

12.5 a. SREBP1- sterol regulatory element binding protein 1 – two transcription factors, SREBP1a and SREBP1c that regulate genes in fatty acid metabolism

b SREBP2 – sterol regulatory element binding protein 2 – a transcription factor that regulates cholesterol metabolism

c. PPAR- peroxisome proliferator–activator receptor – one of several ligand- activated transcription factors that regulate lipid metabolism

d. hypertriglyceridemia – high blood levels of triacylglycerols

e. atheroma – an atherosclerotic lesion in an artery wall; contains macrophages, lipids and cell debris

12.7 a. allyl group – a CH_2=CH-CH_2- group on an organic molecule

b. epoxide – a highly reactive ether in which the oxygen is incorporated into a three-membered ring

c. SAM – S-adenosylmethionine – a methyl donor molecule

d. PAPS – 3'-phosphoadenosine-5'-phosphosulfate – a sulfate donor molecule; used in the syntheses of the sulfatides

e. phase I reaction – a biotransformation reaction involving oxidoreductases and hydrolases that converts hydrophobic substances into more polar molecules

12.8 a. MTTP – microsomal triglyceride transfer protein – a molecular chaperone that facilitates the transfer of lipids onto apoB during the production of apoB-containing lipoproteins

b. abetalipoproteinemia – an autosomal recessive metabolic disorder in which dietary fat absorption is compromised; caused by failure of apoB-lipoproteins to form

c. endogenous lipoprotein pathway – the transport of newly synthesized lipids from the liver via VLDL (very low density lipoprotein)

d. PEPCK-C – the cytoplasmic form of phosphoenolpyruvate carboxykinase; a gluconeogenesis enzyme that plays a role in glyceroneogenesis.

e. triacylglycerol cycle – a mechanism that regulates the level of fatty acids that are available to the body; TGs are constantly being synthesized and degraded to fatty acids and glycerol

12.10 Three differences between fatty acid synthesis and β-oxidation are the following:

(1) The two pathways take place in different cell compartments. Synthesis occurs in the cytoplasm and β-oxidation is within the mitochondria.

(2) The intermediates of fatty acid synthesis and β-oxidation are linked through thioester linkages to ACP and CoASH, respectively.

(3) The electron carrier for fatty acid synthesis is NADPH while those for β-oxidation are NADH and $FADH_2$

12.11 In the short term, hormones alter the activity of preexisting regulatory enzyme molecules. For example, the binding of glucagon inhibits acetyl-CoA carboxylase. Long-term effects of hormones usually involve changes in the pattern of enzyme synthesis in target cells. For example, insulin promotes the synthesis of the enzymes involved in lipogenesis (e.g., acetyl-CoA carboxylase and fatty acid synthase).

12.13 The products of the oxidation of the molecule in question 12 (3-methylhexanoic acid) are two molecules of propionyl CoA, two NADH, one $FADH_2$ and one AMP (which corresponds to a loss of two ATP molecules). Assuming that the propionyl-CoA does not enter the citric acid cycle via its conversion to succinyl-CoA, the total ATP molecules produced are: 5 (from 2NADH) + 1.5 (from $FADH_2$) – 2 (from the conversion of 2-methyl-pentanoic acid to its acyl-CoA activated form) = 4.5 ATP.

The propionyl-CoA may be further oxidized for energy via its conversion to succinyl-CoA with its subsequent entry into the citric acid cycle and the electron transport chain. Listed below is a tally of the ATP molecules produced and their source reaction(s) or pathway. (Conversion of GTP to ATP is assumed.) "Oxaloacetate to oxaloacetate" covers one full turn of the citric acid cycle. "Oxaloacetate to succinyl-CoA" is a partial turn of the citric acid cycle, stopping at succinyl-CoA.

propionyl-CoA to succinyl-CoA:		+1 CO_2	−1 ATP	
succinyl-CoA to oxaloacetate:		+1 ATP	+1 $FADH_2$	+1 NADH
oxaloacetate to oxaloacetate:	−2 CO_2	+1 ATP	+1 $FADH_2$	+3 NADH
oxaloacetate to succinyl-CoA:	−2 CO_2			+2 NADH
total (propionyl-CoA → 3CO_2):	−3 CO_2	+2 ATP	+2 $FADH_2$	+6 NADH
x 2 propionyl-CoA				
total (2 propionyl-CoA → 6 CO_2):		−6 CO_2	**+4 ATP**	+4 $FADH_2$
	+12 NADH			
From the electron transport chain:				(× 1.5 ATP/$FADH_2$)
	(× 2.5 ATP/NADH)			
total ATP from propionyl-CoA and ETC:		**+4 ATP**	**+6 ATP**	**+30 ATP**
		= 40 ATP		

Finally, add the 4.5 ATP from the conversion of 3-methylhexanoic acid to two molecules of propionic acid. Disregarding inorganic products, the complete oxidation of 3-methylhexanoic acid via α-oxidation, β-oxidation, the citric acid cycle, and the ETC yields 7 CO_2 and **44.5 ATP**

12.14 Glyceroneogenesis is an abbreviated version of gluconeogenesis in which glycerol-3-phosphate (required for TG synthesis) is synthesized from substrates other than glucose or glycerol. The key enzymes are pyruvate carboxylase, which converts pyruvate to oxaloacetate, and PEP carboxykinase, which converts oxaloacetate to phosphoenolpyruvate. Instead of glucose, the pathway synthesizes glycerol-3-phosphate from dihydroxyacetone phosphate. Substrates for glyceroneoegenesis are lactate, pyruvate and amino acids such as aspartate and asparagine.

12.16 In the β-oxidation of fatty acids the final reaction (an Cα-Cβ cleavage), catalyzed by thiolase, is described as a thiolytic cleavage. The products of the reaction are acetyl-CoA and an acyl-CoA with two fewer carbon atoms.

12.17 Cells adjust the fluidity of their membrane with membrane remodeling, a process in which phospholipases and acyl transferases alter the fatty acid composition of membrane lipid molecules. The replacement of saturated fatty acids with unsaturated fatty acids increases fluidity.

12.19 Patients with abetalepopteinemia cannot absorb long-chain fatty acids because they lack a protein required for chylomicron synthesis. Medium-chain fatty acids are more easily absorbed and are transported by a different method – transport in the blood by albumin.

12.20 The α-oxidation pathway allows the degradation of branched chain fatty acids such as phytanic acid because it removes methyl groups in the β position of the molecule. The first step is an α-oxidation in which the α-carbon is hydroxlated. α-Hydroxyphytanoyl CoA is converted to pristanal in a decarboxylation reaction that also yields CoA-SH and carbon dioxide. Pristanal is oxidized in an $NAD(P)^+$-requiring reaction to form pristanic acid, which can be further degraded by β-oxidation to yield three acetyl-CoAs, three propionyl-CoAs and one isobutyl-CoA.

12.22 The hydrophobic portions of the molecule are the long hydrocarbon tails of the molecule. The hydrophilic portion of the molecule is the phosphate ester functional group. The hydrocarbon tails are within the bilayer and the phosphate head group is on the surface of the membrane.

12.23 a. glucocorticoid – a steroid hormone that affects carbohydrate, protein and lipid metabolism

b. ketone bodies – acetone, acetoacetate, or β-hydroxybutyrate; produced in the liver from acetyl-CoA

c. biotransformation – reactions that convert hydrophobic molecules into derivatives that contain water-soluble groups in order to improve solubility

d. Phase I reaction - a biotransformation reaction that introduces a functional group into a hydrophobic molecule

e. ACP – acyl carrier protein – a component of fatty acid synthase; a protein that contains a phosphopantetheine prosthetic group to which the acyl intermediates in fatty acid synthesis link via a thioester bond; analogous to CoASH in β-oxidation

12.25 As described (see solution to In-Chapter Question 12.5) each stearic acid molecule generates 120 ATP. Consequently, the three separate stearate products of the hydrolysis of tristearin yield 360 ATP. The glycerol product is transported to the liver where it is used in gluconeogenesis.

12.26 In phase I of biotransformation, reactions catalyzed by oxidoreductases and hydrolases convert hydrophobic substances into more polar molecules. In phase II functional groups on substrate molecules are conjugated with water-soluble substances such as glutathione, glutamate, sulfate or glucuronate, which promote rapid excretion. In animals the term phase III is used to describe this excretory process.

12.28 All the lipid molecules mentioned are originally synthesized from the isoprene units in isopentenyl pyrophosphate molecules. Steroid and terpene molecules are assembled by head-to-tail condensation of these groups

12.29 The functions of each substance are as follows:

a. AMPK - AMP-activated protein kinase – an enzyme that has a major influence of the bodies major metabolic processes

b. sterol carrier protein – to bind squalene during cholesterol synthesis in the cytoplasm

c. lipopotein lipase - an enzyme in the endothelial surface of capillaries that converts TGs in chylomicrons into fatty acids and glycerol

d. GLUT4 – Glucose transport protein in the plasma membrane of insulin sensitive cells

e. hormone sensitive lipase - an enzyme in adipocytes that hydrolyses TGs when glucagon/insulin rations are high

12.31 Conjugation reactions usually improves the solubility of substrate molecules with OH groups that can be esterified with glucuronate or sulfate

12.32 In the small intestine, triacylglycerols mix with bile salts and are emulsified (solubilized). The size of the triacylglycerols and their emulsification with bile salts prevent them from crossing the enterocytes' cell membranes.

12.34 During periods of prolonged starvation, when there is an excess of acetyl-CoA (from the β-oxidation of fatty acids) and very low reserves of glucose, ketone bodies are formed to be metabolized for energy. When the concentration of acetoacetate is high, it decarboxylates to form acetone, which may be detected in the breath.

12.35 Fatty acid biosynthesis occurs in the cytoplasm, whereas fatty acid catabolism takes place in mitochondria and peroxisomes so that these pathways may be regulated independently and energy-wasting futile cycles may be prevented.

12.37 Insulin promotes triacylglycerol synthesis and the storage and uptake of fatty acids. Specifically, insulin inactivates hormone-sensitive lipase to prevent the hydrolysis of fats to glycerol and fatty acids, stimulates the release VLDL from the liver, and activates lipoprotein lipase synthesis and transport to the endothelial cells serving fat and muscle tissue.

12.38 In a long hydrocarbon chain, a *cis*-alkene introduces a kink that partially disrupts the otherwise close packing. *Cis*-fatty acids have lower melting points than saturated fatty acids because less heat energy is required to fully disrupt the packing, which must occur for the melting process to take place. More heat energy is required to disrupt the closer packing in saturated chains, so the melting point is higher for saturated fatty acids. In lipid bilayers that contain a relatively high proportion of *cis*-alkenes in their long hydrocarbon chains, the amphipathic nature of the phospholipids helps to prevent the bilayers from melting *per se*. However, the greater number of *cis*-alkenes does cause an increased degree of fluidity within the membrane bilayer, resulting in more facile lateral motion within the bilayer.

12.40 Refer to Figure 11.18, which illustrates the numbered structure of cholesterol, and copy that structure onto a piece of paper. Each group of carbon atoms that originated as an intact mevalonate unit should be circled as listed below:

- C-2, C-3, C-4 (two methyl groups were lost in the synthesis of 7-dehydrocholesterol from lanosterol)
- C-6, C-5, C-10, C-1, C-19
- C-7, C-8, C-9, C-11 (A methyl shift and/or a demethylation occurred, leaving this mevalonate unit with only four carbons)
- C-15, C-14, C-13, C-12 (C-18 also originated in a mevalonate unit, but is a result of a methyl shift that occurred in the reaction that converted squalene-2,3-epoxide to lanosterol.)
- C16, C-17, C-20, C-21, C-22
- C-23, C-24, C-25, C-26, C-27

12.41 The β-oxidation of oleic acid results in approximately 118.5 ATP equivalents.

$(7FADH_2)(1.5\ ATP/FADH_2) = 10.5\ ATP$

$(8NADH)(2.5ATP/NADH) = 20ATP$

$(9Acetyl\text{-}CoA)(10\ ATP/acetyl\text{-}CoA) = 90\ ATP$

Formation of oleoyl-CoA ffrom oleic acid = – 2ATP

12.43 Instead of cis-double bonds he fatty acids of *trans* fat, created by the partial hydrogenation of vegetable oils, contain *trans*-double bonds. The body can only process fatty acids with the *cis*-isomer. The consumption of *trans* fat-containing food poses a substantial risk for the development of cardiovascular disease because lipoprotein lipase can only bind molecules with the *cis* isomer. As a result, *trans* fat remains in the blood for long periods of time and is, therefore, likely to contribute to arterial plaque formation.

12.44 Fatty acids are activated for β-oxidation and transported into the mitochondrial matrix by conversion to their CoASH derivatives. β-oxidation takes place in the mitochondrial matrix. To cross both the outer and inner mitochondrial membrane the CoASH derivatives are needed. First acyl-CoA synthase catalyzes the formation of the fatty acid acyl-CoA, which is released into the inner membrane space. The acyl group is then transferred to carnitine for transport across the inner membrane into the matrix.

CHAPTER 12: SOLUTIONS TO FILL-IN-THE-BLANK QUESTIONS

12.46 Propionyl-CoA

12.47 Reaction 1 (a dehydrogenation in which one hydrogen is removed from the α- and β-carbons to yield a carbon-carbon double bond) and reaction 3 (the oxidation of the β-carbon's hydroxyl group to yield a carbonyl group)

12.49 Ketone body

12.50 IDL

12.52 Cholesterol

12.53 Eight

12.55 Carnitine

CHAPTER 12: SOLUTIONS TO SHORT-ANSWER QUESTIONS

12.56 In individuals in whom cholesterol synthesis in the liver is higher than normal, cholesterol secreted in bile has a tendency to crystalize because of its insolubility in water.

12.58 Both the SREBPs and the PPARs are transcription factors. Of the three SREBPs (SREBP1a, SREBP1c, and SREBP2), it is SREBP1c that is primarily involved in fatty acid metabolism. Once it is activated by insulin, SREBP1c upregulates the transcription of genes that code for the enzymes of fatty acid and NADPH synthesis. The PPARs are also transcription factors. In liver and adipose tissue under fasting conditions, PPARα stimulates fatty acid catabolism and ketogenesis. PPARγ, expressed in adipose tissue, stimulates glucose uptake and fatty acid and TG synthesis in response to insulin and SREBP1.

12.59 The enzymes that desaturate fatty acids at the Δ9, Δ6, and Δ5 positions of fatty acids use electrons from NADH to activate the oxygen needed to create double bonds. The mechanism whereby these electrons are delivered to the oxygen is composed of two components. Cytochrome b5 reductase accepts a pair of electrons from NADH. Cytochrome b5 accepts electrons one at a time from the reductase and then donates them to the desaturase, where they are used in double-bond formation.

CHAPTER 12: SOLUTIONS TO THOUGHT QUESTIONS

12.61 In the oxidation of butyric acid by the β-oxidation pathway, 1 mole each of $FADH_2$ and NADH and two moles of acetyl-CoA are produced.

12.62 Carnitine is required for the transport of fatty acids into mitochondria where they are oxidized to generate energy. When carnitine levels are low, fat metabolism is impaired. Although glucose metabolism accelerates, an energy deficit occurs. In addition, accumulating acyl-CoA molecules become substrates for competing processes such as peroxisomal β-oxidation and triacylglycerol synthesis.

12.64 Severe dieting (less than 600 calories a day) causes the oxidation of large amounts of fatty acids from the body's TG reserves. Ketone body production increases dramatically with the effect of causing acidosis. Brain cells are damaged by the acidosis and by the ketone bodies themselves, especially the organic solvent acetone.

12.65 Although regular eating is not a panacea, it does provide sufficient carbohydrate to act as a fuel to sustain vital metabolic processes

12.67 When glucagon levels are high, a circumstance that occurs when blood glucose levels are low, glucagon triggers increased fatty acid oxidation.

12.68 The primary effect of the statins is the competitive inhibition of the HMG-CoA reductase, the rate limiting enzyme in cholesterol biosynthesis. The result is lower blood cholesterol levels.

12.70 The molecule shown is phytanic acid . Its oxidation is outlined in figure 12.13.

12.71 Cholesterol is the precursor of the bile salts. If they are not reabsorbed new bile salts are synthesized from the cholesterol pool. This has the result of lowering cholesterol stores in liver, thereby lowering serum cholesterol.

12.73 The ^{14}C label will appear in the mevalonate molecule as indicated by the asterisk.

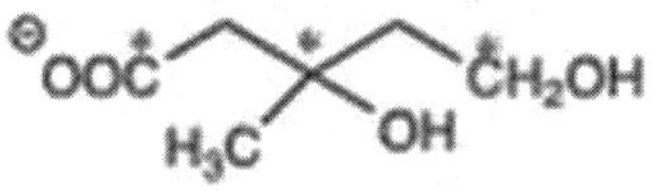

12.74 The ^{14}C label will appear as indicated: *Note*: If ^{14}C-labeled acetyl-CoA is also used to make the β-ketobutyryl-CoA, then a ^{14}C label would appear on the internal alkene carbon in addition to the $–CH_2–O–$ group.

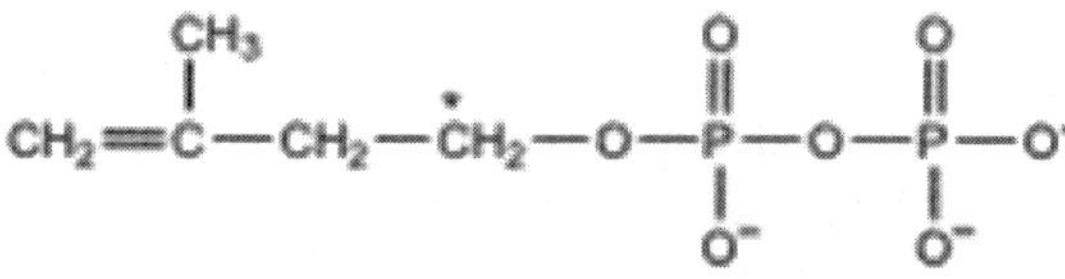

13 Photosynthesis

Brief Outline of Key Terms and Concepts

OVERVIEW

Incorporating CO_2 into organic molecules requires energy and reducing power. In photosynthesis, both of these requirements are provided by a complex process driven by light energy. **REACTION CENTER; PHOTOSYSTEM**

13.1 CHLOROPHYLL AND CHLOROPLASTS

The light energy absorbed by chromophores causes electrons to move to higher energy levels. In photosynthesis, absorption of energy from light drives electron flow.

CHLOROPHYLL; CAROTENOID
CHROMOPHORE; ANTENNA PIGMENT

Photosynthesis occurs in **CHLOROPLASTS,** which consist of: an outer membrane, an inner membrane (encloses the **STROMA**), and the **THYLAKOID MEMBRANE**, which forms an intricate series of flattened vesicles called **GRANA. STROMAL LAMELLAE** connect grana, and the **THYLAKOID LUMEN** is the space within the thylakoid membrane.

THE WORKING UNITS OF PHOTOSYNTHESIS:
PHOTOSYSTEM I (PSI);
PHOTOSYSTEM II (PSII);
CYTOCHROME b_6f COMPLEX;
ATP SYNTHASE (CF_0CF_1ATP SYNTHASE)

13.2 LIGHT

Electrons absorb energy at specific wavelengths and become excited. Electrons release energy (return to ground) by fluorescence, resonance energy transfer, oxidation-reduction, or radiationless decay. This energy drives photosynthesis.

13.3 LIGHT REACTIONS

During the light reactions, H_2O is oxidized to O_2, and the ATP and NADPH required to drive carbon fixation are produced. The **Z SCHEME** is a mechanism that connects PSI and PSII in series. Light-driven photosynthesis begins with PSII.

PHOTOSYSTEM II: O_2 GENERATION
OXYGEN-EVOLVING COMPLEX
WATER-OXIDIZING CLOCK

PHOTOSYSTEM I AND NADPH SYNTHESIS
NONCYCLIC ELECTRON TRANSPORT: Electrons from H_2O are transferred from photosystem II to photosystem I to $NADP^+$ with the production of O_2 and NADPH.

CYCLIC ELECTRON TRANSPORT involves only PSI and generates additional ATP but no NADPH. During the light-independent reactions, CO_2 is incorporated into organic molecules. The first stable product of carbon fixation is glycerate-3-phosphate.

PHOTOPHOSPHORYLATION
Pumping 8 H^+ by the cyt b_6f complex (as a result of absorbing 4 photons) yields 2 ATP.

13.4 LIGHT-INDEPENDENT REACTIONS

THE CALVIN CYCLE is a series of light-independent reactions that incorporates CO_2 into organic molecules such as starch and sucrose (both important energy sources). Three phases of the Calvin cycle reactions are **CARBON FIXATION,** reduction, and regeneration. (Sucrose is also used to translocate fixed carbon throughout the plant.)

PHOTORESPIRATION is a wasteful process in which photosynthesizing cells consume O_2 and release CO_2. Its role in plant metabolism is not understood.

ALTERNATIVES TO C3 METABOLISM
C4 METABOLISM; C4 PLANTS can suppress photorespiration.
CRASSULACEAN ACID METABOLISM

13.5 REGULATION OF PHOTOSYNTHESIS

LIGHT CONTROL OF PHOTOSYNTHESIS
Light is the principal regulator of photosynthesis and affects the activities of regulatory enzymes such as rubisco by means of indirect mechanisms, which include changes in pH, $[Mg^{2+}]$, the **FERREDOXIN-THIOREDOXIN SYSTEM**, and **PHYTOCHROME**.

RIBULOSE-L,5-BISPHOSPHATE CARBOXYLASE (RUBISCO), the most important enzyme in photosynthesis, is highly regulated by allosteric effectors, covalent modification and genetic control (light-activated synthesis of rubisco).

PHOTOSYNTHESIS: WHAT ARE THE ULTIMATE GOALS?

- to convert light energy into chemical energy (ATP)
- to produce reducing equivalents (NADPH)
- to convert CO_2 into sugars: **CARBON FIXATION**

LIGHT ENERGY drives the production of **ATP** and **NADPH**. The ATP and NADPH produced are used to convert **CO_2** into **SUGARS**.

STUDY HINTS

To make sense out of photosynthesis, we need to understand and integrate all of the following topics, and each topic contains new terms and new concepts. The good news is that you are already familiar with a fair amount of this, and there are a number of similarities between systems that you have studied before (especially the electron transport system) and those involved in photosynthesis.

- Chloroplast structure
- How light excites electrons and what those excited electrons can do
- Molecular structures and complexes that are receptive to light (chromophores)
- Electron transfer and the use of the energy released as a result
- ATP as energy currency in the cell; NADPH as reducing power
- Metabolism of carbohydrate molecules and regulation of pathways

STAY FOCUSED ON THE GOAL of each part of photosynthesis that you are studying and how it relates to the ultimate goals listed above. A stumbling block to avoid is to become so mired in the details of one section that you lose sight of its importance. Understand the big picture first and then tackle the specifics.

Example: The *Z Scheme* is just another way of looking at photosynthesis. Compare Figure 13.13 which outlines the Z scheme, to Figure 13.17 which includes a general schematic of the light reactions of photosynthesis.

CHLOROPLAST STRUCTURE: Knowing the location (i.e., stroma, thylakoid lumen, and thylakoid membrane) and how the parts of the chloroplast fit together will really help to give a clear picture of the overall process and will also help to clarify how the different photosystems and the dark reactions fit together. Keep a diagram of the chloroplast in front of you, and when terms such as "thylakoid membrane" and "stroma" pop up, you will have an immediate visual. Figure 13.3 on page 478 is great. Be sure to include *appressed* and *nonappressed* in your diagram.

Example: Since the ATP synthase complex works by allowing H^+ to pass through its pore (and as such, is driven by an $[H^+]$ gradient, just like mitochondrial ATP synthase), it is significant *where* each H^+ is used or released by a particular reaction.

ABBREVIATIONS: Many of the molecules, complexes, and systems have long names, and most of them have abbreviations. To stay afloat in this alphabet soup, keep a running list of abbreviations that you do not know right off the top of your head, and keep this list in front of you as well. *(Examples include PSII, PQ, Yz, MSP, and LHCII.)*

Some molecules have more than one abbreviation. It is valuable to be aware of all of the abbreviations (since different sources/texts have different preferences), but do not let them frustrate the learning process. Have a good concise list on hand. Example: ferredoxin-NADP oxidoreductase is referred to as both FP and FNR, and pheophytin a is shown as both Ph and Phe a.

A PICTURE IS WORTH A THOUSAND WORDS: Make your own diagrams, using the figures in the text as a guide. The drawing and writing process is a powerful learning tool. These figures in your text are particularly good visual summaries:

Figure 13.17	***Membrane Organization of the Light Reactions in Chloroplasts: The Noncyclic Electron Transport Chain and the ATP Synthase Complex*** (p. 491).
Figure 13.18	***The Cyclic Electron Transport Pathway.*** This figure shows the interactions of light with photosystems I and II, their connection, and the roles of the cyt b_6F complex, PQ, and ATP synthase, on location in and around the thylakoid membrane (p. 491).
Figure 13.3	***Chloroplast structure***. Be sure to differentiate between appressed and nonappressed thylakoid membrane in your diagrams (p. 478).
Figures 13.13	***The Z Scheme*** (p. 487).
Figure 13.19	***The Calvin Cycle***: using CO_2, ATP, and NADH to make a 3-carbon sugar. What is the net equation, and what happens at each phase? Note that parts of this cycle resemble the pentose phosphate pathway (p. 494).
Figure 13.21	***Photorespiration*** (p.497).
Figures 13.23-24	***C4 Metabolism. Crassulacean Acid Metabolism*** (p. 499-500).

OVERVIEW

Light energy is converted into chemical energy at **reaction centers**. Each reaction center is a complex of light-absorbing pigments and electron transfer proteins.

13.1 CHLOROPHYLL AND CHLOROPLASTS

SPECIALIZED PIGMENT MOLECULES ABSORB LIGHT ENERGY.

Chromophores that absorb visible light typically have extended chains of conjugated double bonds like the carotenoids. (β-carotene contains 11 conjugated double bonds) In contrast, chlorophylls *a* and *b* and pheophytin *a* have a highly conjugated porphyrin ring system that absorbs visible light. Their long, nonpolar phytol chain embeds (and anchors) in the cell membrane (and contains only one double bond).

Chlorophyll *a*, chlorophyll *b*, and **pheophytin *a*:**
Chlorophylls *a* and *b* have a Mg atom in the porphyrin ring; pheophytin *a* has 2 H atoms. On the porphyrin ring II, chlorophyll *a* and pheophytin *a* have $-CH_3$; chlorophyll *b* has –CHO at the same site.

Lutein and **β-carotene** are the most abundant carotenoids in thylakoid membranes.

CHLOROPLASTS: WHERE PHOTOSYNTHESIS OCCURS IN PLANTS

CHLOROPLAST STRUCTURE *(See Figure 13.3, p. 478, of your text.)*
Chloroplasts have three membranes:

1. an outer membrane that is highly permeable
2. an inner membrane that regulates transport into and out of the chloroplast

   STROMA—the space inside the inner membrane (similar to mitochondrial matrix); contains enzymes (for light-independent reactions and starch synthesis), DNA, and ribosomes

3. **THYLAKOID MEMBRANE**—a third membrane that forms an intricate series of flattened vesicles called **GRANA.** Light-dependent reactions of photosynthesis take place within the thylakoid membrane.

GRANA—stacks of several flattened vesicles (*granum* is singular)

APPRESSED THYLAKOID MEMBRANE—adjacent layers of membrane that fit closely together within each granum; location of most PSII

NONAPPRESSED THYLAKOID MEMBRANE—directly exposed to the stroma; location of most PSI complexes

THYLAKOID LUMEN (SPACE)—internal compartment created by the formation of grana

STROMAL LAMELLAE—tubelike thylakoid membrane that interconnects grana

THE WORKING UNITS OF PHOTOSYNTHESIS:

PSI—PHOTOSYSTEM I

Function: energizes and transfers e^- that are eventually donated to $NADP^+$
Location: primarily in nonappressed thylakoid membrane

PsaA, PsaB—subunits and the largest polypeptides in PSI
PsaAB dimer contains P700, A_0, A_1, and F_4

P700 (SPECIAL PAIR)—a special pair of chlorophyll *a* molecules that absorb light at 700 nm

A_0, A_1, F_4—a series of $1e^-$ carriers

A_0 a specific chlorophyll *a* molecule;

A_1 (phylloquinone, Q, vitamin K_1)—similar in structure to ubiquinone

F_X a 4Fe-4S cluster

F_A, F_B—two 4Fe-4S clusters in an adjacent protein

ANTENNA PIGMENTS—absorb light energy and transfer it to the reaction center; include additional chlorophyll *a* molecules, chlorophyll *b,* and carotenoids

LHCI—LIGHT HARVESTING COMPLEX I—associated with PSI

PSII—PHOTOSYSTEM II

Function: oxidize H_2O to O_2 and donate energized e^- to electron carriers that eventually reduce photosystem I.
Location: primarily in appressed thylakoid membrane (not exposed to stroma)

PSII is a large membrane-spanning protein-pigment complex with at least 23 components. Its reaction center contains D_1/D_2 dimer, cytochrome b_{559}, P680 (a special pair of chlorophyll *a* molecules that absorb light at 700 nm; and several hundred antenna pigments.

O_2-evolving component contains:
MSP (manganese-stabilizing protein)
Mn_3CaO_4 (a cubelike cluster linked to another Mn by a mono-μ-oxo bridge)
Tyrosine residue (Tyr^{161} or Y_z)—located on D_1

Proteins: CP43, CP47

Also associated with PSII:
pheophytin a
plastoquinone (PQ)—similar to ubiquinone; two forms: Q_A, Q_B

LIGHT HARVESTING COMPLEX II (LHCII)—transmembrane protein and a trimer of Lhcb1, Lhcb2, and Lhcb3 (light harvesting complex proteins). Each binds 12-14 chlorophyll a and chlorophyll b molecules and several carotenoids; major component of thylakoid membrane; detachable from PSII.

PQH_2—reduced plastoquinone—maintains balanced absorption of energy between PSI and PSII via a PQ H_2-dependent phosphorylation of LHCII, which (as always) causes a conformational change, which (surprisingly) causes LHCII to detach so it can wander over to PSI. When PQH_2 drops, LHCII loses its phosphoryl group and reassociates with PSII.

CYTOCHROME b_6f COMPLEX is similar to cytochrome bc_1 complex in mitochondria.
Location: throughout the thylakoid membrane.

ATP SYNTHASE = CF_0CF_1ATP SYNTHASE
Structurally and functionally similar to mitochondrial ATP synthase (see Figure 13.8, p. 482. ATP synthase phosphorylates ADP, driven by the transmembrane proton gradient (produced during light-driven electron transport).
Location: thylakoid membrane that is in direct contact with the stroma.

13.2 LIGHT HAS PROPERTIES OF BOTH WAVES AND PARTICLES

VISIBLE LIGHT:	*violet*		*red*
	400 nm	to	700 nm
	shorter wavelength		longer wavelength
	higher frequency		lower frequency
	higher energy		*lower energy*

WAVE PROPERTIES OF LIGHT: LIGHT ACTS LIKE WAVES	PARTICLE PROPERTIES OF LIGHT: PHOTONS LIGHT ACTS LIKE PARTICLES
$\lambda = c/v$ λ = wavelength, v = frequency, c = speed of light	$\varepsilon = hv$ ε = energy of a photon h = Planck''s constant

Molecules absorb light at specific energies that correspond to specific wavelengths. (Complex molecules can absorb light at several wavelengths.)

Chromophores absorb light. When electrons in the chromophores absorb light, they become excited (they move from the ground state to higher energy levels).

HOW AN EXCITED ELECTRON CAN RETURN TO ITS GROUND STATE:

1. **FLUORESCENCE.** Excited electron first relaxes to a lower vibrational (energy) state and then goes back to ground state and emits a photon (of lower energy than the original photon absorbed).

2.* **RESONANCE ENERGY TRANSFER.** Excitation energy is transferred to a neighboring molecule via resonance.

3.* **OXIDATION-REDUCTION.** Transfer of an excited electron to a neighboring molecule makes it a strong reducing agent. The electron returns to its ground state by reducing another molecule.

4. **RADIATIONLESS DECAY.** Excitation energy is given off as heat.

*These two methods are the most important to photosynthesis.

13.3 LIGHT REACTIONS: LIGHT EXCITES ELECTRONS THAT ARE USED TO SYNTHESIZE ATP AND NADPH

OVERALL REACTION: H_2O IS OXIDIZED TO O_2, $NADP^+$ IS REDUCED TO NADPH

$$2NADP^+ + 2H_2O \rightleftharpoons 2NADPH + O_2 + 2H^+ \qquad \Delta E^{\circ\prime} = -1.136 \text{ V}$$

For each mole of O_2 generated:
at least 438 kJ of free energy ($\Delta G^{\circ\prime}$) is needed
at least 8 photons are absorbed to provide 1360 kJ of energy

SUMMARY OF LIGHT REACTIONS

Light is used in photosystem II to excite electrons to a higher energy state. As the electrons are transferred down an energy gradient, protons (H^+) are pumped across the membrane to generate a proton gradient, which is used for ATP synthesis.

The electrons from photosystem II can then be transferred to photosystem I. More light is needed to excite the electrons to an energy state that is high enough to reduce $NADP^+$ to NADPH.

Water serves as a source of electrons when electrons are transferred from photosystem II to photosystem I to $NADP^+$. When H_2O gives up its electrons, the oxygen in H_2O is oxidized to O_2.

Z SCHEME: ELECTRON TRANSFER FROM H_2O TO $NADP^+$

The Z scheme shows where electrons go during photosynthesis. If this path of electrons is shown on a graph where the y-axis is energy, then part of this path resembles the letter *Z*.

Light drives photosynthesis, and the Z scheme shows where and why this is true. From the energy differences between molecules within the photosystems, it is clear that a boost of energy—in the form of a photon of light—is needed at PSII and at PSI to give the electrons enough energy to continue through their path to $NADP^+$.

Overall, ELECTRONS COME FROM H_2O and REDUCE $NADP^+$ TO FORM NADPH.

PATH OF ELECTRONS THROUGH THE PHOTOSYSTEMS IN THE PRESENCE OF LIGHT:

H_2O ↘
PSII ↘ (photosystem II)
PQ ↘ (plastoquinone)
cyt b_6f ↘ (cytochrome b_6f complex)
PC ↘ (plastocyanin)
PSI ↘ (photosystem I)
Fd ↘ (ferredoxin)
FNR ↘ (ferredoxin-NADP oxidoreductase)
$NADP^+$

Here is another way to think about this: begin with the excitation of photosystem I, which results in the transfer of electrons, ultimately to $NADP^+$ to form NADPH. After the electrons are released from PSI, how are they replaced?

P700 in PSI absorbs a photon of light and releases an energized electron. That electron is replaced by an electron from PSII (through PQ, cyt b_6f, and PC, as shown above). The

PSII electron needs to be excited by a photon as well, in order to have enough energy to be transferred. How is the PSII electron replaced? *Water* gives up electrons to form O_2 and H^+.

PHOTOSYSTEM II AND OXYGEN GENERATION

The function of photosystem II is to oxidize water molecules and donate energized electrons to electron carriers, which eventually reduce photosystem I. As the electrons are donated to the electron carriers, protons are pumped out of the thylakoid and ultimately used for ATP synthesis. The water-oxidizing clock is the mechanism by which H_2O is converted into O_2.

REACTIONS OF PHOTOSYSTEM II:

Light energy excites a P680 electron, giving it a large potential energy. Each energized electron is transferred to a series of electron transport carriers that are analogous to the mitochondrial electron transport chain. As electrons are transferred, protons are pumped from the stroma into the thylakoid lumen. This creates a proton gradient that is used to synthesized ATP.

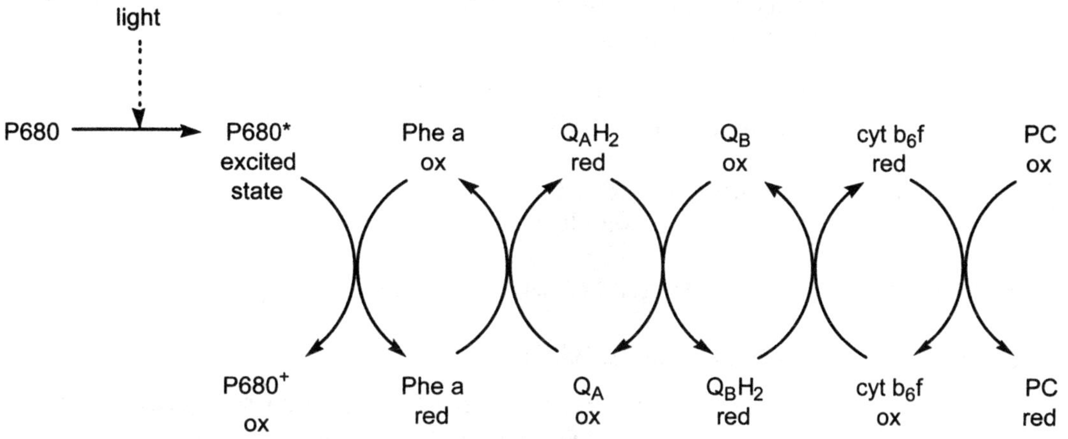

red = reduced state; the carrier marked "red" has the electrons ready to pass to the next carrier
ox = oxidized state; the carrier marked "ox" is ready to accept electrons

The first reaction (first set of arrows pointing down) of P680* is:

$$\underset{\text{excited state}}{\text{P680*}} + \underset{\text{ox}}{\text{Phe a}} \rightarrow \underset{\text{ox}}{\text{P680}^+} + \underset{\text{red}}{\text{Phe a}}$$

The second reaction (second set of arrows, pointing up) is:

$$\underset{\text{red}}{\text{Phe a}} + \underset{\text{ox}}{Q_A} \rightarrow \underset{\text{ox}}{\text{Phe a}} + \underset{\text{red}}{Q_AH_2}$$

THE REDUCED PLASTOCYANIN (PC) TRANSFERS ITS ELECTRONS TO P700 IN PSI.

WHAT HAPPENS TO THE OXIDIZED P680?

These electrons are replaced when H_2O is oxidized.

PQ TRANSFERS ELECTRONS FROM PSII TO cyt b_6f AND PUMPS H^+ TO THE LUMEN.

PQ is plastoquinone, an electron carrier within the thylakoid membrane

$PQ + 2e^-$ (from PSII) $+ 2H^+$ (from stroma) $\rightarrow PQH_2$

$PQH_2 \rightarrow PQ + 2e^-$ (to cyt b_6f) $+ 2H^+$ (to the thylakoid lumen)

CYTOCHROME b_6f COMPLEX TRANSFERS ELECTRONS FROM PQ TO PC

An Fe-S site on the complex transfers electrons from PQ to PC.

PC—plastocyanin—a water-soluble, copper-containing protein in the thylakoid lumen.

PHOTOSYSTEM I USES ELECTRONS FROM PC TO REDUCE $NADP^+ \rightarrow$ NADPH

The ultimate goal of PSI: to reduce $NADP^+$ to NADPH, which requires $2e^-$.

The path of the electrons:

1. from PC (in the lumen) to P700 (in the membrane)
2. reduced P700 absorbs a photon of light that excites and energizes this electron
3. the excited electron is transferred through this series of electron carriers in PSI, in the thylakoid membrane:

   $P700^* \rightarrow A_0 \rightarrow Q \rightarrow F_X \rightarrow F_A, F_B$
   chlorophyll a $\rightarrow$ phylloquinone $\rightarrow$ Fe-S proteins

The next electron acceptor is ferredoxin.
Ferredoxin is mobile, water soluble, and in the stroma.

FERREDOXIN TRANSFERS ITS ELECTRON FROM PSI

EITHER: **TO FNR (TO MAKE NADPH)**

This is the ***NONCYCLIC*** **ELECTRON TRANSPORT PATHWAY.**
FNR (ferredoxin-NADP oxidoreductase) uses two electrons (and a stromal H^+) to reduce $NADP^+$ to NADPH.

OR: **TO PQ (TO PUMP H^+ FOR ATP SYNTHESIS)**

This is the ***CYCLIC*** **ELECTRON TRANSPORT PATHWAY.**

Electrons are cycled *back* to PQ (in the thylakoid membrane), which also takes two H^+ from the stroma to form PQH_2. Remember that PQH_2 then gives its electrons to cyt b_6f and its two H^+ go to the lumen.

So, energized electrons were used to pump additional H^+ across the thylakoid membrane into the lumen *instead of making NADPH.*

The extra H^+ in the lumen produces additional ATP.

The cyclic electron transport pathway drives ATP synthesis without making NADPH. This only occurs in some species (e.g., algae), when the $NADPH/NADP^+$ ratio is high.

ATP Synthase (CF_0CF_1ATP synthase)

ATP synthase phosphorylates ADP, driven by the transmembrane proton gradient (produced during light-driven electron transport), and is located in the thylakoid membrane, in direct contact with the stroma.

Photophosphorylation is the light-driven synthesis of ATP from ADP + P_i.

13.4 Light-Independent Reactions

The Calvin Cycle[1] Incorporates CO_2 into Carbohydrate Molecules

Many Calvin cycle reactions resemble pentose phosphate pathway reactions.

Phases of the Calvin cycle:

1. **Carbon fixation** by the enzyme rubisco (ribulose-1,5-bisphosphate carboxylase) in C3 plants: CO_2 reacts with ribulose-1,5-bisphosphate to form 2 glycerate-3-phosphates (per CO_2). This reaction requires ATP.
2. **Reduction of glycerate-3-phosphate** by NADPH to form glyceraldehyde-3-phosphate
3. **Regeneration of ribulose-1,5-bisphosphate**: Of every six glyceraldehyde-3-phosphates formed, five are regenerated to form three molecules of ribulose-1,5-bisphosphate.

Calvin Cycle Reaction Summary:

3 ribulose-1,5-bisphosphate + 3 CO_2	→	6 glycerate-3-phosphate
6 glycerate-3-phosphate + 6 ATP + 6 NADPH	→	6 glyceraldehyde-3-phosphate + 6 ADP + 6 $NADP^+$ + 6 P_i
5 glyceraldehyde-3-phosphate + 3 ATP	→	3 ribulose-1,5-bisphosphate + 3ADP + 2P_i

Calvin Cycle Net Equation:

3 CO_2 + 6 NADPH + 9 ATP	→	glyceraldehyde-3-phosphate + 6 $NADP^+$ + 9 ADP + 8 P_i

Photorespiration Consumes O_2 and Evolves CO_2

Photorespiration is a wasteful light-dependent process that undermines photosynthesis. In photorespiration, O_2 is consumed and CO_2 is evolved by plant cells while actively engaged in photosynthesis.

The rate of photorespiration depends on the concentrations of CO_2 and O_2, and its function is unknown. Plants that use C4 metabolism and crassulacean acid metabolism (CAM) can counteract the photorespiration process. (See further notes, below.)

[1] The Calvin cycle is also known as the dark reactions, light-independent reactions, the reductive pentose phosphate cycle (RPP cycle), and the photosynthetic carbon reduction cycle (PCR cycle.)

13.5 Regulation of Photosynthesis: Rubisco Control; Effect of Light

Why Regulate Photosynthesis?

- to increase the rate of photosynthesis when light is available;
- to make sure that when the rate of photosynthesis is high, the rates of CO_2 fixation and **sucrose synthesis** is also high.

Photosynthesis depends on temperature, cellular CO_2 concentration, and light.

Light: activates certain photosynthetic enzymes and
deactivates some of the enzymes that degrade sugars.

The key regulatory enzyme in photosynthesis is **RIBULOSE-1,5-BISPHOSPHATE CARBOXYLASE (RUBISCO).** Rubisco catalyzes the fixation of CO_2, which is relatively slow. The best way to increase rubisco's rate is to increase the number of copies of rubisco. Genes that code for rubisco are activated by an increase in light intensity (and appears to involve phytochrome).

Rubisco is also controlled by covalent modification (carbamoylation). The rate of carbamoylation depends on the CO_2 concentration and an alkaline pH.

Light Affects Enzymes by Indirect Mechanisms:

1. **pH** of the stroma increases during photosynthesis (when H^+ is pumped out of the stroma into the thylakoid lumen). This higher pH (~8) activates rubisco and other enzymes.
2. **Mg^{2+}** moves across the membrane into the stroma during the light reactions; this increase in stromal [Mg^{2+}] activates several photosynthetic enzymes (including rubisco).
3. **FERREDOXIN-THIOREDOXIN SYSTEM**
4. **PHYTOCHROME**

Alternatives to C3 Metabolism: C4 Metabolism and CAM (Crassulacean Acid Metabolism)

C4 plants have evolved mechanisms to reduce water loss and photorespiration. C4 metabolism is a mechanism for assimilating CO_2.

When the plant opens its stomata at night, CO_2 enters and reacts to form oxaloacetate. When light is available, photosynthesis makes ATP and NADPH, and CO_2 is released in the bundle sheath cells and converted into sugar. Because the concentration of CO_2 within bundle sheath cells is significantly higher than that of O_2, photorespiration is drastically reduced.

In C4 plants, the light reactions occur in the mesophyll cells, and the Calvin cycle reactions (CO_2 fixation) occur in the bundle sheath cells. This division of labor lets C4 plants concentrate CO_2 in the bundle sheath cells at the expense of ATP hydrolysis.

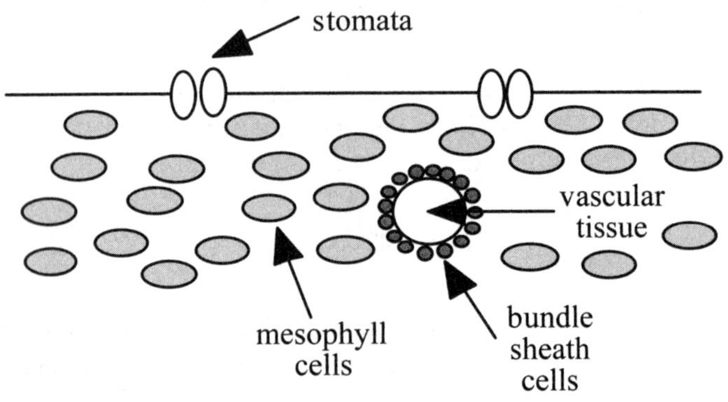

ADVANTAGES OF C4 METABOLISM

- Minimizes photorespiration by increasing CO_2 concentration in cells that have rubisco.
- Reduces the plant's need for water. Stomata must open to allow CO_2 to enter and O_2 to exit the leaf. But water is also lost when the stomata are open. C4 plants can effectively concentrate CO_2, so that the stomata do not need to be open as often. Water loss in C4 plants is only 10-30% compared to C3 plants.

MESOPHYLL CELLS[2]
IN DIRECT CONTACT WITH AIR
LACK RUBISCO
LIGHT REACTIONS MAKE ATP, NADPH RELEASED

BUNDLE SHEATH CELLS
IN CONTACT WITH VASCULAR TISSUE
HAVE RUBISCO
CALVIN CYCLE REACTIONS USE CO_2

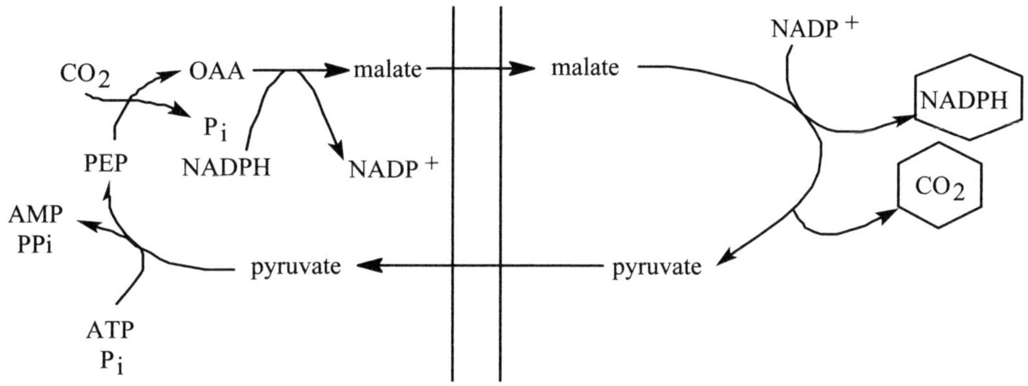

CAM (CRASSULACEAN ACID METABOLISM)

CAM is another type of photosynthetic specialization used by desert plants. These plants grow in high-intensity light without very much water and, in order to survive, have evolved to the point where they are able to temporally separate carbon fixation and ATP/NADPH synthesis. The growth of CAM plants is slowed because the extent of photosynthesis is limited by how much CO_2 is stored as malate during the night.

AT NIGHT:	**DURING THE DAY:**
The stomata open at night when water loss would be at a minimum. CO_2 is stored as malate.	The stomata close. Photosystems I and II work to make ATP and NADPH. The CO_2 that was stored in malate during the night is released, and rubisco uses this CO_2 to make carbohydrates in the Calvin cycle.

[2] In some C4 plants, oxaloacetate is converted to aspartate instead of malate. The Asp travels to the bundle sheath cells, where it is converted back to oxaloacetate, which can then release its CO_2.

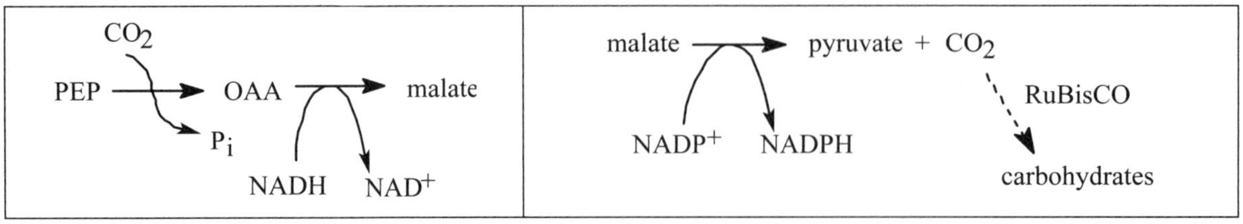

AFTER STUDYING THIS CHAPTER, YOU SHOULD BE ABLE TO:

- Demonstrate an understanding of the overarching goals of photosynthesis and how each pathway studied contributes to those goals.
- Describe the properties of light and how energy from an excited electron is captured in photosynthesis.
- Describe the structure of the chloroplast. State the locations of specific pathways and the property of each location that is advantageous for each pathway.
- Differentiate between the light-dependent and the light-independent reactions.
- Trace electrons through the Z scheme. Identify the points where:
  - energy is captured as reducing power in NADH
  - energy is captured as ATP
  - water is oxidized to O_2
- State the role served by various components of photosystems I and II.
- Describe the similarities and differences between the electron transport system and photosynthesis and between oxidative phosphorylation and photophosphorylation.
- Explain why photorespiration is counterproductive to photosynthesis and describe the mechanisms (pathways) that some plants use to suppress photorespiration.
- Trace carbon and/or oxygen atoms from CO_2 through the Calvin cycle.
- State under which conditions photosynthesis will be activated or inhibited. Specifically, relate the enzyme(s) and/or pathway will be affected, the type of regulation (allosteric, covalent modification, compartmentation, genetic control), and the activating or inhibiting agent.

Use this space to note any additional objectives provided by your instructor.

CHAPTER 13: SOLUTIONS TO REVIEW QUESTIONS

13.1 a. photosynthesis – the trapping of light energy and its conversion to the chemical energy required to incorporate carbon dioxide into organic molecules

b. photosystem – a photosynthetic mechanism composed of light-absorbing pigments

c. reaction-center – the membrane-bound protein complex in a photosynthesizing cell that mediates the conversion of light energy into chemical energy

d. PSI – a large membrane-spanning multisubunit protein-pigment complex that energizes and transfers the electrons that are eventually donated to $NADP^+$

e. PSII – a large membrane-spanning multisubunit protein-pigment complex that oxidizes water molecules and donates energized electrons to electron carriers that eventually reduce PSI

13.2 a. chloroplast – a chlorophyll-containing plastid found in the cells of algae and higher plants; the photosynthetic organelle

b. thylakoid – refers to the intricately folded membrane within chloroplasts that contains the photosynthetic molecular machines

c. stroma – a dense, enzyme-filled substance that surrounds the thylakoid membrane within the chloroplast

d. granum (singular) - the folded portion of the thylakoid membrane; grana (plural) – stacks of folded thylakoid membrane

e. stromal lamellae – a thylakoid membrane segment that interconnects two grana

13.4 a light-dependent reactions – the mechanism whereby electrons are energized by light energy and subsequently used in ATP and NADPH synthesis

b. light-independent reactions – a photosynthetic pathway in which carbon dioxide is incorporated into carbohydrate that can occur in the absence of light; also referred to as the Calvin cycle

c. chlorosome – a membrane-bound vesicle found in the hydrothermal vent green sulfur bacterium *Chlorobium bathomarium* that collects low-energy photons of light

d. P680 – the special pair of chlorophyll a molecules in PSII that absorb light energy at 680 nm

e. P700 – the special pair of chlorophyll a molecules in PSI that absorb light energy at 700 nm

13.5 a. Rieske protein – an iron-sulfur protein in the cytochrome b_6f complex in thylakoid membrane

b. Psa dimer – PsaA and PsaB protein subunits that bind electron carriers in PSI

c. D_1/D_2 dimer – polypeptides that form the protein-pigment complex of the PSII reaction center

d. A_o – a specific chlorophyll a molecule in PSI that accepts an energized electron from P700 and transfers it to A_1, another single electron carrier

e. A_1 – a single electron carrier in PSI identified as phylloquinone (vitamin K)

13.7 a. wavelength – the distance from the crest of one wave to the crest of the next wave

b. chromophore – a molecular component that absorbs light of a specific frequency

c. fluorescence – a form of luminescence in which certain molecules can absorb light of one wavelength and emit light of another wavelength

d. radiationless decay – an excited molecule decays to its ground state by converting the excitation energy into heat

e. resonance energy transfer – excitation energy is transferred to a neighboring chromophore through interaction between adjacent molecular orbitals

13.8 a. C3 metabolism – a photosynthetic pathway in plants that produces glycerate-3-phosphate, a three-carbon molecule, as the first stable product

b. C4 metabolism – a photosynthetic pathway in plants such as corn and sugar cane that produces a four-carbon molecule and avoids photorespiration

c. CAM- Crassulacean acid metabolism – a photosynthetic pathway that produces a four-carbon molecule (malate) in plants that live in hot, dry regions such as deserts

d. phytochrome – a regulatory protein that possesses a light sensitive chromophores that exists in two forms: P_r and P_{fr}; phytochrome activation (P_{fr}) by far red light triggers several signal transduction pathways

e. CA1P – D–carboxyarabinito-1-phosphate – a competitive inhibitor of rubisco that binds to the active site in the dark

13.10 The three primary photosynthetic pigments are the chlorophylls, the carotenoids, and pheophytin. Chlorophyll a absorbs light energy and is involved in light harvesting. Chlorophyll b is a light harvesting pigment that passes absorbed energy to chlorophyll a. the carotenoids function as light-harvesting pigments or protect against overexcitation and ROS. Pheophytin is an electron transfer molecule in PSII.

13.11 Chloroplasts resemble mitochondria in the following ways: (1) they are both similar in size and structure to modern prokaryotes; (2) they both reproduce by binary fission; (3) the genetic information and protein synthesizing capability of both chloroplasts and mitochondria are similar to that of prokaryotes; (4) the ribosomes of chloroplasts and mitochondria are similar in size and function; and (5) they are both thought to have arisen from ancient free living prokaryotes.

13.13 The final electron acceptor in photosynthesis is carbon dioxide, regardless of the $NADPH/NADP^+$ ratio.

13.14 During the light reactions of photosynthesis, light-driven electron transport results in ATP and NADPH synthesis. The net production of the dark or light-independent reactions of photosynthesis is one molecule of glyceraldehyde. See p. 494 for the reactions of the Calvin cycle.

13.16 The molecules of O_2 are generated from H_2O.

13.17 The oxygen evolving system is referred to as a clock because it involves five oxidation-reduction reactions that must be completed in order.

13.19 The net reaction for photosynthesis is

$$3CO_2 + 6NADPH + 9ATP \rightarrow \text{glyceraldehyde-3-phosphate} + 6NADP^+ + 9ADP + 8P_i$$

The fixation of six carbon dioxide into glucose occurs at the expense of 12 NADPH and 18 ATP. The source of the oxygen atoms in glucose is carbon dioxide.

13.20 The phytol chain in a chlorophyll molecule extends into and anchors the molecule into the membrane.

13.22 The term "dark reactions " is misleading because it implies that carbon fixation occurs in the absence of light. Actually, the NADPH and ATP required to drive carbon fixation are only synthesized when light is available. The term light-independent is used instead.

13.23 Rubisco, the enzyme that catalyzes carbon fixation can bind both carbon dioxide and oxygen in its active site. Photorespiration, the reaction of ribulose-1,5-bisphosphate with oxygen, a waste product of the light reactions, converts fixed carbon to carbon dioxide. Photorespiration is favored when plants are exposed to high temperatures and conditions that cause low carbon dioxide and/or high oxygen concentrations.

13.25 The consequences of global warming include rising ocean levels, economic damage from increasingly more powerful hurricanes, drought and flood-caused food and water shortages, and the increased prevalence of insect-borne diseases.

13.26 The criteria for sustainable biofuel production are: (1) economically viable production of large quantities of energy, (2) noncompetition with food production (e.g., no use of arable land), and (3) neutral effects on the environment.

13.28 Among the most notable metals present in photosynthesizing systems are magnesium, manganese, iron, and copper. Magnesium stabilizes the porphyrin ring in chlorophyll molecules without adding a redox element site during the reduction of water to oxygen. Manganese acts as a redox center in the transfer of electrons from the quinone electron carriers to the terminal electron acceptors in photosynthesis. Finally, copper acts as the terminal electron acceptor in PSII before the electrons are transferred to P700 in PSI.

13.29 The oxygen-evolving complex of PSII exists in five transient oxidation states (S_0 through S_4), collectively referred to as a clock. Because oxygen evolution occurs only when the S_4 state has been reached, several light bursts are required.

13.31 Carbon dioxide fixation occurs in the stroma of the chloroplasts.

13.32 The chloroplast contains the thylakoid membranes in both appressed (stacked) and unappressed format. The ATPase is oriented in the membrane so that ATP synthesis is always exposed to the stromal compartment. LCHII and PSII are richly concentrated in the appressed regions to maximize light collection and electron transfer. PSI, which should not receive the excitation energy directly from PSII, is physically separated from it in the unappressed regions. Electron replacement of PSI and PSII is mediated by mobile carriers, so physical separation is not a problem.

13.34 The first reaction in both photosynthesis and photorespiration is catalyzed by rubisco, which has both carboxylase and oxidase activity. CO_2 and O_2 compete for the enzyme's active site. Higher levels of CO_2 enable it to out-compete O_2, and photosynthesis will occur at the expense of photorespiration.

13.35 The radiolabeled ^{14}C in $^{14}CO_2$ becomes the first carbon of one of two glyceraldehyde-3-phosphate molecules produced after step 3 of the Calvin cycle. From there, the ^{14}C may be incorporated into ribulose-1,5-bisphosphate or used in the biosynthesis of starch, sucrose or other metabolites.

13.37 Since O_2 is consumed in the process of photorespiration, and C4 plants are able to avoid photorespiration, exposure to $^{15}O_2$ should not produce any molecules that contain ^{15}O-

labeled atoms. Although C4 plants utilize rubisco, which catalyzes the first reaction in both the Calvin cycle and in photorespiration, C4 plants maintain a much higher concentration of CO_2 than O_2. This higher concentration, combined with rubisco's greater affinity for CO_2, results in CO_2 outcompeting any O_2 that might be present. Should CO_2 be removed from the air surrounding the C4 plant prior to exposure to $^{15}O_2$, then it's possible that photorespiration may occur until ATP and NADH are depleted. In this case, the first products of photorespiration, glycolate-2-phosphate and glycerate-3-phosphate, would contain ^{15}O-labeled atoms.

13.38 When corn has been exposed to $^{14}CO_2$, the radioactive label will first be found in oxaloacetate, at the carbon atom located in the carboxylate group adjacent to the CH_2.

13.39 When hydrogen sulfide is the source of hydrogen atoms, the end products of photosynthesis are glucose and elemental sulfur. (The carbon and oxygen atoms in glucose are from carbon dioxide molecules.)

13.40 Dinitrophenol is an uncoupler molecule that destroys the proton gradient needed for ATP synthesis. Photosynthesis requires ATP. Therefore, without ATP photosynthesis comes to a halt.

CHAPTER 13: SOLUTIONS TO FILL-IN-THE-BLANK QUESTIONS

13.41 The mitochondrial matrix-like stroma of chloroplasts

13.43 Thylakoid

13.44 Antenna

13.46 Radiationless decay

13.47 Fluorescence

13.49 Oxygen

13.50 Crassulacean acid metabolism

CHAPTER 13: SOLUTIONS TO SHORT-ANSWER QUESTIONS

13.52 In the early days of oxygenic photosynthesis, atmospheric oxygen levels were quite low. As a result, there was no selection pressure for an extended period of time for improvement in Rubisco's capacity to distinguish between CO2 and O2.

13.53 A blackbody is a substance that absorbs radiant energy and emits light energy. The hotter the substance becomes, the shorter the wavelength of light that it emits. Geothermal energy is sufficiently hot that hydrothermal vents act as black-bodies. GSB1 is a photosynthetic organism that can absorb sufficient amounts of the low-energy light emitted at hydrothermal vents to sustain its biochemical processes because it has an extremely efficient light-harvesting antenna.

13.55 A chromophore is a region in a molecule in which energy is absorbed by exciting an electron from its ground state into a higher energy level. Chromophores that absorb wavelengths in the visible electromagnetic spectrum typically have chains of conjugated double bonds that are often linked to aromatic rings. Note that the wavelengths that give a chromophore-containing molecule its color are those that are reflected. For example, the chlorophyll absorbs blue-violet and red wavelengths and reflects green wavelengths.

CHAPTER 13: SOLUTIONS TO THOUGHT QUESTIONS

13.56 Biofuels are an improvement over fossil fuels because they are renewable and carbon neutral (the combustion of biofuels is offset by the carbon dioxide they remove from the atmosphere).

13.58 The electrons energized by light absorption are used to generate NADPH. This NADPH is used in large quantity to fix CO_2 into carbohydrate molecules. If CO_2 is not available, NADPH accumulates and the chlorophyll molecules do not have electron transfer available as a way to return to the relaxed state. One avenue available is to release the energy as a photon, i.e., to fluoresce.

13.59 Conjugation is a system of alternate double and single bonds. When light with sufficient energy strikes the p electrons of a conjugated system (or any double bond), an electron is promoted from the ground state to a higher energy state, referred to as the excited state. Conjugation lowers the energy difference between the ground and the excited state; hence photons of lower energy are capable of achieving this transition.

13.61 Oxidative phosphorylation and photophosphorylation use many of the same molecules in their reactions and both are linked to an electron transport system. However, chloroplasts use light energy to drive redox reactions while mitochondria use the energy of chemical bonds to drive redox reactions. In contrast to mitochondrial inner membrane, thylakoid inner membrane is permeable to magnesium and chloride ions. Therefore, the electrochemical gradient across the thylakoid membrane consists mainly of a proton gradient.

13.62 High oxygen concentrations promote photorespiraton.

13.64 Chloroplasts possess DNA similar to that of modern cyanobacteria as well as prokaryotic-like protein synthesizing machinery. In addition, they multiply by binary fission as do bacteria.

13.65 Energy generated from large scale farming of photosynthetic algae does not require arable land and does not degrade the environment.

13.67 Because of the capacity of C4 plants to avoid the process of photorespiration, herbicides that promote photorespiration do not affect these organisms.

13.68 The herbicides kill marine photosynthesizing organisms, thereby depressing worldwide oxygen production.

13.70 The spectrum of light that reaches the surface of the earth is richer in the blue region than in the ultraviolet. Also, pigments that absorb in the ultraviolet would be damaged or destroyed more easily by the more powerful ultraviolet.

13.71 C4 plants do not have as much photorespiration. As a consequence the total energy that they expend in the fixation of carbon dioxide is less. Therefore. they are more efficient and can out compete the C3 plants for carbon.

13.73 Triazine herbicides promote photorespiration.

13.74 The ratio of energies is: $\dfrac{-hc/\mathbf{700}}{-hc/\mathbf{1000}} = 1.48$

14 Nitrogen Metabolism I: Synthesis

Brief Outline of Key Terms and Concepts

OVERVIEW

Nitrogen cycle
NAA—nonessential amino acid
EAA—essential amino acid
BCAA—branched-chain amino acid
transamination reactions
transaminases; aminotransferases

14.1 NITROGEN FIXATION

THE NITROGEN FIXATION REACTION

NITROGENASE COMPLEX—dinitrogenase and dinitrogenase reductase (Fe protein); contains P cluster (8Fe-7S) and a MoFe cofactor; regulated by transcriptional control of nif (nitrogen fixation genes).

NITROGEN ASSIMILATION

Nitrogen assimilation is the incorporation of inorganic nitrogen compounds into organic molecules.

14.2 AMINO ACID BIOSYNTHESIS

AMINO ACID METABOLISM OVERVIEW

Amino acid pool; positive nitrogen balance; negative nitrogen balance (Kwashiorkor)
Transport of amino acids into cells

REACTIONS OF AMINO GROUPS

TRANSAMINATION REACTIONS transfer amino groups from one carbon skeleton to another; transaminases require PLP in a ping-pong reaction mechanism. Important α-keto acids/α-amino acids pairs:
α-ketoglutarate/Glu pair; oxaloacetate/Asp pair; pyruvate/Ala pair
REDUCTIVE AMINATION is the synthesis of amino acids by incorporating free NH_4^+ or the amide nitrogen of glutamine or asparagine into α-keto acids. Ammonium ions are also incorporated into cellular metabolites by the amination of glutamate to form glutamine.

SYNTHESIS OF THE AMINO ACIDS

Six families of amino acids:
glutamate, serine, aspartate, pyruvate, the aromatics, and histidine.
The nonessential amino acids are derived from precursor molecules available in many organisms. The essential amino acids are synthesized from metabolites produced only in plants and some microorganisms.

14.3 BIOSYNTHETIC REACTIONS INVOLVING AMINO ACIDS

ONE-CARBON METABOLISM

Important carriers of single carbon atoms:
—THF (tetrahydrofolate, the biologically active form of folic acid)
—SAM (*S*-adenosylmethionine)

GLUTATHIONE (GSH)

GSH, the most common intercellular thiol, is involved in many cellular activities. In addition to reducing sulfhydryl groups, GSH protects cells against toxins and promotes the transport of some amino acids.
γ-GLUTAMYL CYCLE; BLOOD-BRAIN BARRIER; GLUTATHIONE-S-TRANSFERASES

NEUROTRANSMITTERS

NEUROTRANSMITTERS may be excitatory or inhibitory. **BIOGENIC AMINES** are amino acid derivatives and include γ-aminobutyric acid (GABA), catecholamines [dopamine (D), norepinephrine (NE), and epinephrine], serotonin, and histamine. DOPA is 3,4-dihydroxyphenylalanine; BH_4 is tetrahydrobiopterin; and PNMT is phenylethanolamine-*N*-methyltransferase.

NUCLEOTIDES

Nucleotides are the building blocks of the nucleic acids. They also regulate metabolism and transfer energy.
PURINE and **PYRIMIDINE NUCLEOTIDES** are synthesized in both *de novo* and salvage pathways. **NUCLEOSIDES**; *syn* versus *anti*; **IMP** (inosine-5′-monophosphate)
DEOXYRIBONUCLEOTIDES; **RNRI** (ribonuclease reductase I)

HEME

Heme has an iron-containing porphyrin ring system and is synthesized from glycine and succinyl-CoA. Protoporphyrin IX, the precursor of heme, is also a precursor of the chlorophylls.

BIOCHEMISTRY IN PERSPECTIVE:

GASOTRANSMITTERS (NO, CO, H_2S); **ONLINE**: Parkinson's disease and dopamine; Pb poisoning

Overview

Our Source of Nitrogen: Amino Acids

BCAA	Branched chain amino acids	Leu, Ile, Val
EAA	Essential amino acids must be provided by the diet.	Ile, Leu, Lys, Met, Phe, Thr, Trp, Val
NAA	Nonessential amino acids can be synthesized from available metabolites.	Ala, Arg*, Asn, Asp, Cys, Glu, Gln, Gly, His*, Pro, Ser, Tyr
		*essential for infants

> *Hint:*
>
> *Now would be a great time to brush up on the amino acid R groups! It'll be a tremendous help to you if their structures instantly pop into your mind as you study.*

14.1 Nitrogen Fixation

The Nitrogen Fixation Reaction

Nitrogen fixation ($N_2 \rightarrow 2NH_3$) is a reduction that requires energy: at least 16 ATP/N_2.

Nitrogenous complex (*must be anaerobic—protected from O_2*) includes:

Dinitrogenase (or Fe-Mo protein): $N_2 + 8H^+ + 8e^- \rightarrow 2NH_3 + H_2$

Dinitrogenase reductase (or Fe protein): binds ATP; ATP hydrolyzes to ADP and P_i; this causes conformational changes that help the e^- transfer to dinitrogenase

Path of electrons:

NADH (or NADPH) → ferredoxin → dinitrogenase reductase → dinitrogenase

NH_3 then travels from the bacteria to the host cell to be used in glutamine synthesis.

Nitrogen fixation allows plants and animals to synthesize many *N*-containing biomolecules such as proteins and nucleic acids. Only a few prokaryotes can fix nitrogen.

Nitrogen Assimilation

Plants receive their nitrogen via symbiotic relationships with N_2-fixing prokaryotes or by absorbing NH_3 and NO_3^- synthesized by soil bacteria (or provided by artificial fertilizers). Animals take in organic nitrogen mainly as amino acids. The liver determines the fate of ingested amino acids.

14.2 Amino Acid Biosynthesis

Amino Acid Metabolism Overview

Amino acid pool are the amino acids available for metabolic processes. Amino acids from the degradation of dietary and tissue proteins enter the pool; excreted nitrogenous end products such as urea and uric acid leave the pool. Amino acids enter cells via membrane-bound transport proteins; some are Na^+-transport dependent.

Nitrogen Balance: Nitrogen Intake = Nitrogen Loss

Positive nitrogen balance:	nitrogen intake > nitrogen loss. Protein synthesis exceeds degradation (e.g., in growing children, pregnant women, recuperating patients).
Negative nitrogen balance:	nitrogen intake < nitrogen loss. Nitrogen can't be replenished fast enough; leads to malnutrition (Kwashiorkor)

Reactions of Amino Groups

Transamination Reactions: The Transfer of Amino Groups Makes the Synthesis of New Amino Acids Possible

Amino groups are transferred **from an α-amino acid to an α-keto acid.** (Remember that "α-acid" places these groups *right next to* a carboxylic acid.) Because transamination reactions are readily reversible, they play an important role in both the synthesis and degradation of the amino acids.

Transamination example: Remember the alanine cycle?

pyruvate glutamate alanine α-ketoglutarate

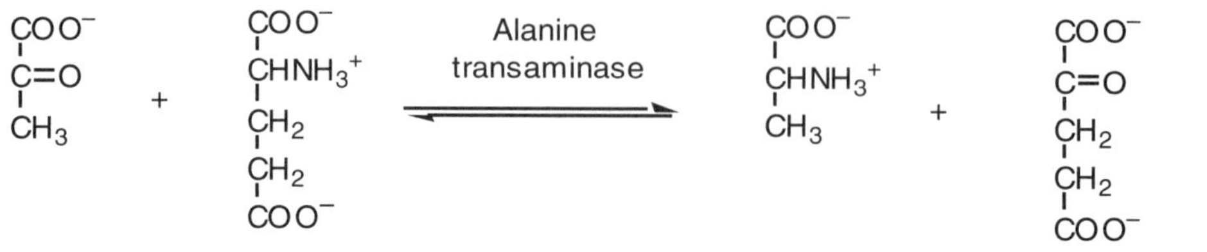

AMINOTRANSFERASES:

Aminotransferases require the coenzyme pyridoxal-5′-phosphate (PLP) (from pyridoxine/vitamin B_6) that forms a Schiff base (–C=N–) between its aldehyde C and the N of the amino acid's $–NH_3^+$.

Two types of aminotransferases:

- Specific for the type of α-amino acid that donates the α-amino group
- Specific for the α-keto acid that accepts the α-amino group.

BIMOLECULAR PING-PONG MECHANISM: The first substrate has to leave the active site before the second one can enter.

1. An amino acid enters, leaves its amino group behind (with the PLP), and leaves as an α-keto acid.
2. A different α-keto acid enters, the reverse reactions occur for it to take the amino group (from the PLP), and it leaves as an amino acid.

IMPORTANT TRANSAMINATION PAIRS

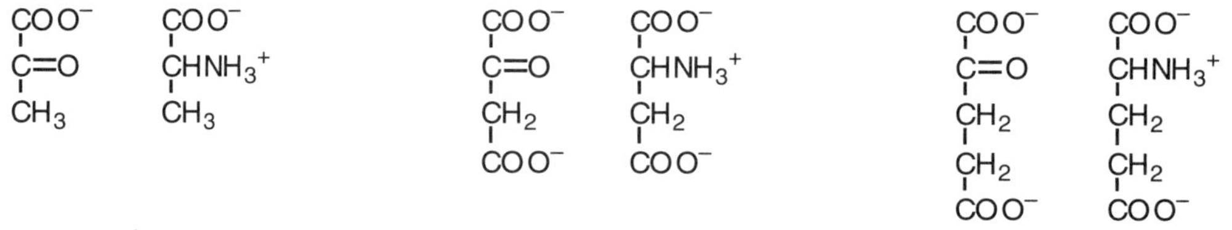

pyruvate/alanine oxaloacetate/aspartate α-ketoglutarate/glutamate

Note that α-ketoglutarate and oxaloacetate are citric acid cycle intermediates.

Direct Incorporation of Ammonium Ions into Organic Molecules

REDUCTIVE AMINATION OF α-KETO ACIDS

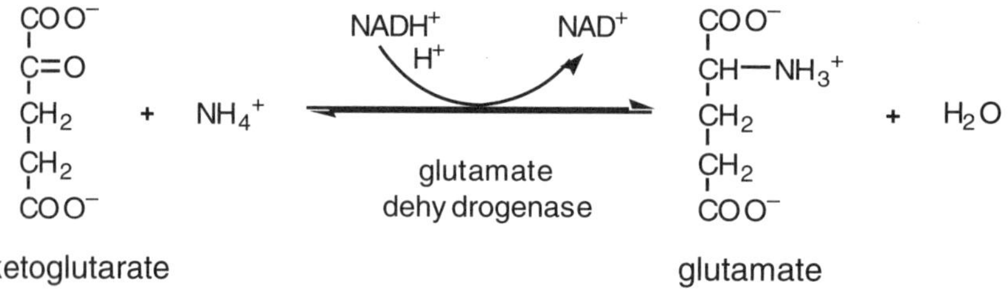

α-ketoglutarate glutamate

(This reaction typically functions to get rid of NH_4^+, but can also be reversed if excess ammonia is present.)

FORMATION OF THE AMIDES OF ASP AND GLU (TO MAKE ASN AND GLN)

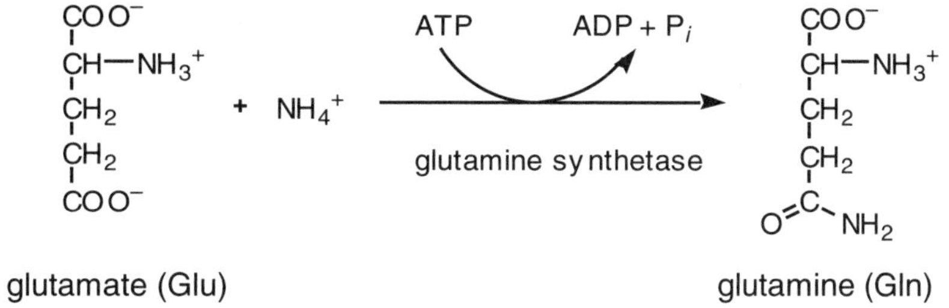

The brain uses this reaction to get rid of NH_4^+. Plants couple the reaction above with the following reaction, using two electrons from ferredoxin or NADPH, for a net of 1 Glu per NH_4^+.

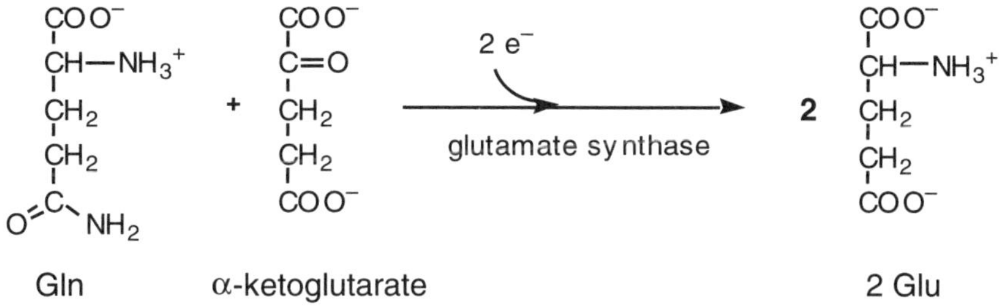

Synthesis of the Amino Acids

To be able to synthesize an amino acid, its α-keto acid precursor must be independently synthesized by the organism. In *de novo* pathways, amino acids are synthesized from metabolic intermediates (and not only by transamination).

AMINO ACIDS IN THE SAME FAMILY ARE SYNTHESIZED FROM A COMMON PRECURSOR.

Family	Precursor (Parent)	Amino Acids
Glutamate family	α-Ketoglutarate	Glutamate Glutamine Proline (cyclized derivative of Glu) Arginine
Serine family	Glycerate-3-phosphate	Serine Glycine Cysteine

FAMILY	**PRECURSOR (PARENT)**	**AMINO ACIDS**
Aspartate family	Oxaloacetate (OAA)	Aspartate (from OAA + Glu transamination) Asparagine (from Asp + Gln transamination) Lysine Methionine Threonine Isoleucine*
Pyruvate family	Pyruvate	Alanine (from pyruvate + Glu transamination) Valine Leucine Isoleucine*
Aromatic family	Phosphoenolpyruvate Erythrose-4-phosphate (shikimate pathway; chorismate intermediate)	Phenylalanine Tyrosine (from chorismate OR from hydroxylation of Phe) Tryptophan
Histidine	PRPP (phosphoribosyl-pyrophosphate)	Histidine

* Isoleucine is listed in two families because the carbons in isoleucine are derived from both pyruvate and oxaloacetate.

As an exercise, go through Figures 14.9–14.11 in the text to be sure you can follow how a given amino acid is made from each parent molecule. Make your own diagrams of each family's reactions and trace specific atoms through the pathways.

14.3 BIOSYNTHETIC REACTIONS INVOLVING AMINO ACIDS

ONE-CARBON METABOLISM: THE TRANSFER OF ONE-CARBON GROUPS

ONE-CARBON GROUPS AND THEIR RELATIVE OXIDATION LEVELS:

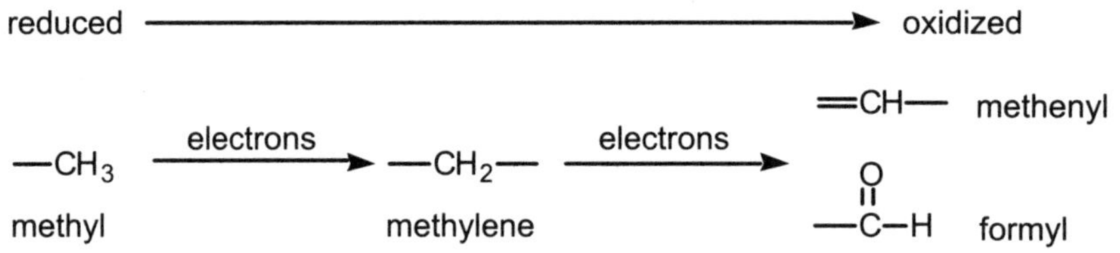

IMPORTANT CARRIERS/DONORS OF ONE-CARBON GROUPS:

TETRAHYDROFOLATE (THF) is the biologically active form of **FOLIC ACID**. Once the additional C is attached to THF, the oxidation state can be interconverted (see below). Sources of C groups include Gly, Ser, His, and formate (from Trp).

S-ADENOSYLMETHIONINE (SAM) contains an activated methyl thioether group that transfers its methyl group.

FOLIC ACID

TETRAHYDROFOLATE (THF)

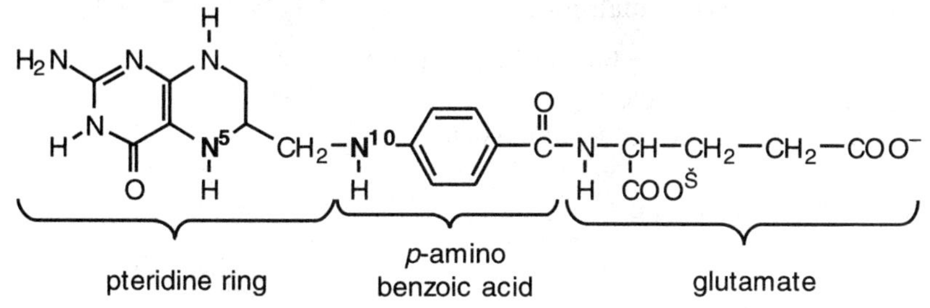

THF: INTERCONVERSION OF ONE-CARBON GROUP OXIDATION STATES

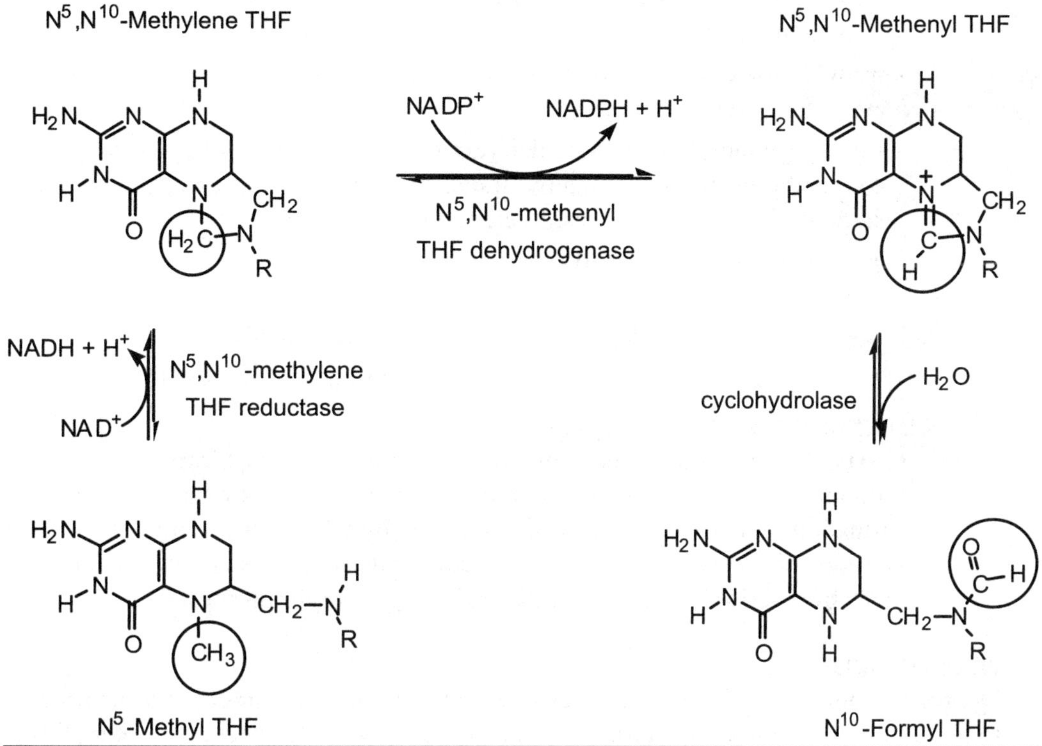

S-ADENOSYLMETHIONINE (SAM)

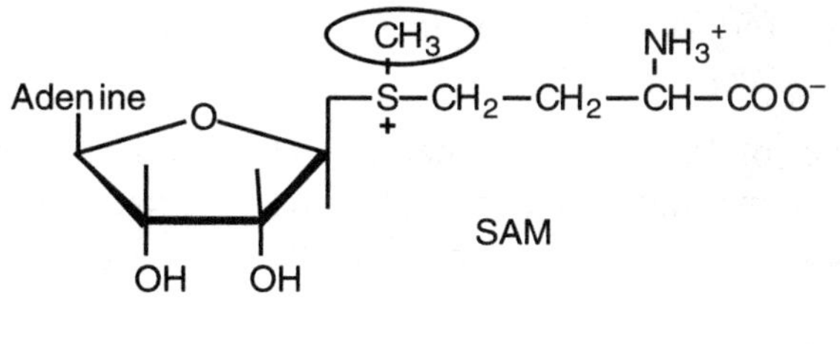

SAM = S-ADENOSINE (FROM ATP) + METHIONINE

Methyl transferases can transfer the circled $-CH_3$ to an acceptor. Without the circled CH_3, this molecule is SAH, *S*-adenosylhomocysteine, which can be used to regenerate methionine.

GLUTATHIONE: AN IMPORTANT REDUCING AGENT FORMED FROM GLU, CYS, AND GLY

Important functions of GSH (glutathione):

- reduces molecules in various biosyntheses (DNA, RNA, certain eicosanoids, etc.)
- protects cells from radiation, O_2 toxicity, and environmental toxins
- promotes amino acid transport

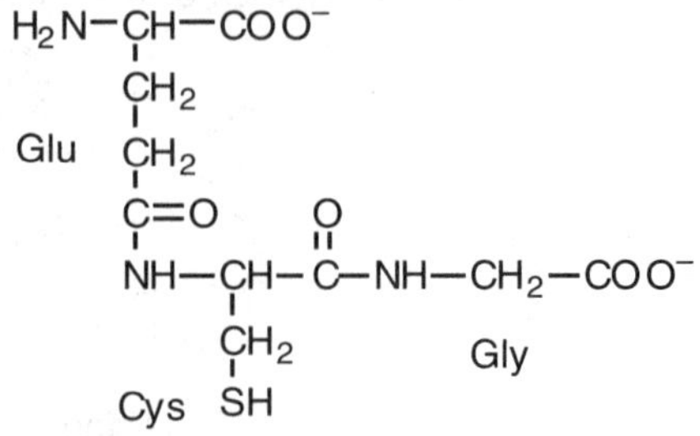

GSH TRANSPORT OUT OF CELLS FUNCTIONS TO:

—transfer sulfur (cysteine) between cells,

—protect the plasma membrane from oxidative damage,

—provide for active transport of several amino acids (in the brain, intestine, pancreas, liver, and kidney)

—form γ-glutamyl amino acid derivatives in a process that begins with GSH transfer to membrane bound γ-glutamyl transpeptidases. This initiates the γ-glutamyl cycle in brain, intestine, pancreas, liver, and kidney cells.

Functions of the γ-glutamyl cycle:

—maintain cellular GSH levels,

—stimulate the Na^+-dependent transport of amino acids

—contribute to the regulation of amino acid transport across the blood-brain barrier

GLUTATHIONE-S-TRANSFERASES

GSH helps to protect cells from environmental toxins by forming GSH conjugates with these foreign molecules in order to prepare them for excretion. GSH conjugate formation may be spontaneous or catalyzed by GSH-S-transferases (**LIGANDINS**). Before their excretion in urine, GSH conjugates are usually converted to mercapturic acids by a series of reactions initiated by γ-glutamyltranspeptidases.

NEUROTRANSMITTERS

Neurotransmitters are signal molecules released from neurons. Excitatory neurotransmitters (e.g., Glu, acetylcholine) open Na^+ channels. Inhibitory neurotransmitters (e.g., Gly) open Cl^- channels, inhibiting the formation of an action potential. **BIOGENIC AMINES** are amino acid derivatives and include:

GABA, γ-AMINOBUTYRIC ACID, is produced when glutamate decarboxylase decarboxylates glutamate.

THE CATECHOLAMINES, derivatives of Tyr, include dopamine (D), norepinephrine (NE, noradrenaline), and epinephrine (adrenaline). The enzyme tyrosine hydroxylase catalyzes the rate-limiting first step of catecholamine synthesis, producing L-DOPA (3,4-dihydroxyphenylalanine).

SYNTHESIS OF CATECHOLAMINES: | Type of reaction:

Tyrosine + O_2 + BH_4 $\xrightarrow{\text{Tyrosine hydroxylase}}$ H_2O + BH_2 + L-DOPA — Hydroxylation

(Rate-limiting reaction)

BH_4 (tetrahydrobiopterin) activates O_2.

L-DOPA + H^+ $\xrightarrow{\text{DOPA decarboxylase}}$ CO_2 + Dopamine — Decarboxylation

requires pyridoxal phosphate

$$\text{Dopamine} + O_2 \xrightarrow[\text{Ascorbic Acid}]{\text{Dopamine-}\beta\text{-hydroxylase,}} H_2O + \text{Norepinephrine} \qquad \text{Hydroxylation}$$

$$\text{Norepinephrine} + \text{SAM} \xrightarrow{\text{PNMT}} \text{SAH} + \text{Epinephrine} \qquad \text{Methylation}$$

PNMT is phenylethanolamine-*N*-methyltransferase.

SEROTONIN is a derivative of tryptophan (hydroxylated and then decarboxylated). Note the similarities between the syntheses of dopamine and serotonin.

$$\text{Tryptophan} + O_2 + BH_4 \xrightarrow{\text{Tryptophan hydroxylase}} H_2O + BH_2 + \text{5-Hydroxytryptophan}$$

$$\text{5-Hydroxytryptophan} + H^+ \xrightarrow[\text{decarboxylase}]{\text{5-Hydroxytryptophan}} CO_2 + \text{Serotonin}$$

requires pyridoxal phosphate

HISTAMINE

$$\text{Histidine} + H^+ \xrightarrow{\text{Histidine decarboxylase}} CO_2 + \text{Histamine}$$

requires pyridoxal phosphate

NUCLEOTIDES: STRUCTURE AND BIOSYNTHESIS

First, let's look at the purine and pyrimidine bases. The general structures are as follows:

Purine: 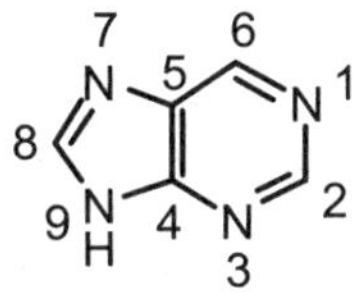

Pyrimidine: 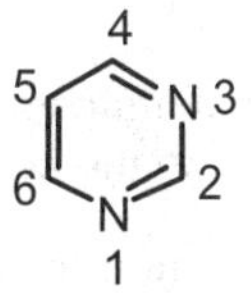

PURINE STRUCTURES

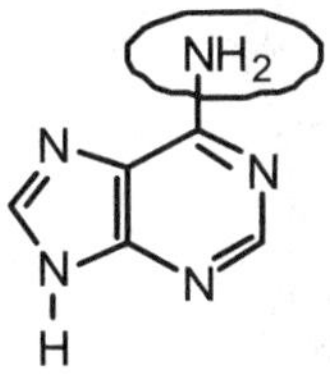

Adenine

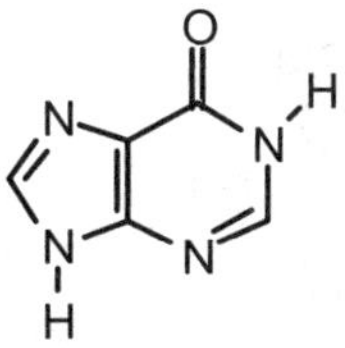

Hypoxanthine

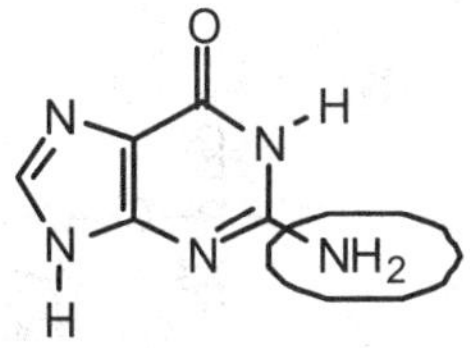

Guanine

Hypoxanthine is not a component of DNA or RNA, but it is the base in IMP, inosine-5′-monophosphate, an intermediate in purine biosynthesis. To convert hypoxanthine to either adenine or guanine, the circled amino groups must replace either a C=O or an H.

PYRIMIDINE STRUCTURES

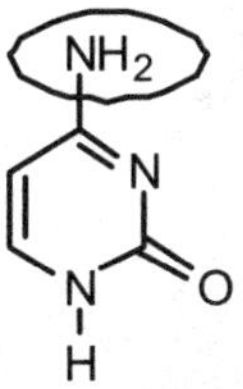

Cytidine
(DNA and RNA)

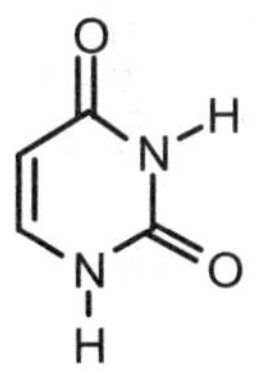

Uracil
(RNA only)

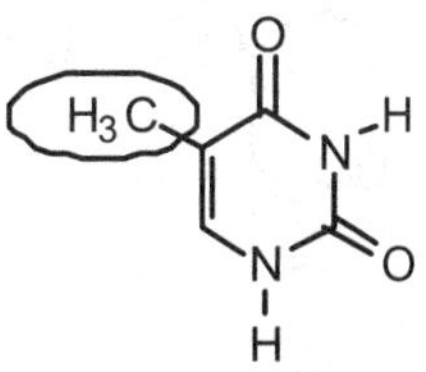

Thymine
(DNA only)

Uracil monophosphate (UMP) is an intermediate in pyrimidine biosynthesis; molecules containing cytosine and thymine are derived from UMP. Again, the groups that differ from uracil are circled. (Note that cytosine, like adenine, also has an extra double bond in the ring.)

NUCLEOSIDES AND NUCLEOTIDES

Nucleo***S***ide = ***S***ugar + base (purine or pyrimidine)

Nucleo***T***ide = The ***T***otal ***P***ackage: sugar + base + ***P***hosphate, ***T***oo.

The carbons in the pentose are numbered beginning with the anomeric carbon, which is consistent with what you learned in Chapter 7. So as not to confuse the pentose carbons with those of the bases, the pentose carbon numbers carry a prime, such as 3′ and 5′.

The sugar in nucleotides can either be ribose (for RNA) or deoxyribose (for DNA). The base is connected to the anomeric carbon. The phosphate is connected to the 5′ carbon. In deoxyribose, the 2′-OH is replaced by an H.

Since the bases do not have any chiral carbons, the bases can be drawn facing the left or right. For example, these two structures of adenine are the same:

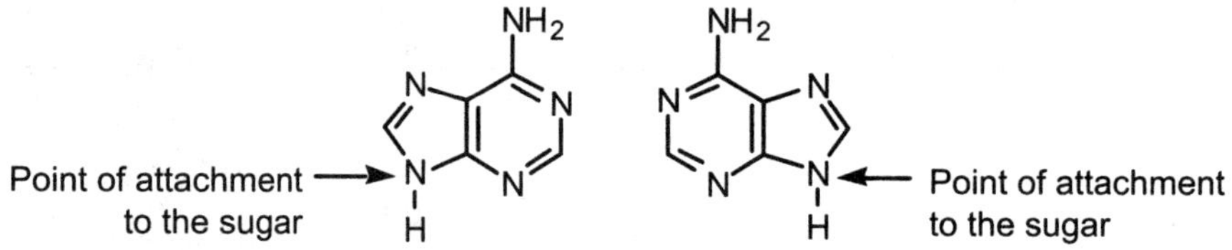

However, once the bases are attached to the pentose, how they are drawn will represent either *syn* or *anti* conformations. Purines occur as both *syn* and *anti*, but pyrimidines are typically *anti* (the base's C=O interferes with the pentose).

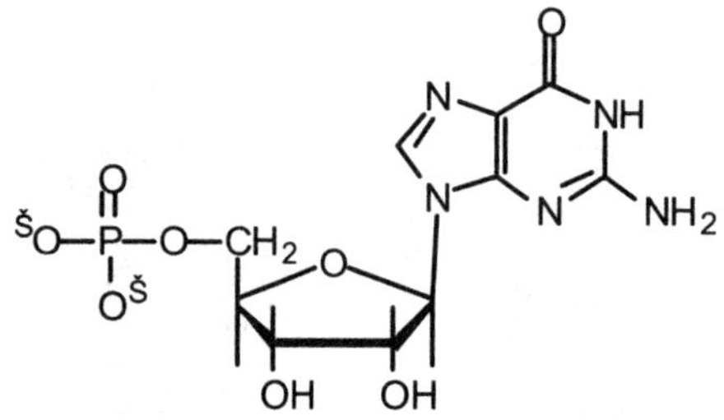

Anti-guanosine-5′-monophosphate

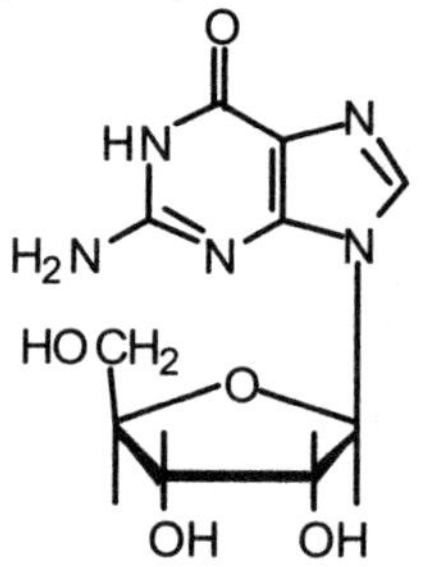

Syn-guanosine-5′-monophosphate

NUCLEOSIDES (RIBOSE + BASE)

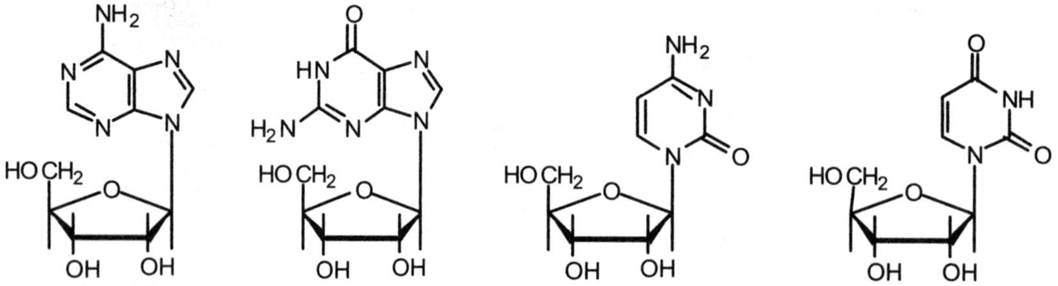

adenosine guanosine cytidine uridine

DEOXYNUCLEOSIDES (DEOXYRIBOSE + BASE)

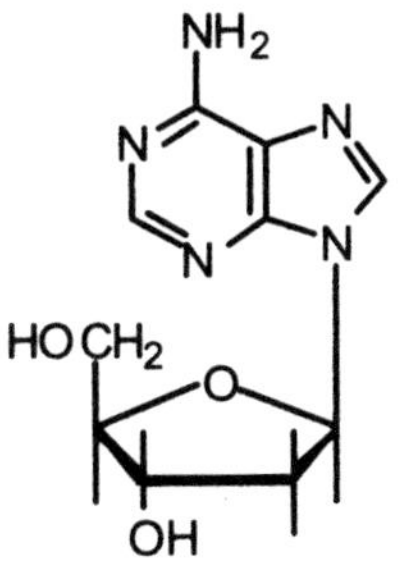
deoxyadenosine

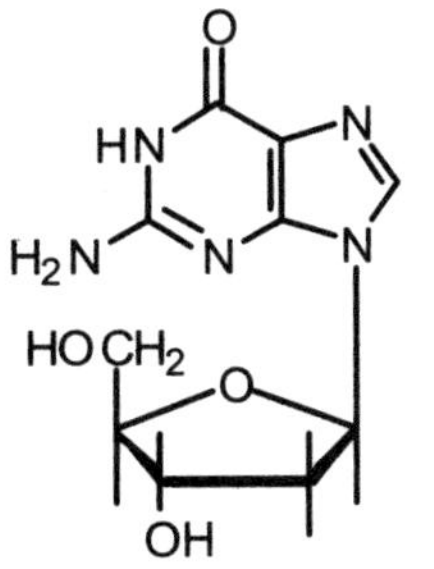
deoxyguanosine

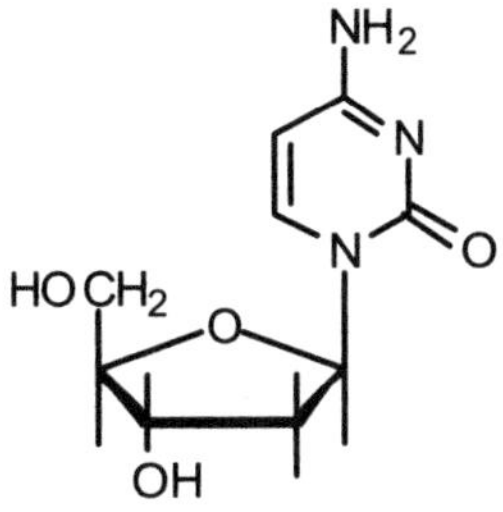
deoxycytidine

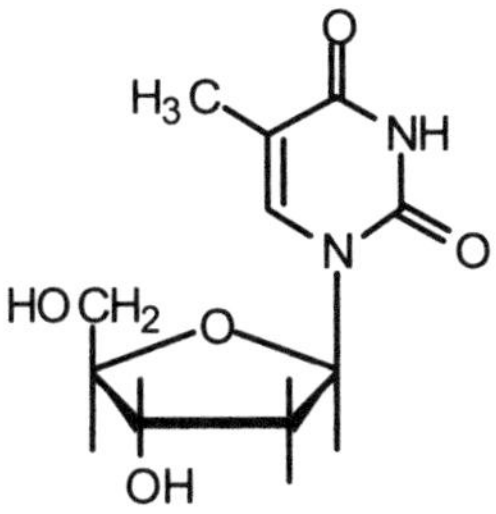
deoxythymidine

NUCLEOTIDES (RIBOSE + BASE + PHOSPHATE)

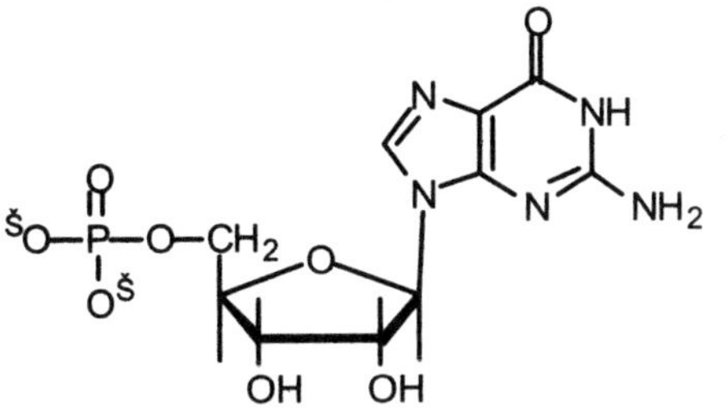
guanosine-5′-monophosphate
GMP

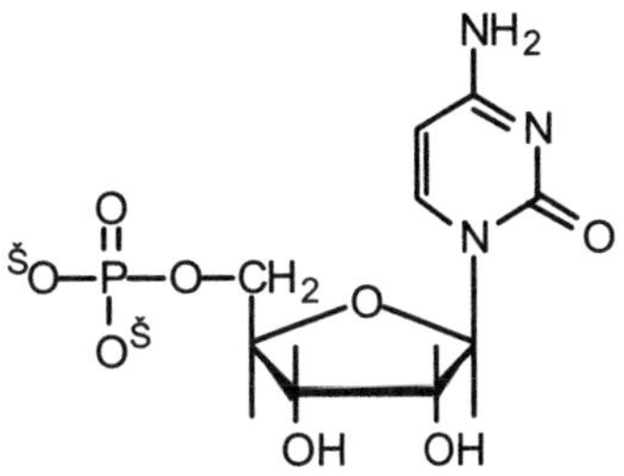
cytidine-5′-monophosphate
CMP

For the deoxynucleotides, the OH at the 2′ carbon is replaced by an H, and the abbreviations are designated by a small "d" (e.g., dGMP, dCMP).

As an exercise, draw the structures of the nucleotides dAMP, UMP, and dTMP. Check the structures of the bases with those in Figure 14.23 (p. 540).

NUCLEOTIDE BIOSYNTHESIS: THE REAL CHALLENGE LIES IN MAKING THE BASE.

Nucleotides can be synthesized in *de novo* pathways (meaning "from scratch") or in "salvage" pathways. Many nucleic acids, such as RNA, are "turned over," meaning they are synthesized and degraded. Rather than having to synthesize the purine nucleotide *de novo*, the cell can recycle the bases. This makes sense when considering how much energy (ATP equivalents) is expended in *de novo* biosynthesis.

For *de novo* biosynthesis of . . .

. . . purines: the base is built onto ribose-5-phosphate.

. . . pyrimidines: the base is made first and *then* attached to ribose-5-phosphate.

Where does each atom in the purine or pyrimidine ring come from?

PURINES

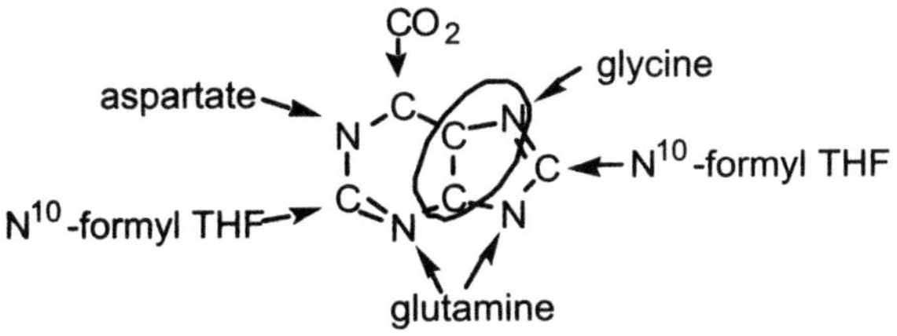

PYRIMIDINES

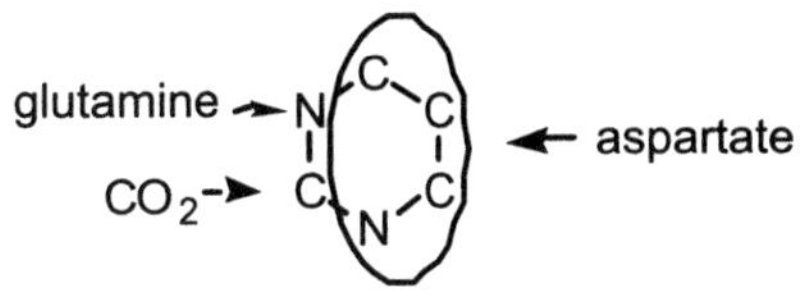

PURINES: RINGS ARE BUILT ONTO AN ACTIVATED RIBOSE (PRPP).

PURINE *DE NOVO* PATHWAYS:

1. Ribose-5-phosphate is activated by converting it to PRPP:

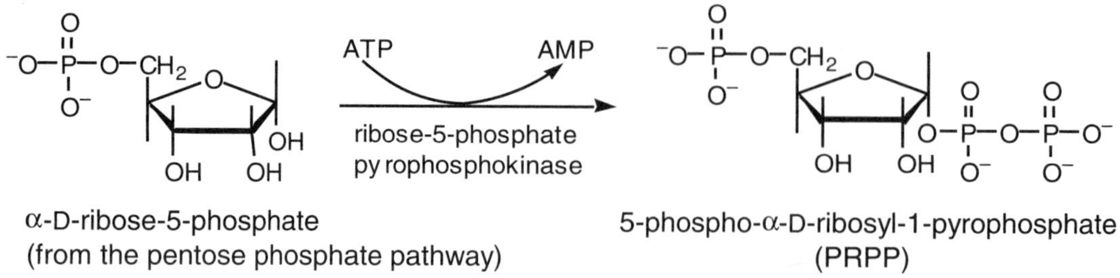

   Note that the anomeric carbon of ribose-5-phosphate and PRPP are in the α-configuration. This changes in the next step.

2. PRPP's pyrophosphate group is displaced by the amide nitrogen of glutamine to form 5-phospho-β-D-ribosylamine (and glutamate).

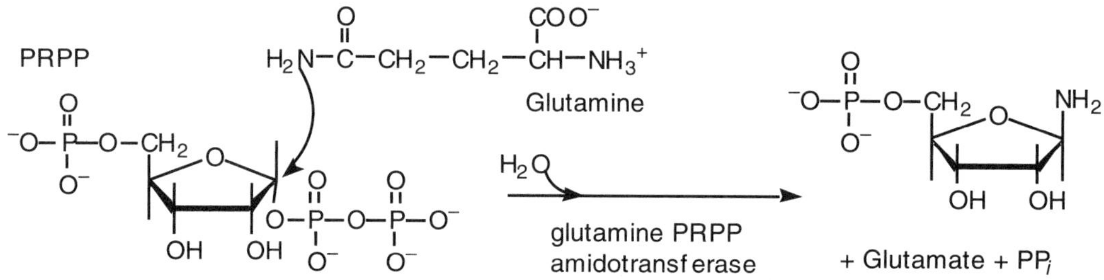

3. Nine further reactions add various groups and close the rings to form IMP. The sources of these groups are listed in the order that they react: Gly, N^{10}-formyl-THF, Gln (5-membered-ring closure), CO_2, Asp (loss of fumarate), N^{10}-formyl-THF (ring closure to form IMP).

4. Further reactions change IMP into either AMP or GMP.

PURINE SALVAGE PATHWAYS USE PRPP AND THE FREE BASE → NUCLEOTIDES

Hypoxanthine-guanine phosphoribosyltransferase (HGPRT) catalyzes nucleotide synthesis using PRPP and either hypoxanthine or guanine:

Guanine + PRPP → GMP + PP_i

Adenine + PRPP → AMP + PP_i Adenine phosphoribosyltransferase

PYRIMIDINES: THE BASE IS CREATED FIRST, THEN ATTACHED TO THE SUGAR

1. Carbamoyl phosphate is formed from glutamine and HCO_3^-. This reaction uses one ATP for the phosphate and a second ATP to drive the reaction. It's catalyzed by the cytoplasmic enzyme carbamoyl phosphate synthetase II, a key regulatory enzyme in mammals. It's inhibited by UTP (the product) and stimulated by purine nucleotides (so that the numbers of pyrimidines and purines made are balanced).

2. Carbamoyl phosphate then condenses with aspartate (by aspartate transcarbamoylase).

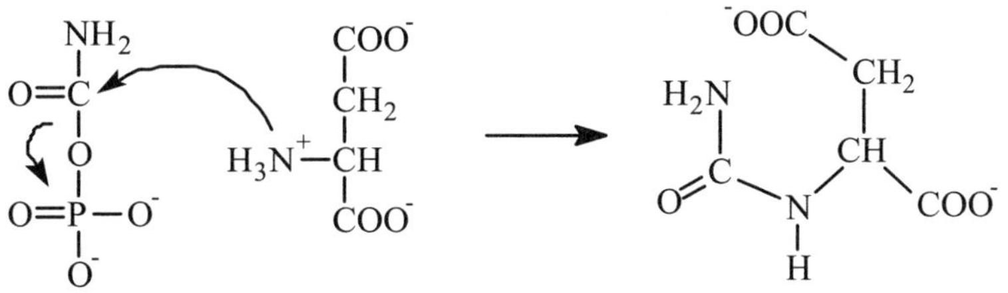

carbamoyl phosphate + aspartate → carbamoyl aspartate + P_i

3. The ring is closed by dihydroorotase to form dihydroorotate.
4. Dihydroorotate is then oxidized (hydrogens are removed to form a double bond) by the catalyst dihydroorotate dehydrogenase to form orotate.
5. The ring is attached to PRPP by orotate pyrophosphoribosyl transferase to form orotidine-5′-monophosphate (OMP).
6. A CO_2 is removed to form uridine monophosphate (UMP) (catalyzed by OMP decarboxylase).

UMP serves as a precursor for the other pyrimidine nucleotides.

UMP is phosphorylated twice by ATP to form UTP.

$$\text{UMP} + \text{ATP} \rightarrow \text{UDP} + \text{ADP}$$

$$\text{UDP} + \text{ATP} \rightarrow \text{UTP} + \text{ADP}$$

The amide nitrogen from glutamine replaces the carboxyl group on UTP to form cytidine triphosphate (CTP).

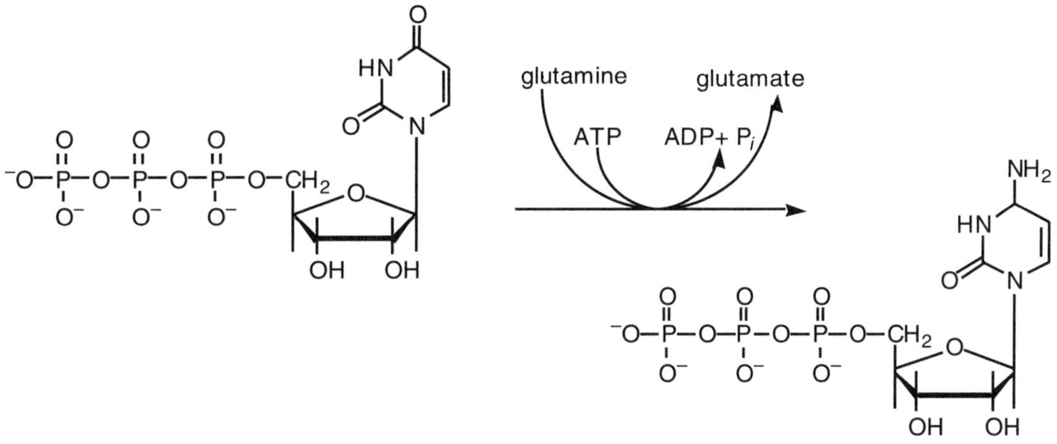

DEOXYRIBONUCLEOTIDES: REDUCTION OF RIBONUCLEOTIDE DIPHOSPHATES

Ribonucleotide reductase replaces the 2′-OH with an H, converting a ribonucleotide *di*phosphate into a deoxyribonucleotide diphosphate.

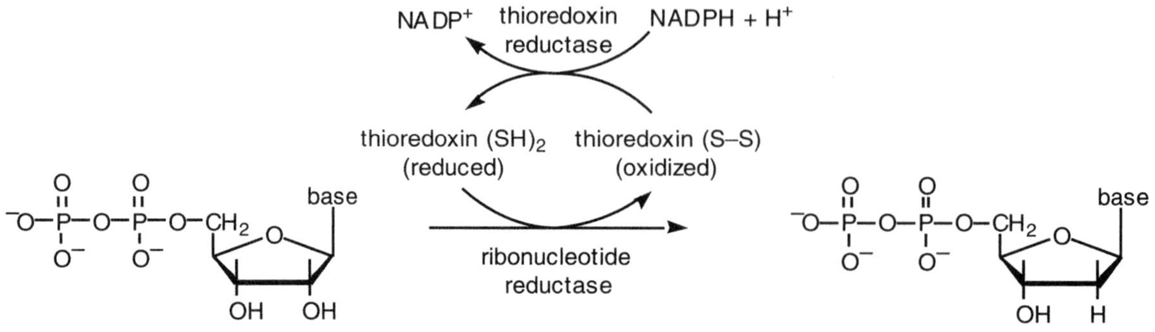

WHAT ABOUT THYMIDINE?

The only nucleotide that's missing is deoxythymidine monophosphate (dTMP).

Note that the only difference between uracil and thymine is a methyl group. The transformation of UMP to dTMP is UMP → UDP → dUDP → dUMP → dTMP.

Ribonucleotide reductase catalyzes UDP → dUDP. Conversion to dUMP is a simple dephosphorylation. Adding the methyl group to dUMP requires N5,N10-methylene THF and the enzyme thymidylate synthase to obtain dTMP:

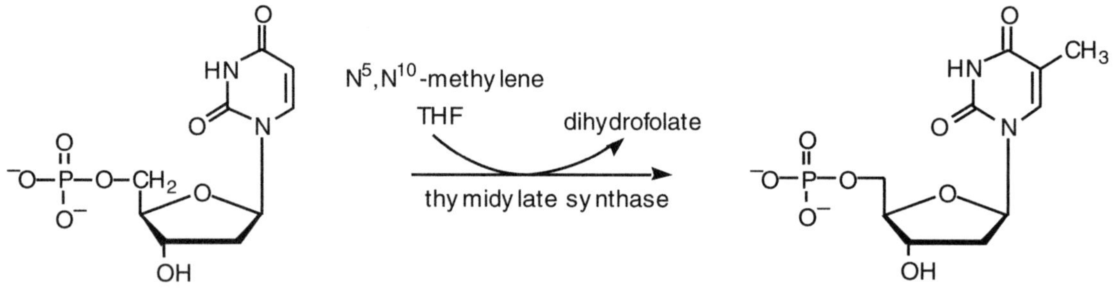

dUMP dTMP

NADPH donates its electrons (and hydrogens) to the dihydrofolate produced to regenerate THF (tetrahydrofolate).

Since cancer cells grow rapidly, they require large amounts of deoxynucleotides for DNA synthesis. Many antitumor or anticancer agents block these steps.

For example, methotrexate and aminopterin block the reduction of DHF to THF, and fluorodeoxyuridine monophosphate (FdUMP) inhibits the methylation of dUMP to form dTMP.

CHAPTER 14: SOLUTIONS TO REVIEW QUESTIONS

14.1 a. nitrogen fixation – conversion of molecular nitrogen (N_2) into a reduced biologically useful form (NH_3) by nitrogen-fixing microorganisms

b. nitrogenase – the nitrogenase complex – the multisubunit enzyme that converts nitrogen into ammonia

c. nitrogen assimilation – the incorporation of inorganic nitrogen compounds into organic molecules

d. amino acid pool – the amino acid molecules that are immediately available in an organism for use in metabolic processes

e. nitrogen cycle – the biogeochemical cycle in which nitrogen atoms flow through the biosphere

14.2 a. nonessential amino acid – an amino acid that can be synthesized by the body

b. essential amino acid – an amino acid that cannot be synthesized by the body

c. branched chain amino acid – one of a group of essential amino acids (leucine, isoleucine, and valine) with branched carbon skeletons

d. nitrogen balance – the condition in which nitrogen intake into the body equals nitrogen loss

e. transamination – one of several reactions in amino acid metabolism in which α–amino groups are transferred from an α-amino acid to an α-keto acid.

14.4 a. Vitamin B_{12} – cobalamin – a complex cobalt-containing molecule that is required for the N^5-methylTHF–dependent conversion of homocysteine to methionine

b. blood-brain barrier – a physical barrier between the cells of the brain and blood that prevents many polar substances and ions from being transported from brain capillaries into brain cells

c. neurotransmitter – a molecule released at a nerve terminal that binds to and influences the function of other nerve cells or muscle cells

d. serotonin – a neurotransmitter derived from tryptophan

e. pernicious anemia – a disease caused by a deficiency of vitamin B_{12}; symptoms include low red blood cell counts, weakness and neurological disturbances

14.5 a. excitatory neurotransmitter – a molecule released from a neuron that opens sodium channels and promotes the depolarization of the membrane of another cell

b. inhibitory neurotransmitter – a neurotransmitter molecule that opens chloride channels and makes the membrane potential of the postsynaptic cell more negative

c. retrograde neurotransmitter – a neurotransmitter molecule that once released from the presynaptic cells binds to the presynaptic cell to potentiate its activity

d. dopamine – the catecholamine neurotransmitter synthesized by the decarboxylation of L-DOPA

e. epinephrine – the catecholamine neurotransmitter synthesized from norepinephrine by a SAM-requiring methylation reaction

14.7 a. GSH- glutathione (reduced) – γ-glutamylcysteinylglycine; the most common intracellular reducing agent.

b. mercapturic acid – a GSH conjugate of a xenobiotic molecule

c. CSE – γ-cystathionase – an enzyme in the transsulfuration pathway

d. CBS – cystathionine-β-synthatase - an enzyme in the transsufuration pathway

e. γ-glutamyl cycle – a pathway that facilitates the import of amino acids;

γ-glutamyltranpeptidase catalyzes the reaction between an extracellular amino acid and GSH to yield an intracellular γ-glutamylamino acid and glycine and cysteine

14.8 a. SAM – S-adenosylmethionine – the major methyl donor in one-carbon metabolism; contains an activated methyl thioether group

b. SAH – S-adenosylhomocysteine – the product of reactions in which SAM is the methyl donor

c. THF – tetrahydrofolate – the biologically active form of folic acid; carrier of methyl, methylene, methenyl, and formyl groups in one-carbon metabolism

d. homocysteine – the demethylated precursor of methionine

e. one-carbon metabolism – a set of reactions in which single carbon atoms are transferred from one molecule to another

14.10 Transamination reactions represent a method of conserving valuable nitrogen reserves. The reaction is reversible and can be used to convert α-keto acids produced by metabolic reactions to α-amino acids that may be in short supply. Surplus amino acids present in larger amounts than required by current metabolic needs are used as the source of the amino groups.

14.11 While nitrogen incorporation into organic molecules is thermodynamically favored, the conditions in the biosphere (temperature, pressure, and pH) are such that the reactions have low kinetic probability. Only a few organisms possess the machinery to surpass the energy barrier to nitrogen compound synthesis

14.13 Glutamate is synthesized from a-ketoglutarate by two means (1) transamination, catalyzed by the aminotransferases (pyridoxal phosphate is required coenzyme) and (2) direct amination, catalyzed by glutamate dehydrogenase. NADPH provides the reducing power for this reaction

14.14 The Schiff base forms when pyridoxal phosphate reacts with an amino acid, which loses a proton to form a carbanion. The free electrons of the carbanion are in conjugation with the positively charged nitrogen of the pyridinium ion. As electrons flow to the positively charged nitrogen, the double bond system reorganizes itself to quench the charge on the nitrogen.

14.16 Neurotransmitters are either excitatory or inhibitory. Excitatory neurotransmitters (e.g., glutamate and acetylcholine) promote the depolarization of a postsynaptic cell. Inhibitory neurotransmitters (e.g., glycine) inhibit action potentials in postsynaptic cells, i.e., they make the membrane potential more negative.

14.17 a. Glutamine is produced in the following reactions:

α-Ketoglutarate + NADH + NH_3 + H^+ → Glutamate

Glutamate + NH_3 + ATP → Glutamine + ADP + P_i

b. Serine is produced in the following series of reactions

Glycerate-3-phosphate + NAD^+ → 3-Phosphohydroxypyruvate + NAD + H^+

3-Phosphohydroxypyruvate+Glutamate → 3-Phosphoserine + α-Ketoglutarate

3-Phosphoserine + H_2O → Serine + P_i

c. Arginine is produced in the following series of reactions

Glutamate + Acetyl-CoA → N-acetylglutamate + CoASH

N-acetylglutamate → → → L-ornithine

L-ornithine → → L Arginine

d. Glycine is produced in the following series of reactions:

3-Phosphoglycerate + NAD^+ → 3-Phosphohydroxypyruvate + NADH + H^+

3-Phosphohydroxypyruvate + Glutamate → 3-Phosphoserine + α-Ketoglutarate

3-Phosphoserine + H_2O → Serine + P_i

Serine + THF → Glycine + N^5,N^{10}-methylene THF + H_2O

e. Serine + L-Homocysteine → Cystathione + H_2O

Cystathione + H_2O → Cysteine + α-Ketobutyrate + NH_4^+

14.19 The two most prominent one-carbon carriers are THF (tetrahydrofolate, the biologically active derivative of folic acid) and SAM (S-adenosylmethionine). THF plays important roles in the synthesis of several amino acids and the nucleotides. SAM contains an activated methyl thioether group, and is a major methyl donor in the synthesis of numerous biomolecules, such as phosphatidylcholine, epinephrine, and carnitine. Refer to Figure 14.18 (The Tetrahydrofolate and S-Adenosylmethionine Pathways) for an example of processes in which both THF and SAM participate.

14.20 Glutathione is involved in the synthesis of DNA, RNA and the eicosanoids. It is also utilized as a reducing agent that protects cells from radiation and oxygen, and as a conjugating agent for environmental toxins. Glutathione is also believed to play a role in amino acid transport.

14.22 No. Perhaps surprisingly, the larger purine does not have similar steric interactions with the pentose. Even though the pyrimidine has only one ring, its carbonyl group interferes sterically with the pentose ring to hinder the formation of the *syn* conformation. In contrast, the shape and functionality of the purine rings are such that steric interactions with the pentose are minimal when in the *syn* conformation, in spite of the overall larger size of the purines.

14.23 Pyrimidine nucleosides occur predominantly in the *anti* conformation. Steric hindrance between the pentose sugar and the carbonyl oxygen at C-2 of the pyrimidine ring prevents free rotation around the N-glycosidic bond.

14.25 Vegans are at risk for vitamin B_{12} deficiency because of their diet. If significant amounts of antibiotics are also consumed, normal intestinal microorganisms, some of which synthesize vitamin B_{12}, may be eliminated. With the loss of this last source of the vitamin, the body becomes so depleted that the symptoms of pernicious anemia will manifest themselves.

14.26 In the γ-glutamyl cycle glutathione is excreted from the cell. γ-Glutamyltranspeptidase converts GSH to a γ-glutamylamino acid and Cys-Gly. The γ-glutamylamino acid is transported into the cell where it is converted to 5-oxoproline and the free amino acid. 5-oxoproline is eventually reconverted to GSH. The Cys-Gly is transported into the cell where it is hydrolyzed to cysteine and glycine. The location of the γ-glutamyltransferase on the plasma membrane facilitates the transport process because its function is linked to transport. The transported glutamylamino acid is immediately converted to 5-oxoproline and the free amino acid, therefore driving the reaction in favor of transport (uptake of amino acids).

14.28 In ping-pong reactions, the first substrate must leave the active site before the second can enter. In the reaction of alanine with α-ketoglutarate to produce pyruvate and glutamate the following steps take place: (1) the alanine enters the active site and transfers the amino group to pyridoxal phosphate, (2) water enters the reaction site and hydrolyses the Schiff base to produce pyridoxamine phosphate and pyruvate, (3) pyruvate diffuses from the active site, (4) α-ketoglutarate, the second substrate, enters the reaction site and forms a Schiff base with the pyridoxamine phosphate, (5) water hydrolyzes the Schiff base to give pyridoxal phosphate and glutamate, and (6) glutamate diffuses out of the active site.

14.29

a) Reaction of pyridoxyl phosphate with alanine:

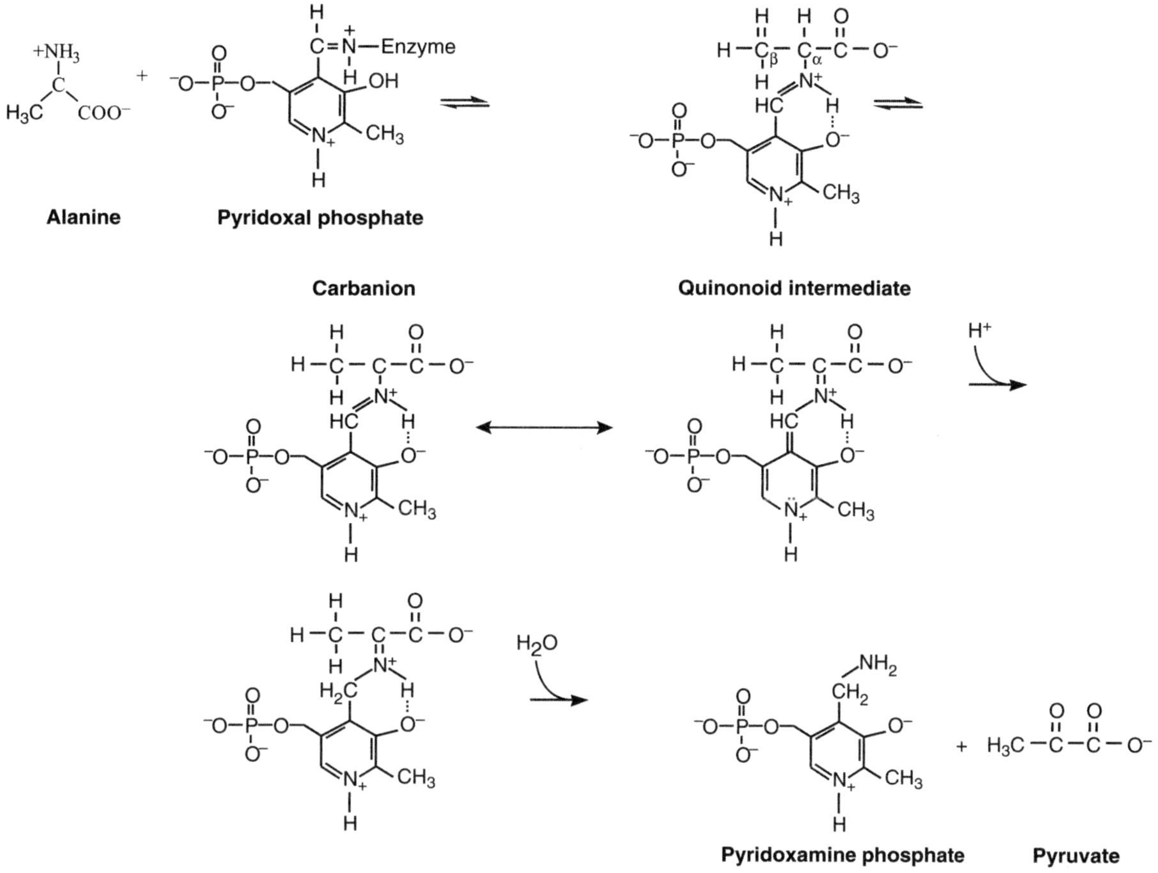

b) Reaction of pyridoxamine phosphate with α-ketoglutarate

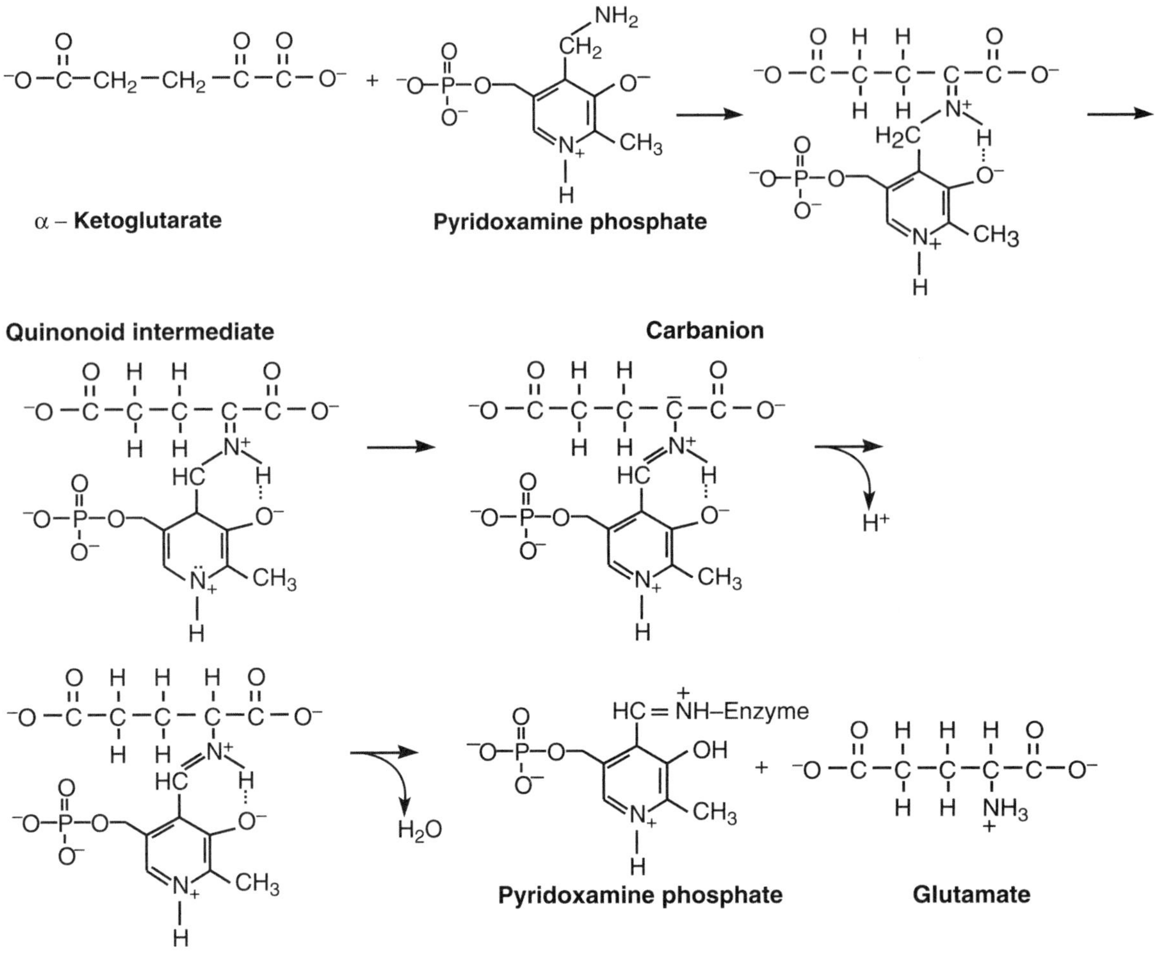

14.31 a. γ-Aminobutyric acid is an inhibitory neurotransmitter.

b. Tetrahydrobiopterin is an essential cofactor in the hydroxylation of certain amino acids.

c. Oxaloacetate is a citric acid cycle intermediate and the α-keto acid in transamination reactions involving aspartate.

d. Phosphoribosylpyrophosphate is the source of the ribose moiety in nucleotide synthesis pathways.

e. S-adenosylmethionine is a methyl transfer agent.

14.32 Nitrogen fixation is highly endergonic. One possibility of a naturally occurring method of fixing nitrogen is via lightning. The tremendous amount of energy in a lightning bolt splits N_2 molecules, which allows them to react with oxygen to form nitric oxide. Nitric oxide dissolves in, and/or reacts with water in the atmosphere to form nitrates that eventually "drop" to the earth as rain.

14.34 The energy provided by ATP hydrolysis in the conversion of glutamate to glutamine makes the reaction energetically favorable. Note that this reaction also provides a means by which brain cells can decrease ammonia concentration.

14.35 In the absence of melanin, skin is vulnerable to sunburn and other forms of light-induced skin damage. A lack of melanin is caused by a defective enzyme in the pathway that converts tyrosine, a precursor of L-DOPA, into melanin.

CHAPTER 14: SOLUTIONS TO FILL-IN-THE-BLANK QUESTIONS

14.37 High energy requirements for nitrogen fixation

14.38 Nitrogen assimilation

14.40 One-carbon metabolism

14.41 Biogenic

14.43 Purine

14.44 Pyrimidine

14.46 Glutathione

CHAPTER 14: SOLUTIONS TO SHORT-ANSWER QUESTIONS

14.47 Lesch-Nyhan syndrome is caused by excessive synthesis of uric acid, the result of a deficiency of the purine salvage pathway enzyme HGPRT (hypoxanthine-guanine phosphoribosyltransferase). Symptoms of this X-linked disease include mental retardation, self-mutilation, involuntary movements, gout, and renal failure.

14.49 The physiological roles of CO include neuromodulation (neurotransmitter release regulation) in functions such as learning and memory and cardiac protection during hypoxia.

14.50 NDMA (*N*-methyl-D-aspartate) receptors are a type of glutamate receptor that, when activated, opens a membrane channel that allows the inward flow of Ca^{2+} and other ions. They are believed to play a role in synaptic plasticity.

CHAPTER 14: SOLUTIONS TO THOUGHT QUESTIONS

14.52 Atmospheric nitrogen contains a nonpolar triple bond that must be broken to reduce nitrogen to ammonia. Also diatomic oxygen is a diradical. The inherent instability of radicals results in a molecule that is much more reactive (e.g., ready to accept electrons in a redox reaction). than the stable nitrogen triple bond.

14.53 Purine and pyrimidine bases are recycled in salvage pathways. Instead of being degraded to form precursors for catabolic pathways they are, instead, oxidized and excreted as nitrogenous waste products.

14.55 Microorganisms synthesize folic acid from its precursors. This pathway is blocked if the para aminobenzoic acid analogue sulfanilamide is present since it is a competitive inhibitor. Without folate processes such as DNA, RNA and protein synthesis cannot occur. Hence growth ceases. Note that sulfanilamide does not kill susceptible infectious microorganisms. It stops their growth, thus limiting their numbers so that the immune system can effectively destroy them.

14.56 Running is a physical activity that requires the rapid metabolism of both fatty acids and glucose. The most rapidly utilized source of glucose is blood glucose. Although certain amino acids can be absorbed easily in the intestine and used as substrates in gluconeogenesis in the liver, this process is slower than the immediate absorption of glucose into the blood.

14.58 The hydroxyl radical extracts a hydrogen atom from glutathione to form water in a reaction catalyzed by glutathione peroxidase.

$$\text{GSH} + \cdot\text{OH} \rightarrow \text{GS}\cdot + H_2O$$

Two molecules of oxidized glutathione then react to form GSSG. GSH is regenerated from GSSG by glutathione reductase using NADPH as the source of hydrogen atoms.

$$\text{GSSG} + \text{NADPH} + H^+ \rightarrow 2\text{GSH} + \text{NADP}^+$$

14.59 Radiolabel both the carbon and nitrogen of glycine, aspartate and glutamine. If the amino acid is used in the ring assembly, then both carbon and nitrogen should bear the label. If nitrogen exchange takes place only the carbon will be labeled. In addition isolating each intermediate allows the origin of each atom to be traced.

14.61 Breaking the bond perpendicular to the pyridinium ring generates a *p* orbital that can then interact with the p system of the pyridinium ring. The resulting delocalization of the charge stabilizes the carbanion.

14.62 The labeled carbon in proline is indicated by an arrow in this structure:

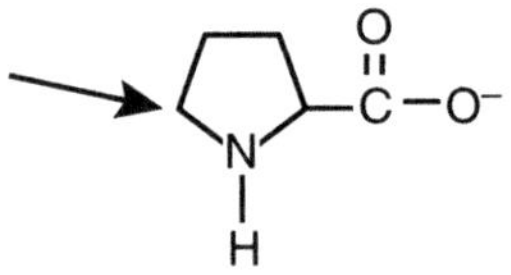

15 Nitrogen Metabolism II: Degradation

Brief Outline of Key Terms and Concepts

OVERVIEW

AMMONOTELIC; UREOTELIC; URICOTELIC

15.1 PROTEIN TURNOVER

Animals are constantly synthesizing and degrading nitrogen-containing molecules such as proteins and nucleic acids. **PROTEIN TURNOVER** is believed to provide cells with metabolic flexibility, protection from accumulations of abnormal proteins, and the timely destruction of proteins during developmental processes.

Protein motifs common in proteins that tend to have a shorter **HALF-LIFE** include the **PEST** sequence and, in cyclins, the cyclin destruction box. Other markers include basic or bulky hydrophobic N-terminal residues or oxidized residues.

Most cellular proteins are degraded by the **UBIQUITIN PROTEASOMAL SYSTEM.** In **UBIQUINATION**, **UBIQUITIN** (a small, highly conserved protein) is covalently linked to worn-out or damaged proteins or short-lived regulatory proteins. Most proteins are then transferred to **PROTEASOMES** where they are cleaved into peptide fragments. The fragments are then degraded by cytoplasmic **PROTEASES** to amino acids, which enter the **AMINO ACID POOL.**

AUTOPHAGY-LYSOSOMAL SYSTEM

Cell components such as organelles and long-lived proteins are degraded by a process that involves lysosomal enzymes.

15.2 AMINO ACID CATABOLISM

Amino acids are classified as **KETOGENIC** or **GLUCOGENIC** on the basis of whether their carbon skeletons are converted to fatty acids or to glucose. Several amino acids can be classified as both ketogenic and glucogenic because their carbon skeletons are precursors for both fat and carbohydrates.

DEAMINATION

In general, amino acid degradation begins with deamination. Most deamination is accomplished by **TRANSAMINATION** followed by **OXIDATIVE DEAMINATION** to produce NH_4^+. Although most oxidative deaminations are catalyzed by glutamate dehydrogenase, other enzymes also contribute to NH_4^+ formation.

UREA SYNTHESIS

Ammonia is prepared for excretion by the enzymes of the **UREA CYCLE.** Aspartate and CO_2 also contribute atoms to urea.

CONTROL OF THE UREA CYCLE

Short-term control is via substrate availability. Carbamoyl phosphate synthetase I is allosterically activated by *N*-acetylglutamate, the production of which is activated by Arg. Transcription of urea cycle enzymes are activated by glucagon and repressed by insulin.

CATABOLISM OF AMINO ACID CARBON SKELETONS

The α-amino acids are degraded to form acetyl-CoA, acetoacetyl-CoA, pyruvate, α-ketoglutarate, succinyl-CoA, and oxaloacetate.

15.3 NEUROTRANSMITTER DEGRADATION

The degradation of neurotransmitters is critical to the proper functioning of information transfer in animals. The amine neurotransmitters such as the catecholamines, are among the best-researched examples.

15.4 NUCLEOTIDE DEGRADATION

NUCLEASES degrade the nucleic acids to **OLIGONUCLEOTIDES**. (**DEOXYRIBONUCLEASES** degrade DNA; **RIBONUCLEASES** degrade RNA.) **PHOSPHODIESTERASES** convert oligonucleotides to mononucleotides. **NUCLEOTIDASES** remove phosphoryl groups to convert nucleotides to nucleosides. **NUCLEOSIDASES** hydrolyze nucleosides to form free bases and ribose or deoxyribose. **NUCLEOSIDE PHOSPHORYLASES** convert ribonucleosides to free bases and ribose-1-phosphate.

PURINE CATABOLISM

Cellular purines are converted to uric acid. Many animals (other than primates) produce enzymes that can degrade uric acid further.

PYRIMIDINE CATABOLISM

Pyrimidine bases are degraded to either β-alanine (UMP, CMP, dCMP) or β-amino-isobutyrate (dTMP).

OVERVIEW

HOW DO VARIOUS ANIMAL SPECIES DISPOSE OF NITROGENOUS WASTE?

The nitrogen in amino acids is removed by deamination reactions (transamination and oxidative deamination) and converted to ammonia, which must be detoxified and/or excreted as fast as it is made. The form in which nitrogen is excreted depends on the availability of water.

AMMONOTELIC: Many aquatic animals that can excrete ammonia, which dissolves in the surrounding water and is quickly diluted.

UREOTELIC: Terrestrial animals (such as mammals) convert ammonia into urea because they must conserve body water. Urea can be excreted without a large loss of water.

URICOTELIC: Animals such as birds, certain reptiles, and insects that convert ammonia to uric acid because they have more stringent water conservation requirements. In many animals, uric acid is also the nitrogenous waste product of purine nucleotide catabolism.

15.1 PROTEIN TURNOVER

Protein turnover is the continuous degradation and resynthesis of proteins.

WHAT ROLE DOES PROTEIN TURNOVER PLAY IN CELLULAR METABOLISM?

Metabolic flexibility is afforded by relatively quick changes in the concentrations of key regulatory enzymes, peptide hormones, and receptor molecules.
Protein turnover also protects cells from accumulation of abnormal or damaged proteins. Growth and development processes depend upon timely destruction of proteins as much as on protein synthesis.

UBIQUITIN PROTEASOMAL SYSTEM (UPS)

UPS is a rapid, efficient, and elaborate mechanism to degrade proteins.

HOW ARE PROTEINS TARGETED FOR DEGRADATION?

The mechanisms by which proteins are targeted for destruction by ubiquitination or other degradative processes are not fully understood. However, the following features of proteins appear to signal their destruction:

- **N-TERMINAL RESIDUES**. Very basic or bulky hydrophobic N-terminal residues tend to be very short-lived. N-terminal residues that are nonbulky and/or contain sulfur or –OH are more stable, with longer half-lives.
- **PEPTIDE MOTIFS. PEST SEQUENCES:** Pro-Glu-Ser-Thr. Proteins that have extended sequences containing proline, glutamate, serine, and threonine (PEST, the amino acid one-letter abbreviations) have fairly short half-lives (e.g., less than 2 hours). **CYCLIN DESTRUCTION BOX:** In cyclins, this set of homologous 9-residue sequences near the N-terminus marks these proteins for rapid ubiquination.
- **OXIDIZED RESIDUES.** Oxidized amino acid residues (i.e., residues that are altered by oxidases or attacked by ROS) promote protein degradation.

PROTEASOMES are massive proteolytic molecular machines that cleave proteins into peptide fragments with 7-8 amino acid residues. **CYTOPLASMIC PROTEASES** degrade these fragments into amino acids, which enter the **AMINO ACID POOL,** where they are available to be incorporated into new protein molecules.

UBIQUITIN is a small, highly conserved 76-residue protein.

UBIQUINATION, the attachment of ubiquitin to worn-out, damaged, or short-lived regulatory proteins, occurs in several stages and involves three enzyme classes:

E1 (ubiquitin-activating enzyme)
E2 (ubiquitin-conjugating enzyme)
E3 (ubiquitin ligase; binds both E2 and protein)

15.2 AMINO ACID CATABOLISM CAN BE DIVIDED INTO THREE STAGES:

1. Deamination (removal of the α-amino group and transport to the liver)
2. Urea synthesis (to excrete nitrogen; occurs only in the liver)
3. Conversion of the carbon skeleton to one of six metabolic intermediates

DEAMINATION: TRANSAMINATION AND OXIDATIVE DEAMINATION

The first step in amino acid catabolism is almost always to remove the α-amino group by transamination, in which the α-amino group is transferred to α-ketoglutarate to form glutamate (Glu).

TRANSAMINATION:

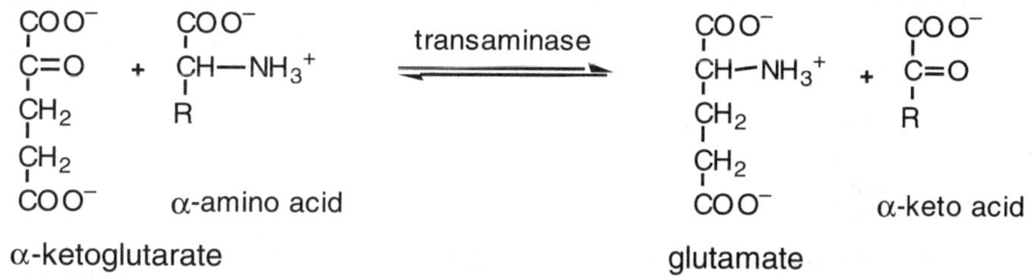

EXCESS NH_4^+ IS CARRIED TO THE LIVER BY GLUTAMINE OR ALANINE:[1]

MUSCLE (ALANINE CYCLE): ("α-kg" = α-ketoglutarate)

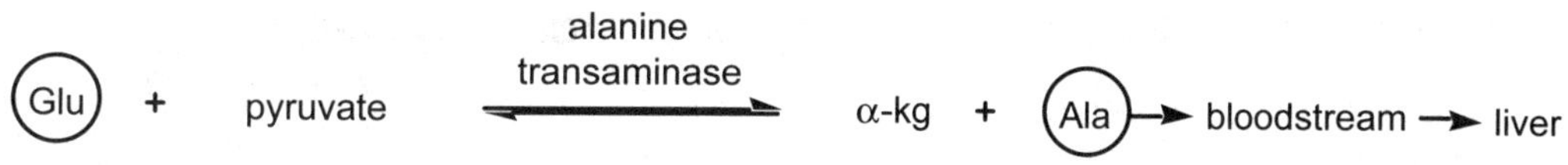

MOST OTHER TISSUES: OXIDATIVE DEAMINATION,[2] THEN GLUTAMINE SYNTHETASE

In most tissues (except the liver), two reactions (and two separate glutamates) are needed to transport NH_4^+ through the blood in the form of glutamine (Gln).

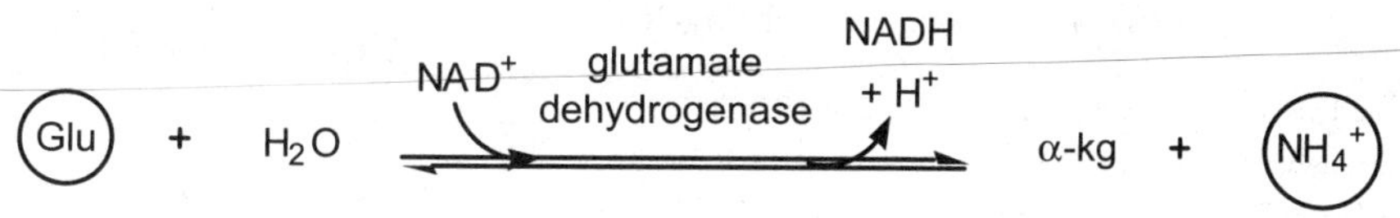

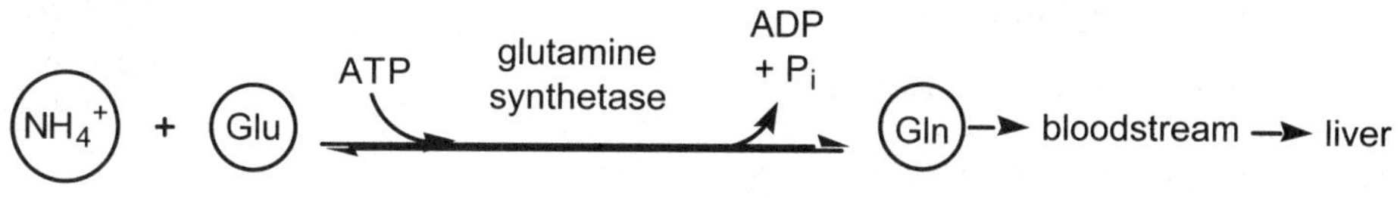

---

[1] Why does glutamate not carry nitrogen to the liver? One possible reason is that glutamate is also a neurotransmitter in the brain. If glutamate levels in the blood are elevated, this could cause the brain to short circuit. In fact, this is an explanation for the headaches that some people experience as a result of consuming MSG (monosodium glutamate), a seasoning ingredient sometimes found in Asian cuisine.

[2] We saw the reverse of this reaction, reductive amination, in Chapter 14.

LIVER: FREES NH_4^+ FROM Gln (FROM OTHER TISSUES) AND Ala (FROM MUSCLE)

$$\text{Ala} + \alpha\text{-kg} \xrightleftharpoons{\text{alanine transaminase}} \text{pyruvate} + \text{Glu}$$

Gln loses its NH_4^+ to regenerate Glu (in a reaction catalyzed by glutaminase). Glu is then oxidatively deaminated to release NH_4^+.

$$\text{Gln} + H_2O \xrightleftharpoons{\text{glutaminase}} \text{Glu} + NH_4^+$$

$$\text{Glu} + H_2O \xrightleftharpoons[\text{NAD}^+ \rightarrow \text{NADH} + H^+]{\text{glutamate dehydrogenase}} \alpha\text{-kg} + NH_4^+$$

OTHER NH_4^+-GENERATING REACTIONS

1. Amino acid oxidases (in the liver and kidney)
2. Serine and threonine dehydratases (because serine and threonine cannot be transaminated): Ser → pyruvate; Thr → α-ketobutyrate
3. Bacterial urease
4. Adenosine deaminase

THE UREA CYCLE OCCURS IN LIVER CELLS

THE OVERALL REACTION OF UREA SYNTHESIS:

$$CO_2 + NH_4^+ + \text{Asp} + 3\text{ATP} + 2H_2O \rightarrow \text{urea} + \text{fumarate} + 2\text{ADP} + 2P_i + \text{AMP} + PP_i + 5H^+$$

THE UREA CYCLE: REACTION NOTES (SEE NEXT PAGE FOR DIAGRAM)

1. Carbamoyl phosphate synthetase I (mitochondrial matrix)	$NH_4^+ + HCO_3^- + 2ATP \rightarrow$ Carbamoyl phosphate + 2ADP + P_i + $3H^+$
Similar to a reaction in pyrimidine biosynthesis Uses 2 ATP: one to activate HCO_3^-, and one to phosphorylate carbamate	
2. Ornithine transcarbamoylase (mitochondrial matrix)	Carbamoyl phosphate + Ornithine → Citrulline + P_i
Carbamoyl phosphate has a high phosphate group transfer potential. Releasing its P_i drives this forward. Ornithine is an amino acid; in fact, it looks just like lysine but with one less $-CH_2$ group.	Citrulline → transport to cytoplasm
3. Arginosuccinate synthase (cytoplasm)	Citrulline + Asp → Arginosuccinate
Cleavage of PP_i (by pyrophosphatase) drives this forward. The α-amino of Asp adds the second N of urea. Uses 2 ATP equivalents, since 1ATP → AMP (not ADP)	
4. Arginosuccinate lyase (cytoplasm)	Arginosuccinate → Arg + Fumarate
Asp''s N is left behind when fumarate is cleaved from argininosuccinate to form Arg. Fumarate's carbons came from Asp.	

5. Arginase (cytoplasm) hydrolyzes Arg Ornithine → transport back to mitochondrion Urea → transport (bloodstream) to kidneys → excretion	Arg + H_2O → Ornithine + Urea

THE UREA CYCLE

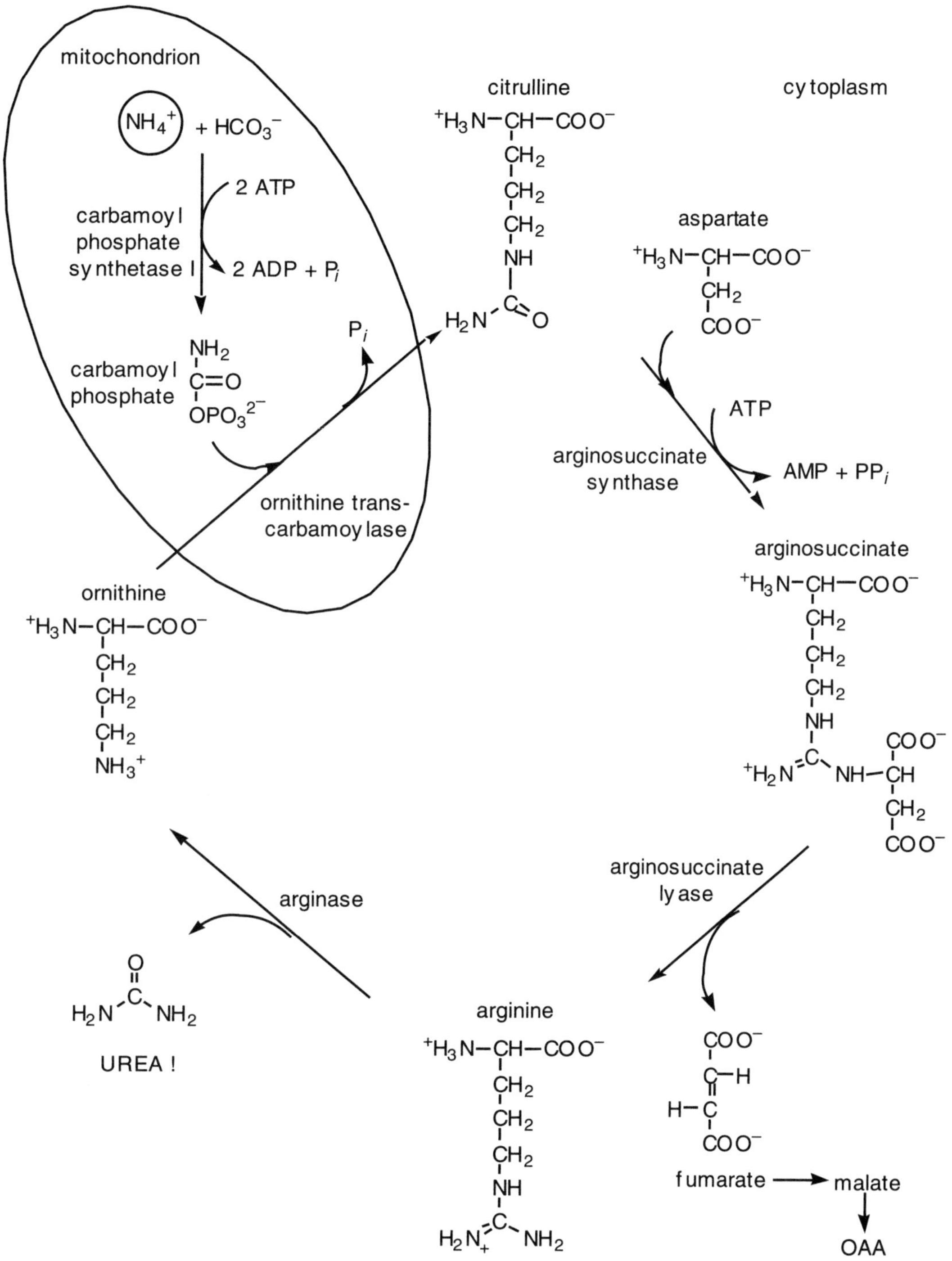

KREBS BICYCLE: THE UREA CYCLE—CITRIC ACID CYCLE CONNECTION

What happens to the fumarate? Where can those four carbons go?

Fumarate supplied by the urea cycle can enter the mitochondrion and the citric acid cycle: fumarate → malate → oxaloacetate (OAA). From OAA, three paths are possible, depending (of course) on the metabolic needs of the cell.

- Transamination OAA + α-Amino acid → α-Keto acid + Asp → Urea cycle
- Gluconeogenesis: OAA → PEP → etc. → Glucose-6-phosphate → Glucose
- Citric acid cycle: OAA + Acetyl-CoA → Citrate → etc. → Energy

CONTROL OF THE UREA CYCLE

Short-term control is via substrate concentrations (availability). So, it makes sense that urea synthesis is stimulated when the diet is high in protein or during starvation, when muscle protein is degraded for energy. At these times, more NH_4^+ would need to be excreted.

Carbamoyl phosphate synthetase I is allosterically activated by *N*-acetylglutamate, an indicator of Glu concentration (Glu + acetyl-CoA → *N*-acetylglutamate). The production of *N*-acetylglutamate is activated by arginine.

The transcription of urea cycle enzymes is activated by glucagon and glucocorticosteroids and repressed by insulin.

CATABOLISM OF AMINO ACID CARBON SKELETONS TO METABOLIC INTERMEDIATES

The carbon skeletons are eventually degraded to one of six intermediates. The degradation of the amino acid carbon skeletons is classified in terms of their converted end products. Acetyl-CoA and acetoacetyl-CoA are grouped together.

KETOGENIC AMINO ACIDS are degraded to form acetyl-CoA or acetoacetyl-CoA, which can be converted to either fatty acids or ketone bodies. Remember that no net synthesis of glucose can occur with acetyl-CoA or acetoacetyl-CoA.

GLUCOGENIC AMINO ACIDS are degraded to form intermediates that can be used in gluconeogenesis. The intermediates are pyruvate, α-ketoglutarate, succinyl CoA, fumarate, and oxaloacetate. These intermediates can then be catabolized by the citric acid cycle or converted to fatty acids, ketone bodies, or glucose.

Keep in mind the following reactions that can result in some overlap or confusion between groups:

pyruvate → acetyl-CoA

acetoacetate → acetoacetyl-CoA → acetyl-CoA

citric acid cycle reactions

PRODUCT OF AMINO ACID CATABOLISM	AMINO ACID	PATHWAY NOTES
Acetyl-CoA (KETOGENIC AMINO ACIDS)	Lys, Trp, Tyr, Phe, Leu	Acetoacetyl-CoA is an intermediate common to all five amino acids. acetoacetyl-CoA → acetyl-CoA Phe → Tyr → acetoacetate → acetoacetyl-CoA

PRODUCT OF AMINO ACID CATABOLISM	AMINO ACID	PATHWAY NOTES
Pyruvate	Ala, Cys, Thr, Gly, Ser	Thr → Gly → Ser → pyruvate (Acetyl-CoA is not an intermediate between these five amino acids and pyruvate.)
α-Ketoglutarate	Gln, Arg, Pro, His, Glu	Glu is an intermediate common to the other four amino acids. Glu → α-ketoglutarate
Succinyl-CoA	Met, Ile, Val, Thr	Propionyl-CoA is an intermediate common to all four amino acids. In the **TRANSSULFURATION PATHWAY,** the sulfur atom in Met becomes the sulfur atom of cysteine in two reactions that involve SAM.
Oxaloacetate	Asp, Asn	
Fumarate	Tyr	Catabolism of Tyr forms both acetoacetate and fumarate.

15.3 NEUROTRANSMITTER DEGRADATION

WHAT ROLE DOES THE DESTRUCTION OF NEUROTRANSMITTERS PLAY IN THE FUNCTIONING OF NEURONS AND MUSCLE CELLS?

Neurotransmitter degradation in a timely and efficient fashion is required for maintaining precise information transfer in the nervous system. Degradation of neurotransmitters essentially results in the inactivation of the signal.

Catecholamines (epinephrine, norepinephrine, dopamine): inactivated by transport out of the synaptic cleft followed by monoamine oxidase (MAO) oxidation; also inactivated by methylations by catechol-*O*-methyltransferase (COMT); epinephrine is catabolized in nonneural tissue

15.4 NUCLEOTIDE DEGRADATION

Digestion:

Nucleic acids + H_2O → Oligonucleotides	Nucleases (DNases, RNases)
Oligonucleotides + H_2O → Nucleotides	Phosphodiesterases
Nucleotide + H_2O → Nucleoside + P_i	Nucleotidases
Nucleoside + P_i → Base + Ribose-1-phosphate	Nucleosidases

Purines are degraded to uric acid.

Pyrimidines are degraded to NH_4^+, CO_2, and either β-alanine or β-aminoisobutyrate.

PURINE CATABOLISM

(**HINT**: Review nucleoside/nucleotide nomenclature.)

Dephosphorylation of AMP, IMP, XMP, and GMP by 5′-nucleotidase forms the nucleosides adenosine, inosine, xanthosine, and guanosine.

Adenosine is deaminated to form inosine and then the ribose is removed. (Note that adenine is not an intermediate.) AMP may also be deaminated first (to form IMP) and then dephosphorylated (to form inosine). Inosine's ribose is removed to form the free base hypoxanthine, which is then oxidized to xanthine.

Guanosine loses its ribose to form the guanine, which is deaminated to form xanthine.

Finally, xanthine is oxidized to uric acid by xanthine oxidase.

Animals other than primates can degrade uric acid into other nitrogen products such as urea and NH_4^+.

Look at the structure of uric acid. Note that the purine rings remain intact.

Pyrimidine Catabolism

As an exercise, compare the outline of pyrimidine catabolism (following purine catabolism) with Figure 15.18 in your text (p. 578). Write in the groups that are added and removed (NH_4^+, P_i) and the names of the enzymes (or copy this figure onto a separate sheet of paper). Drawing careful comparisons between the molecular structures will help you to understand and learn this pathway. Note that the pyrimidine ring can be degraded, whereas the purine ring cannot.

CHAPTER 15: SOLUTIONS TO REVIEW QUESTIONS

15.1 a. protein turnover – the continuous synthesis and degradation of an organism's proteins

b. proteasome – multienzyme complex that degrades proteins linked to ubiquitin

c. ubiquitin – a protein that is covalently attached by enzymes to proteins that are to be degraded

d. ubiquitination – the covalent attachment of ubiquitin to proteins that are to be degraded

e. autophagy – a cellular degradation pathway in which cell components are degraded by the hydrolytic enzymes within lysosomes; also referred to as macroautophagy

15.2 a. autophagosome – formed during autophagy; an isolation membrane, which sequesters cytoplasmic components and then seals itself

b. endosome – a membrane bound vesicle with functions in endocytosis and the autophagy lysosomal system

c. amphisome – a membrane bound vesicle in the autophagy lysosomal system that is formed from the fusion of the autophagosome with an endosome

d. lysosome – a saclike organelle that contains hydrolytic enzymes capable of degrading most biomolecules

e. autophagolysosome – a membranous vesicle formed from the fusion of an amphisome and a lysosome; lysosomal enzymes proceed to degrade the biomolecules and organelles contained in the amphisome

15.4 a. urea cycle – a cyclic pathway in which waste ammonia molecules, carbon dioxide, and the amino nitrogen of aspartate molecules are converted to urea

b. Krebs bicycle – a biochemical pathway in which the aspartate required in the urea cycle is generated from oxaloacetate, an intermediate in the citric acid cycle

c. carbamoyl phosphate – the first product in urea biosynthesis; formed from ammonium ions and bicarbonate in the mitochondrial matrix

d. ureotelic organism – an organism that utilizes urea synthesis for the disposal of ammonia

e. ornithine translocase – transports ornithine into the mitochondrial matrix in exchange for citrulline and protons

15.5 a. glucogenic – describes amino acids that are degraded to pyruvate or a citric acid cycle intermediate; these amino acids are used as substrates in the synthesis of glucose in gluconeogenesis

b. ketogenic – describes amino acids degraded to form acetyl-CoA or

acetoacetyl-CoA

c. N-acetylglutamate – an allosteric regulator of carbamoyl phosphate synthetases I, the liver mitochondrial enzyme that catalyzes the first reaction in urea synthesis; a sensitive indicator of the cell's glutamate concentration

d. hyperammonemia – a potentially fatal elevation of the concentration of ammonium ions in the blood

e. BH_4 – tetrahydrobiopterin – a folic acid–like molecule that is an essential cofactor in numerous reactions, which include the hydroxylation of aromatic rings.

15.7 a. chaperone-mediated autophagy (CMA) – a receptor-mediated process in which specific proteins that are bound to a chaperone complex are unfolded and then translocated to a lysosome where they are degraded.

b. microautophagy–a degradative process in which small amounts of cytoplasm are directly engulfed by lysosomes.

c. macroautophagy – a lysosomal pathway for bulk degradation of cytoplasmic components; also referred to as autophagy.

d. ubiquitin proteasomal system – a degradative mechanism in which protein destruction is initiated by covalent modification with ubiquitin; subsequent degradation occurs within a proteolytic molecular machine referred to as a proteasome

e. autophagy lysosomal system – a cellular degradation system involving lysosomes that is used to degrade long-lived proteins and organelles

15.8 a. albinism – a condition in which the missing or defective enzyme tyrosinase results in the absence of the black pigment melanin; affected individuals are extremely sensitive to sunlight.

b. maple syrup urine disease – a deficiency in the enzyme complex branch-chain α-keto acid dehydrogenase results in the accumulation of toxic α-keto acids derived from leucine, isoleucine, and valine; symptoms include convulsions, brain damage and mental retardation.

c. alkaptonuria – deficiency in the enzyme homogentisate oxidase results in accumulation of its substrate homogentisate; symptoms include arthritis and uneven skin pigmentation.

d. methylmalonic acidemia – deficiency in methylmalonyl-CoA mutase cause symptoms similar to those of maple syrup urine disease; may also be caused by adenosylcobalamin deficiency

e. phenylketonuria – deficiency of phenylalanine hydroxylase; causes accumulation of phenylalanine in blood, which disrupts amino acid transport into the brain; symptoms include brain damage and mental retardation

15.10 a. oligonucleotide – a short nucleic acid segment that contains fewer than 50 nucleotides

b. nuclease – an enzyme that hydrolyzes nucleic acid molecules to form oligonucleotides to produce nucleotides

c. phosphodiesterase – an enzyme that hydrolyses oligonuclotides to produce nucleosides

d. nucleosidase – an enzyme that hydrolyses nucleosides to yield free bases and ribose or deoxyribose

e. nucleotidase – an enzyme that hydrolyzes nucleotides to yield nucleosides and phosphate

15.11 a. uric acid – the end waste product of purine nucleotide catabolism in some animals including humans

b. allantoin – the excretory product of purine nucleotide metabolism in many mammals; formed from uric acid by uric oxidase

c. allantoate – the excretory product of purine metabolism in bony fish; formed from allantoin by allantoicase

d. urate oxidase – the enzyme that converts uric acid to allantoin

e. allantoicase – the enzyme that splits allantoic acid to glyoxylate and urea.

15.13 The major nitrogen-containing excretory molecules are ammonia, urea, uric acid, allantoin, and allantoate.

15.14 The process of protein turnover promotes metabolic flexibility, protects a cell from the accumulation of abnormal proteins, and is a key feature of organismal developmental processes.

15.16 The metabolic products of amino acid degradation are acetyl-CoA, acetoacetyl-CoA, pyruvate, α-ketoglutarate, succinyl-CoA, fumarate, and oxaloacetate, ammonia is also a product.

15.17 Aspartate is converted to oxaloacetate via transamination, then catabolized by the citric acid cycle.

15.19 a. Ketogenic, b. Ketogenic, c. Glycogenic, d. Glycogenic, e. Glycogenic, f. Both.

15.20 Glutamate is converted to α-ketoglutarate by glutamate dehydrogenase or by transamination, and is then catabolized by the citric acid cycle.

15.22 In the muscle, pyruvate undergoes a transamination reaction and is converted to alanine. Alanine is then transferred to the liver, where it is reconverted to pyruvate by transferring its amino group to α-ketoglutarate to form glutamate. Glutamate is then oxidatively deaminated to form α-ketoglutarate and ammonia. The pyruvate and α-ketoglutarate are degraded by the citric acid cycle.

15.23 Alanine transfers its amino group to α-ketoglutarate to form pyruvate and glutamate. The glutamate is then oxidatively deaminated to form ammonia and α-ketoglutarate. The pyruvate and α-ketoglutarate are degraded by the citric acid cycle

15.25 The NADH formed during the conversion of fumarate to aspartate results in the synthesis of approximately 2.5 ATP. Therefore, the net ATP requirement for urea synthesis (4 ATP – 2.5 ATP) is approximately 1.5 moles of ATP per mole of urea.

15.26 The term Krebs bicycle refers to two interlocking cyclic reaction pathways. The aspartate-arginosuccinate shunt of the citric acid cycle is responsible for regenerating the aspartate needed for the urea cycle from fumarate. The molecule the two cycles have in common is arginosuccinate.

15.28 In ubiquitination, the major mechanism of protein degradation, ubiquitin (a small hsp) is covalently attached to lysine residues of a protein with structural features such as oxidized amino acid residues that mark it for destruction. Once the protein is ubiquitinated it is degraded by proteases in ATP-requiring reactions.

15.29 For the degradation of lysine see Figure 15.7.

15.31 The following compounds yield uric acid when degraded: DNA, FAD and NAD^+, all of which contain the purine ring.

15.32 Tyrosine is catabolized in five reactions to acetoacetate and fumarate. The resulting acetoacetate is then converted to acetoacetyl-CoA, and then to acetyl-CoA. The resulting acetyl-CoA and pyruvate are finally catabolized by the TCA cycle.

15.34 Nitrogen enters the urea cycle as NH_4^+ and aspartate. The two nitrogen atoms in urea most likely originated in the α-amino groups of amino acids. The pathways from α-amino acids to ammonium and aspartate are outlined below, with 1) and 2) providing most of the nitrogen in the urea cycle.

1) From muscle cell amino acids to NH_4^+ in the liver: In muscle cells, a transamination reaction with α-ketoglutarate produces glutamate (and an α-keto acid). Glutamate in muscle cells may transfer its amino group to pyruvate via another transamination to produce alanine, or it may be oxidatively deaminated to generate NH_4^+, which reacts with another glutamate to form glutamine. Alanine and glutamine transport the nitrogen atom from α-amino acids to the liver, where glutamate is regenerated: a) from alanine via transamination to α-ketoglutarate, or b) from glutamine via hydrolysis of the amide, which also forms a free NH_4^+. The glutamate produced from both alanine and glutamine is oxidatively deaminated to form NH_4^+.
2) From liver cell amino acids to aspartate: An α-amino acid and oxaloacetate react via transamination to form an α-keto acid and aspartate.
3) From serine and threonine to NH_4^+ in the liver: Since Ser and Thr cannot undergo transamination, they are deaminated by serine dehydratase and threonine dehydratase.
4) From urea present in the blood to NH_4^+ in the liver: Urea in the bloodstream diffuses into the intestinal lumen, where it is hydrolyzed by bacteria with urease to form ammonia, which diffuses back into the blood for transport to the liver.
5) From amino acids in the liver and kidney to NH_4^+ via the action of L-amino acid oxidases.
6) From the C-6 amino group of adenine to NH_4^+ via the action of adenosine deaminase.

15.35 Ketogenic amino acids that are degraded to form acetyl-CoA or acetoacetyl-CoA cannot be converted into glucose because (1) acetyl-CoA cannot be converted to pyruvate because the pyruvate dehydrogenase-catalyzed reaction is irreversible and (2) when acetyl-CoA enters the citric acid cycle, two carbon atoms are lost as carbon dioxide; thus no carbon atoms, which could have contributed to the formation of oxaloacetate (and eventually glucose), are added.

15.37 Foods high in purines include liver and sardines. Increased purine levels lead to increased uric acid synthesis. Increased uric acid levels increase the possibility of sodium urate crystal deposits in and around joints and within the kidney. These deposits cause the symptoms of gout.

15.38 The nitrogen atom in position 3 on the uracil ring eventually ends up in urea molecules. (This nitrogen is between the two carbonyl carbons in uracil.)

15.40 A characteristic odor in the urine is the primary symptom of maple syrup urine disease. If untreated, symptoms include vomiting, convulsions, severe brain damage, mental retardation, and, often, death within an infant's first year. Maple syrup urine disease is caused by a deficient branched chain α-keto acid dehydrogenase complex, which converts α-keto acids to their acyl-CoA derivatives. Large quantities of the α-keto acids derived from branched-chain amino acids – leu, ile, and val – accumulate in the blood.

15.41 The molecules that cause the urine odor that is characteristic of the maple syrup urine disease are the α-keto acids derived fro leucine, isoleucine and valine.

15.43 Tyrosine catalyzes the conversion of DOPA to dopaquinone, a highly reactive molecule that is the precursor of the pigment molecule melanin. In the absence of tyrosine melanin can't be synthesized, thus resulting in albinism.

15.44 Humans do not synthesize the enzymes urate oxidase and allantoinase, which catalyze the conversion of uric acid to allantoate, the first product of purine degradation that does not contain a ring.

CHAPTER 15: SOLUTIONS TO FILL-IN-THE-BLANK QUESTIONS

15.46 Ketogenic

15.47 Glucogenic

15.49 Humoral

15.50 Aspartate/NH_4^+

15.52 Urea

15.53 Pyrimidine

CHAPTER 15: SOLUTIONS TO SHORT-ANSWER QUESTIONS

15.55 Taurine monochloride is synthesized in white blood cells, where it modulates the inflammatory process by downregulating the synthesis of nitric oxide and proinflammatory proteins such as tumor necrosis factor α.

15.56 In the absence of functional PNP, high levels of purine nucleotides (e.g., dGTP) apparently impair T cells.

15.58 Methymalonic acidemia is caused by the accumulation of methylmalonate, an intermediate in the conversion of propionyl-CoA, a product of the catabolism of the amino acids methionine, isoleucine, and valine, to succinyl-CoA. The causes of this malady include the inheritance of a defective gene for the enzyme methylmalonyl-CoA mutase. Treatment consists largely of a low-protein diet. Cases caused either by the enzyme's low affinity for adenosylcobalamin or by a vitamin B12 deficiency are treated with vitamin B12 injections.

15.59 Named for Sir Hans Krebs, the discoverer of the citric acid cycle and the urea cycle (along with Kurt Henseleit), the Krebs bicycle illustrates the linkage of these two metabolic cycles. The urea cycle intermediate citrulline reacts with aspartate to form arginosuccinate, which is then cleaved to form arginine and the citric acid cycle intermediate fumarate. Fumarate is then converted to oxaloacetate, which is the α-keto acid precursor of aspartate. Refer to Figure 15.4.

CHAPTER 15: SOLUTIONS TO THOUGHT QUESTIONS

15.61 Once your diagram of the glycolysis and citric acid pathways is drawn, note that the linkage between the citric acid cycle and the purine nucleotide cycle is via fumarate, a citric acid cycle intermediate and aspartate, the product of oxaloacetate transamination.

15.62 As the concentration of glutamate (as well as its deamination product ammonia) rises, the enzyme N-acetylglutamate synthase catalyzes the synthesis of N-acetylglutamate, an activator of carbamoyl phosphate synthetases I. The latter enzyme catalyzes the first committed step in urea synthesis.

15.64 These amino acids are intermediates in the urea cycle. Therefore, their addition stimulates the formation of urea.

15.65 a. Methylene (-CH_2-) groups

b. Methyl groups

c. Methyl groups

d. Methyl groups

15.67 Urea and uric acid, both of which are less toxic than ammonia, require significantly less water for their excretion than does ammonia. As a result the use of these molecules facilitates water conservation in land animals. Osmotic pressure is an important factor. Since urea and uric acid contain two and four nitrogen atoms, respectively, less water is required (per nitrogen atom) for their excretion than for ammonium ion.

15.68 If parkin, an E3 (ubiquitin ligase) enzyme, is defective or inoperative the capacity of dopaminergic nerve cells to degrade proteins is compromised. The pace of protein turnover will slow and abnormal proteins will accumulate. Eventually such cells will die.

15.70 The nitrogen at ring position 3 of uracil is released as ammonia, a substrate of the urea cycle when β-ureidopropionate, an intermediate formed during the degradation of uracil, is converted to β-alanine.

15.71 Primates lack urate oxidase, which converts uric acid to allantoin, a more soluble molecule. As a result uric acid crystals can build up and gout results.

15.73 The α-keto acids react with ammonia to produce amino acids thus sparing α-ketoglutarate, a citric acid cycle intermediate needed to generate energy.

15.74 Three ATPs are directly required for the synthesis of each molecule of urea. Since the conversion of fumarate to oxaloacetate in the citric acid cycle yields 2.5 molecules of ATP, urea synthesis is a net energy requiring process.

16 Integration of Metabolism

Brief Outline of Key Terms and Concepts

16.1 OVERVIEW OF METABOLISM

Multicellular organisms require sophisticated regulatory mechanisms to ensure that all their cells, tissues, and organs cooperate and use **HORMONES** to convey information between cells. **ENDOCRINE HORMONES** are secreted from cells that are distant from target cells.

TARGET CELLS; SIGNAL TRANSDUCTION; SECOND MESSENGERS; ENZYME CASCADE

To ensure proper control of metabolism, the synthesis and secretion of many mammalian hormones are regulated by a complex cascade mechanism ultimately controlled by the central nervous system.

Hormones: peptide or polypeptide, steroid, and amino acid derivatives

16.2 HORMONES AND INTERCELLULAR COMMUNICATION

PEPTIDE HORMONES

HYPOTHALAMUS; **PITUITARY**

Protection against overstimulation by hormones: **FEEDBACK INHIBITION**, **DESENSITIZATION**, **DOWNREGULATION** (e.g., **INSULIN RESISTANCE**)

Types of cell-surface receptors:

- **G-PROTEIN-COUPLED RECEPTORS (GPCRs)**
- **RECEPTOR TYROSINE KINASES (RTKs)**

GPCRs

- **G PROTEINS** and the action of the α-subunit
- **GUANINE NUCLEOTIDE EXCHANGE FACTOR (GEF)**
- **SECOND MESSENGERS:** cAMP; cGMP; the phosphatidylinositol cycle, DAG, PIP_3, Ca^{2+}

RECEPTOR TYROSINE KINASES (RTKs) activate phosphorylation cascades when they undergo autophosphorylation triggered by the binding of signal molecules. **RTKs** do not directly involve the generation of a second messenger.

GROWTH FACTORS

EPIDERMAL GROWTH FACTOR (EGF)
PLATELET-DERIVED GROWTH FACTOR (PDGF)
SOMATOMEDINS
INSULIN-LIKE GROWTH FACTOR
CYTOKINES: INTERLEUKIN-2; INTERFERONS; TUMOR NECROSIS FACTORS (TNF)

STEROID AND THYROID HORMONE MECHANISMS

Nonpolar steroid and thyroid hormones diffuse through the lipid bilayer of cell membranes and bind to intracellular receptors. The hormone-receptor complex then binds to a DNA sequence, a **HORMONE RESPONSE ELEMENT (HRE)**. The binding of a hormone-receptor complex to an HRE enhances or diminishes the expression of specific genes.

16.3 METABOLISM IN THE MAMMALIAN BODY: THE FEEDING-FASTING CYCLE

A variety of organs contribute via hormones and neurotransmitters to the acquisition and use of nutrients. The goal is to maintain balance between energy acquisition and energy expenditure. The hypothalamus contains the critical neural circuits that control appetite and satiety.

THE FEEDING PHASE

Food is consumed, digested, absorbed, and transported to the organs to be used or stored.

POSTPRANDIAL; CHYLOMICRON REMNANTS

THE FASTING PHASE

During fasting, several metabolic strategies maintain blood glucose levels.

POSTABSORPTION; POSTABSORPTIVE STATE

FEEDING BEHAVIOR

ARCUATE NUCLEUS (ARC) of the hypothalamus
NUCLEUS TRACTUS SOLITARIUS (NTS)

BIOCHEMISTRY IN PERSPECTIVE

- Diabetes mellitus
- Obesity and the metabolic syndrome

OVERVIEW

It is all about control—balancing anabolism and catabolism based upon the organism's energy needs, energy stores (e.g., ATP, glycogen, and triacylglycerols), and nutrient availability—and integrating these metabolic processes for optimal results.

Information transfer within an organism is critical for effective, efficient, and timely control and is achieved by coordination of the nervous system with chemical signals (usually hormones). Descriptions of information transfer systems may vary somewhat depending upon the focus and depth of the discussion. Here are the major points that had been covered in previous chapters. The description in Section 16.1 (below) is more detailed.

PROCESS that organisms use to receive and interpret information (**SIGNAL TRANSDUCTION**):
- **RECEPTION**—ligand binds to receptor
- **TRANSDUCTION**—ligand binding triggers a conformational change in the receptor that results in the conversion of the extracellular signal into an intracellular message
- **RESPONSE**—a cascade of events that involve covalent modifications; possible results: changes in enzyme activities and/or gene expression, cytoskeletal rearrangements, cell movement, or cell cycle progression (growth or division).

COMPONENTS of a simple system for information transfer:
- **PRIMARY SIGNAL**—the first messenger, usually a hormone; additional examples: neurotransmitter, growth factor, or cytokine
- **TARGET**—a specific **RECEPTOR**, usually bound to a membrane
- **TRANSDUCER SYSTEM**—converts the signal to a cellular response

16.1 OVERVIEW OF METABOLISM

STEADY STATE is when the rate of anabolic processes equals (approximately) the rate of catabolic processes, and as a result, the organism does not grow or change appreciably.

COMPONENTS AND PROCESS OF AN INTERCELLULAR COMMUNICATION SYSTEM VIA CHEMICAL SIGNALS:

PRIMARY SIGNAL (the first messenger, usually a hormone)

ENDOCRINE HORMONES, secreted by specialized glandular cells, fall into three categories: peptide or polypeptide hormones, steroid hormones, and amino acid derivatives.

GROWTH FACTORS are hormone-like polypeptides and proteins. Unlike hormones, they are not synthesized by specialized glandular cells, and they specifically regulate growth, differentiation, and proliferation of various cells. **CYTOKINES** are proteins produced by blood-forming cells and immune system cells and may stimulate or inhibit cell growth or proliferation.

TRANSPORT to the **TARGET CELL**

RECEPTION: A signal molecule binds to its specific **RECEPTOR**.

SIGNAL TRANSDUCTION: Upon hormonal binding to a receptor, a conformational change transmits the signal across the membrane. As a result, second messengers may be released inside the cell. (Examples of second messengers include cAMP, cGMP, DAG, IP_3, Ca^{2+}, and the inositol phospholipid system.) Most second messengers that bind to an enzyme cause a conformational transition that switches the enzyme from an inactive to an active form and often triggers an enzyme cascade that amplifies the signal.

ENZYME CASCADE: Once activated by the second messenger, the enzyme modifies multiple copies of a number of *different* target enzymes. These newly activated target enzymes may also modify multiple copies of *another* set of target proteins.

CELLULAR RESPONSE: Cellular functions are altered by changes in the activities of existing enzymes, rearrangements of the cytoskeleton, and/or altered gene expression.

16.2 HORMONES AND INTERCELLULAR COMMUNICATION

PEPTIDE HORMONES

SYNTHESIS AND SECRETION OF PEPTIDE HORMONES are regulated by the endocrine hormone cascade system that is ultimately controlled by the central nervous system. An overview of mammalian hormones is provided in an online Biochemistry in Perspective essay on mammalian hormones and the hormone cascade system.

MECHANISMS THAT PROTECT AGAINST OVERSTIMULATION BY HORMONES

FEEDBACK INHIBITION—the hypothalamus and the anterior pituitary are controlled by the target cells they regulate.

DESENSITIZATION—target cells decrease the number of cell surface receptors (or inactivate receptors) in response to changes in stimulation.

DOWN-REGULATION—the reduction in cell surface receptors (by endocytosis) in response to stimulation by specific hormones. The receptors may eventually be recycled back to the cell surface or degraded. An example is **INSULIN RESISTANCE**, in which a decrease in functional insulin receptors causes one type of diabetes.

CELL-SURFACE RECEPTORS

GPCRs—G-PROTEIN-COUPLED RECEPTORS

RTKs—RECEPTOR TYROSINE KINASES (next page)

GPCRs—G-PROTEIN COUPLED RECEPTORS

GPCRs are composed of seven membrane-spanning helices, with an N-terminal segment as part of the ligand-binding site and a C-terminal segment that interacts with G proteins on the cytoplasmic side of the membrane.

G PROTEINS (HETEROTRIMERIC GTP-BINDING PROTEINS)

Located on the cytoplasmic side of the cell membrane, G proteins transduce ligand binding to GPCRs into intracellular signals. G proteins contain three subunits (α, β, γ). While inactive, the $\beta\gamma$ complex binds to and inhibits the α subunit, which binds GDP.

GPCR ACTIVITY AND G PROTEIN ACTIVATION:

1. Ligand binding to the GPCR

   Examples of GPCR ligands: hormones (e.g., glucagon, TSH, the catecholamines, and the endocannabinoids), neurotransmitters (e.g., glutamate, dopamine, and GABA), neuropeptides (e.g., vasopressin and oxytocin), odorants and tastants, and light (rhodopsin).

2. Signal initiation

   Binding of the ligand causes a conformational change in the receptor, and the G protein binds to the occupied receptor. **GEF (GUANINE NUCLEOTIDE EXCHANGE FACTOR)** mediates the GDP/GTP exchange:

$$\text{GDP-}\alpha\beta\gamma + \text{GTP} \xrightarrow{\text{GEF}} \text{GTP-}\alpha + \beta\gamma + \text{GDP}$$

The activated GTP α-subunit dissociates from the βγ complex and moves over the cytoplasmic surface of the membrane to activate an enzyme that synthesizes a specific second messenger, which initiates an enzyme cascade.

3. Primary signal termination: GTP hydrolysis: GTP-α → GDP-α

   GDP-α recombines with GPCR-βγ.

4. Secondary signal termination: removal of the second messenger.

EXAMPLES OF SECOND MESSENGERS: cAMP, cGMP, DAG, IP_3, Ca^{2+}, AND THE INOSITOL PHOSPHOLIPID SYSTEM.

cAMP $ATP \xrightarrow{\text{Adenylate cyclase}} cAMP + PP_i$

The G-protein $\mathbf{G_S}$ stimulates (and $\mathbf{G_i}$ inhibits) adenylate cyclase.

Upon ligand binding, G_S binds to the occupied receptor.

GDP-α_S → GTP-α_S. The GTP-α_S subunit (the activated α subunit of G_S) detaches from βγ and activates adenylate cyclase, which synthesizes cAMP.

The cAMP diffuses into the cytoplasm, where it binds to and activates PKA (cAMP-dependent protein kinase). PKA then phosphorylates key regulatory enzyme to change its activity.

Primary signal termination: GTP-α_S → GDP-α_S

Secondary signal termination: cAMP + H_2O → AMP, catalyzed by phosphodiesterase.

THE PHOSPHATIDYLINOSITOL CYCLE, DAG, and $\mathbf{Ca^{2+}}$ mediate the action of hormones and growth factors. (See Figure 16.7, page 591 in your text.)

DAG (DIACYLGLYCEROL) activates protein kinase C, which phosphorylates specific regulatory enzymes.

IP_3 (INOSITOL-1,4,5-TRIPHOSPHATE): Its receptor is a Ca^{2+} channel in the calcisome (SER). When activated, the channel opens and calcium ions flow through, and the action of Ca^{2+}-regulated proteins is affected.

THE PHOSPHATIDYLINOSITOL PATHWAY

The binding of certain hormones to their receptor activates the α-subunit of a G protein. The α-subunit then activates phospholipase C, which cleaves IP_3 from PIP_2, leaving DAG in the membrane.

$$PIP_2 \xrightarrow{\text{Phospholipase C}} IP_3 + DAG$$

DAG, acting with phosphatidylserine (PS) and Ca^{2+}, activates protein kinase C, which then phosphorylates key regulators. IP_3 binds to receptors on the SER, opening Ca^{2+} channels. Then Ca^{2+} moves into the cytoplasm and activates additional targets.

cGMP $GTP \xrightarrow{\text{Guanylate cyclase}} cGMP + PP_i$

TWO TYPES OF MEMBRANE-BOUND GUANYLATE CYCLASE:

ANF (atrial natriuretic factor)—lowers blood pressure—activates membrane-bound guanylate cyclase to produce cGMP, which activates the phosphorylating enzyme protein kinase G.

BACTERIAL ENTEROTOXIN—causes diarrhea—binds to and activates another type of membrane-bound guanylate cyclase.

RTKs—Receptor Tyrosine Kinases

Like GPCRs: RTKs are a family of transmembrane receptors with an external domain that binds specific ligands. Upon ligand binding, a conformational change in the receptor protein transduces the signal to the cell's interior.

Unlike GPCRs: RTKs have a cytoplasmic catalytic domain with tyrosine kinase activity. Upon ligand binding, the tyrosine kinase domain autophosphorylates, initiating a phosphorylation cascade.

The Insulin Receptor: A Well-Researched Example of an RTK

Structure: a transmembrane glycoprotein with four subunits connected by disulfide bridges. Two α-subunits extend out of the cell to bind insulin; two β-subunits extend through the membrane and contain a tyrosine kinase domain on the cytoplasmic side.

A Simplified Model of Insulin Signaling (See Figure 16.9 in your text.)[1]

Insulin binds to its RTK, causing autophosphorylation and activation of its tyrosine kinase domain, which triggers a phosphorylation cascade that modulates various intracellular proteins.

This activated insulin receptor then phosphorylates several proteins, such as **insulin receptor substrate 1 (IRS1)**.

$$\text{IRS1}_{\text{(inactive)}} \xrightarrow{\text{Insulin Receptor (activated)}} \text{IRS1}_{\text{(phosphorylated, active)}}$$

This newly phosphorylated IRS1 then binds to and activates several proteins, including phosphatidylinositol-3-kinase, which catalyzes:

$$\text{PIP}_2 \xrightarrow{\text{Phosphatidylinositol-3-kinase}} \text{PIP}_3$$

PIP_3 activates PIP_3-dependent protein kinase, which initiates a phosphorylation cascade that leads to the activation of several kinases (e.g., PKB and PKC), which in turn trigger pathways that lead to changes in the expression of several genes.

In general, insulin induces processes that promote uptake and storage of nutrients. Some of the end results of insulin-receptor binding:

- Inhibits hormone-sensitive lipase in adipocytes;
- Induces transfer to the cell's surface of several types of protein that affect the uptake of nutrients (examples: receptors for LDL and IGF-II; glucose transporter GLUT4, to the plasma membrane of adipocytes and muscle cells (and facilitated by activated PKB).

Growth Factors

Epidermal Growth Factor (EGF)—a **mitogen** (stimulates cell division) for epidermal and gastrointestinal lining cells; triggers cell division when it binds to plasma membrane EGF receptors (transmembrane tyrosine kinases)

1

PIP_2	phosphatidylinositol 4,5-bisphosphate	PKB	protein kinase B
PIP_3	phosphatidylinositol 3,4,5-trisphosphate	IRS1	insulin receptor substrate 1
RTKs	receptor tyrosine kinases		

PLATELET-DERIVED GROWTH FACTOR (PDGF)—secreted by blood platelets during clotting; stimulates mitosis in fibroblasts and other cells during wound healing; promotes collagen synthesis in fibroblasts

SOMATOMEDINS—secreted by the liver into the bloodstream; mediate the growth-promoting actions of GH; include **INSULIN-LIKE GROWTH FACTORS I** and **II (IGF-I, IGF-II)**; bind to cell surface receptors that are also tyrosine kinases

CYTOKINES

INTERLEUKIN-2 (IL-2)—secreted by T cells after they have been activated by binding to a specific antigen-presenting cell; stimulates cell division so that numerous identical T cells are produced; regulates the immune system, promotes cell growth and differentiation.

INTERFERONS are growth inhibitors.
Type I protects cells from viral infection.
Type II inhibits cancer cell growth and also has several immunoregulatory effects.
TUMOR NECROSIS FACTORS (TNF), toxic to tumor cells, suppress cell division.

How do hormones, growth factors, and cytokines differ? *(See p. 245 for the answer.)*

STEROID AND THYROID HORMONES

Steroid and thyroid hormones switch genes on or off, changing the pattern of proteins that an affected cell makes.

STEROID HORMONES are lipid soluble,

- Steroid hormones require transport proteins to travel through the bloodstream. Upon reaching the target cell, the hormone dissociates from its transport protein.
- A steroid hormone can diffuse across a cell membrane into a target cell and bind to a receptor protein in the cytoplasm (or in the nucleus). (Water-soluble hormones cannot do that, so they bind to a receptor on the cell membrane surface.)

If in the cytoplasm, the hormone-receptor complex moves to the nucleus.

In the nucleus, it binds to specific sites on DNA. Specifically, the hormone-receptor complex binds to the base sequence of an **HRE** (**HORMONE RESPONSE ELEMENT** = specific sites on DNA) via zinc finger domains, **HRE** – hormone receptor complex binding changes a cell's pattern and rate of gene transcription, ultimately affecting protein synthesis).

Since several HREs can bind to the same hormone-receptor complex, the expression of numerous genes can be altered simultaneously.

16.3 METABOLISM IN THE MAMMALIAN BODY: THE FEEDING-FASTING CYCLE

Consuming food intermittently is possible because of elaborate mechanisms for storing and mobilizing energy-rich molecules derived from food. Regulation of opposing pathways ensures that they do not occur simultaneously.

THE FEEDING PHASE: FOOD IS CONSUMED, DIGESTED, AND ABSORBED . . .

Absorbed nutrients are transported to various organs where they are either used or stored. During the feeding phase, hormones such as gastrin, secretin, and cholecystokinin stimulate the secretion of various digestive enzymes or aids such as bicarbonate and bile. This phase is regulated by interactions between enzyme-producing cells of the digestive organs, the nervous system, and several hormones.

Lipids

Chylomicrons transport lipid molecules from the small intestine, through the lymph and the bloodstream, to target tissues. **Chylomicron remnants** (chylomicrons after most triacylglycerols have been removed) are taken up by the liver, where they are reused or degraded. Elevated fatty acids in the blood promote lipogenesis in adipose tissue.

Glucose

As glucose moves from the small intestine to the liver via the blood, β-cells in the pancreas are stimulated to release insulin. Insulin release triggers several processes that ensure nutrient storage (see table below). Insulin also influences amino acid metabolism. For example, insulin promotes the transport of amino acids into cells and stimulates protein synthesis in most tissues.

The Fasting Phase

Early Postaborptive State

Glucagon is released as blood glucose and insulin levels return to normal. Glucagon prevents hypoglycemia by promoting glycogenolysis and gluconeogenesis in the liver.

Prolonged Fast

Strategies to maintain blood glucose levels include the following:

- Norepinephrine stimulates lipolysis in adipose tissue, releasing fatty acids that provide an alternative energy source.
- Glucagon increases gluconeogenesis by using amino acids from muscle.

Starvation

Metabolic changes occur; further fatty acids from adipose tissue are converted into ketone bodies in the liver.

Feeding-Fasting Cycle: Effects of Insulin and Glucagon

	Postprandial State	**Postabsorptive State**
	After a meal **High blood nutrient levels**	**After a fast** **Low blood nutrient levels**
	Insulin **Promotes nutrient storage**	**Glucagon** **Prevents hypoglycemia**
Blood	Lowers blood glucose	Raises blood glucose
Liver	Glucose → Glycogen Fatty acids, glycerol → Fats	Glycogen → Glucose Gluconeogenesis → Glucose
Muscle	Amino acids → Protein Glucose → Glycogen Glucose uptake	Protein → Amino acids (Ala, Gln)
Adipose tissue	Fatty acids, glycerol → Fats Glucose uptake	Fats → Fatty acids, glycerol

FEEDING BEHAVIOR

Feeding behavior is the complex mechanism by which animals, including humans, seek out and consume food. Appetite and satiety in humans are regulated by hormonal and neural signals from peripheral organs (e.g., the GI tract and adipose tissue).

Neuronal pathways of the autonomic nervous system (such as the vagus nerve) continuously supply the brain with information related to the status of the body's internal organs. Also, the mammalian brain links appetite systems to taste, olfaction, and reward systems to create a powerful drive that ensures survival.

Neural circuits that control appetite are in the hypothalamus and in the brain stem:

ARC (*arcuate nucleus*)—location in the hypothalamus of the primary neurons that control feeding behavior

NTS (*nucleus tractus solitarius*)—location in the brain stem that integrates information with appetite-regulating signals from the GI tract.

REGULATION OF FEEDING BEHAVIOR: FACTORS THAT AFFECT APPETITE[2]

	INHIBIT APPETITE *Stop; you are full!*	**STIMULATE APPETITE** *Eat!*
Release of peptide hormones	PYY (cells in the GI tract) CCK (cells in the GI tract) Insulin (pancreatic β-cells) Leptin (adipose tissue)	Ghr (cells in the stomach and small intestine) (Also: falling leptin levels caused by weight loss)
Appetite-regulating neurons in the **ARC** (in the hypothalamus)	Activated POMC neurons by leptin (insulin to a lesser extent) Inhibition of NPY/AgRP neurons by leptin and PYY	Activated NPY/AgRP neurons (by Ghr) Inhibition of POMC neurons
AMPK mediates when hormonal signals differ from nutrient signals. *(That looks really good, but I am not hungry.)*	Inhibition of AMPK occurs when hormones such as leptin and insulin bind to their cell-surface receptors, resulting in the inhibition of NPY/AgRP neurons.	Activation of AMPK (by molecules like Ghr) triggers signal transduction events that cause neurons to fire and release NPY and AgRP neurotransmitter molecules that bind to cell-surface receptors in neurons in ARC (and other hypothalamic regions)
Sensory input	Nauseating sight, smell, taste	Appetizing sight, smell, taste

Signals from AgRP/NPY and POMC neurons are communicated via second-order neurons to other parts of the hypothalamus and then other brain centers. Signals from these centers are then sent to the NTS, which integrates this information with appetite-regulating signals from the GI tract and other organs.

[2] NPY (neuropeptide Y); AgRP (agouti-related peptide); PYY (peptide YY)
POMC (pro-opiomelanocortin) neurons produce α-MSH (α-melanocyte-stimulating hormone).

Chapter 16: Solutions to Review Questions

16.1 a. second messenger – a molecule that mediates the actions of some hormones

b. desensitization – a process in which target cells adjust to changes in stimulation by hormone molecules by decreasing the number of cell surface receptors or by inactivating those receptors

c. target cell – a cell that responds to the binding of a hormone or growth factor to a receptor protein

d. insulin resistance – the insensitivity of tissues to insulin: a common cause is the downregulation of insulin receptors

e. adenylate cyclase – the enzyme that converts ATP to cyclic AMP (cAMP), the second messenger molecule

16.2 a. ketosis – elevated concentrations of ketones in the blood

b. ketoacidosis – elevated concentrations of ketone bodies in the blood; occurs in type II diabetics as the result of unrestrained fatty acid oxidation

c. osmotic diuresis – a process in which solutes in the uninary filtrate causes excessive loss of water and electrolytes

d. GLUT4 – an insulin sensitive glucose transporter found in the plasma membrane of muscle and adipose tissue cells

e. body mass index – BMI – a measure of a person's body composition; based on weight and height

16.4 a. hyperinsulinemia – high blood levels of insulin

b. dyslipidemia – high blood levels of total cholesterol and triacylglycerol and low HDL levels

c. hyperglycemia – blood glucose levels that are higher than normal

d. glucosuria – the presence of glucose in the urine

e. hyperosmolar hyperglycemic non ketosis – severe dehydration in non-insulin dependent diabetics; caused by persistently high blood glucose levels

16.5 a. Insulin-like growth factor (IGF) – a protein in humans that mediates the growth-promoting actions of growth hormone; has insulin-like properties (e.g., promotes glucose transport and fat synthesis)

b. interferon – one of a group of glycoproteins that have nonspecific antiviral activity that inhibits the synthesis of viral RNA and proteins and regulates the growth and differentiation of immune system cells.

c. interleukin – one of a group of cytokines that regulate the immune system in addition to promoting cell growth and differentiation

d. hormone response element – a specific DNA sequence that binds hormone- receptor complexes; the binding of a hormone–receptor complex either enhances or diminishes the transcription of a specific gene

e. histocompatability antigen – one of a group of proteins on the surface of most of the body's cells that play an important role in determining how the immune system will react to foreign substances or cells

16.7 a. postprandial – the phase in the feeding-fasting cycle immediately after the meal; blood nutrient levels are relatively high

b. postabsorptive – the phase in the feeding-fasting cycle in which nutrient levels are low

c. ARC – arcuate nucleus – the primary neurons in the hypothalamus that regulate feeding behavior

d. NPY – neuropeptide Y – an appetite stimulating peptide synthesized in the arcuate nucleus of the hypothalamus

e. POMC – pro-opiomelanocortin – a precursor polypeptide that in the POMC cells in the hypothalamus is converted to α-MSH (α-melanocyte stimulating hormone) which suppresses appetite

16.8 a. mitogen – a substance that stimulates cell division

b. phorbol ester – a tumor promoting molecule found in croton oil; mimics the actions of DAG

c. enteric - a term referring to the intestine, e.g., enterocytes of the small intestine absorb nutrient molecules and deliver them to blood and lymph

d. SUA – serum uric acid level – the concentration of uric acid in blood

e. endocrine – refers to hormones secreted into the bloodstream that act on distant target cells.

16.10 a. liver, b. liver, c. intestine, d. brain, e. adipose tissue

16.11 a. triiodothyronine (T_3) - general stimulation of many celluar reactions

b. insulin - promotes general anabolic effects, including glucose uptake by some cells, protein synthesis and lipogenesis.

c. glucagon-promotes glycogenolysis and lipolysis.

d. growth hormone - promotes general anabolic effects in many tissues

e. cholcystokinin - stimulates secretion of digestive enzymes and bile

16.13 The major recognized second messengers are (1) cAMP (generated from ATP by adenylate cyclase) stimulates changes in cellular activities by activating several protein kinases, (2) cGMP (generated from GTP by guanylate cyclase) activates protein kinase G, (3) diacylglycerol (DAG, a product of phospholipase C) activates protein kinase C, (4) inositol-1,4,5-triphosphate (IP_3, a product of phospholipase C) triggers the release of Ca^{2+} from the calcisome and (5) calcium ions, which regulate the activities of numerous cellular activities when they bind to calcium-dependent regulatory proteins.

16.14 Phorbol esters, found in croton oil, activate protein kinase C, an action that stimulates cell growth and division. However, unlike DAG, the molecule that they mimic, phorbol esters continue to activate protein kinase C for a prolonged time. This circumstance provides the affected cell with an advantage over unstimulated cells. Phorbol esters may transform a cell previously exposed to a carcinogenic initiating event into a cancerous cell whose unrestrained proliferation creates a tumor.

16.16 In uncontrolled diabetes, large amounts of glucose are excreted in the urine. Excessive urine flow caused by the large amounts of water that are excreted along with the glucose dehydrates the body. Dehydration then usually triggers the thirst response.

16.17 a. vasopressin – maintains blood pressure and water balance

b. PYY- an appetite stimulating molecule produced by cells in the small intestine and colon; inhibits NPY/AgRP neurons

c. leptin – induces satiety and is secreted into the bloodstream primarily by adipose tissue

d. ghrelin – stimulates appetite (food intake)

e. adiponectin – enhances glucose-stimulated insulin secretion and cellular responses to insulin

16.19 Anabolic steroid hormones change the expression of a specific set of genes that code for proteins (e.g., enzymes) that would increase protein synthesis in skeletal muscle (among other metabolic changes).

16.20 a. IGF-1 - insulin-like growth factor 1 – a polypeptide that mediates the growth-promoting actions of growth hormone

b. PGDF – platelet-derived growth factor – stimulates mitosis during wound healing; secreted by blood platelets

c. epidermal growth factor – stimulates cell division of a large number of epithelial cells, including epidermal and gastrointestinal lining cells

d. IRS-1 – insulin receptor substrate 1 - one of several proteins that are directly phosphorylated by the tyrosine kinase activity of the insulin receptor.

e. tumor necrosis factor (TNF) - growth inhibitor that suppresses cell division; toxic to tumor cells, and may have a role in regulating several developmental processes.

16.22 During the initial phase of a prolonged fast, blood glucose and insulin levels fall, and glucagon release is triggered. Glucagon acts to prevent hypoglycemia by promoting glycogenolysis and gluconeogenesis. The amino acids derived from muscle protein are a major source of the carbon skeleton substrates in gluconeogenesis.

16.23 The metabolism of glutamine and glutamate generates ammonia which leaves the kidney in the urine, taking with it a proton. This process, along with the active transport of protons down a sodium gradient and into the kidney tubules, helps to maintain the blood pH at 7.4.

16.25 a. mTOR – mammalian target of rapamycin – a serine/threonine kinase that integrates the input of pathways regulated by several hormones (e.g., insulin) and growth factors (e.g., IGF-1 and IGF -2)

b. SREBP-1c – sterol regulatory element 1c – a transcription factor that regulates genes involved with fatty acid metabolism

c. PDK1 - PIP_3-dependent kinase – an enzyme that activates several kinases such as PKB and PKC

d. TSH – (thyrotropin; thyroid-stimulating hormone) – stimulates the synthesis of thyroid hormone

e. interleukin – regulates the immune system by stimulating the division of T-cells in response to an antigen; promotes cell growth and differentiation

16.26 The most common sites on proteins that are phosphorylated during signal transduction cascades are the hydroxyl groups of serine and tyrosine.

16.28 Obesity contributes to the onset of diabetes by promoting tissue insensitivity to insulin. Specifically, this process begins as enlarged adipocytes release free fatty acids into the bloodstream and are taken up by other cells, causing the disruption of signal transduction pathways. Consequences include: lipotoxicity, high blood insulin levels, excess glucose production (liver), inhibited insulin-mediated glucose uptake by muscle, and further release of fatty acids from adipocytes that have developed insulin resistance. Adipose tissue in the obese develops a low-level chronic inflammation, and as a result, adipocytes experience ER stress and oxidative stress, and release inflammatory cytokines that increases insulin resistance further. All of these factors contribute to the onset of diabetes.

16.29 In addition to carbohydrate metabolism, insulin also impacts lipid and protein metabolism by triggering several processes that are anabolic and/or ensure nutrient storage. These processes include: stimulating fat synthesis in the liver and fat storage in adipocytes, promoting amino acid uptake by cells (especially liver and muscle cells), stimulating protein synthesis in most tissues, decreasing lipolysis, and promoting satiety by inhibiting NPY/AgRP neurons and activating POMC neurons in the hypothalamus.

16.31 Hormones can be classified as peptides or polypeptides (composed of amino acids), amino acid derivatives, or as steroids (derived from cholesterol).

16.32 Steroid transport proteins include corticosteroid-binding globulin, sex hormone-binding protein, and albumin.

CHAPTER 16: SOLUTIONS TO FILL-IN-THE-BLANK QUESTIONS

16.34 Downregulation

16.35 Receptor tyrosine kinases

16.37 Atrial natriuretic factor

16.38 Pancreatic β cells

16.40 Glucosuria

16.41 Osmotic diuresis

16.43 Ketosis

CHAPTER 16: SOLUTIONS TO SHORT-ANSWER QUESTIONS

16.44 Insulin receptor substrate 1 (IRS-1) is one of the substrates of activated insulin receptor phosphorylation reactions. Once it is phosphorylated, IRS-1 activates several proteins including PI3K (phosphatidylinositol 3-kinase), which subsequently phosphorylates PIP_2, a minor component of the cell membrane, yielding PIP_3. Once PIP_3 has bound to PDK1 (PIP_3-dependent protein kinase), the activated protein proceeds to phosphorylate several kinases, including PKB. Activated PKB stimulates glycogen synthesis by contributing to the activation of glycogen synthase. It does so by inactivating GSK3, one of several kinases that inhibit glycogen synthase.

16.46 Both leptin deficiency and Prader-Willi syndrome (unregulated ghrelin synthesis) are genetic disorders of body-weight regulation. In both cases, patients are obese because of insatiable hunger.

16.47 In addition to promoting weight loss, adiponectin decreases the risk of type II diabetes by improving glucose-stimulated insulin release and insulin sensitivity.

CHAPTER 16: SOLUTIONS TO THOUGHT QUESTIONS

16.49 Flies are well known for their preference for sugar. If urine has detectable sugar in it, flies will be attracted.

16.50 Release of fatty acids by enlarged adipocytes and their subsequent nonspecific insertion in cell membranes causes the disruption of signal transduction pathways such as the insulin receptor pathway.

16.52 Increased mobilization of fatty acids provides an alternate energy source for muscle, thereby sparing glucose for the brain. In addition, glucagon stimulates gluconeogenesis, a pathway that utilizes amino acids derived from muscle.

16.53 The recognition by the conscious centers in the brain that danger is imminent results in a discharge of epinephrine from the sympathetic nervous system and the adrenal medulla. The large amounts of glucose and fatty acids that flood into blood as a consequence of epinephrine's stimulation of glycogenolysis and lipolysis have several effects on the body. One effect, high blood glucose levels, provides the energy required for rapid decision-making processes in the brain. In addition, large quantities of glucose and fatty acids are required for strenuous physical activity if a decision is made to run away from the danger.

16.55 In uncontrolled diabetes mellitus, massive breakdown of fat reserve results in the production of large amounts of ketone bodies. Two of the ketone bodies (acetoacetic acid and β-hydroxybutyric acid) are weak acids. The release of hydrogen ions from large numbers of these molecules overwhelms the body's buffering capacity.

16.56 To ensure proper control of metabolism, powerful hormones are synthesized in small quantities. Hormones also elicit responses in only specific target cells. They are metabolized quickly to ensure the precision of metabolic regulation.

19.58 The steroid molecule is covalently bound to the matrix in a chromatographic column. The extract suspected of containing a steroid binding protein is then passed through the column. Any proteins remaining on the column are eluted by changing the salt concentration of the eluting buffer. After isolation and purification such proteins can be examined specifically for binding activity to the steroid.

16.59 While muscle does not synthesize fatty acids it does oxidize them. The product of the carboxylation of acetyl-CoA is malonyl-CoA. When cellular ATP and NADH levels are high, malonyl-CoA has the effect of inhibiting the oxidation of fatty acids because it inhibits carnitine acyl transferase I, the enzyme involved in the transport of fatty acids into the mitochondria.

16.61 The fatty acid components of triacylglycerol molecules cannot be converted into glucose molecules in animals because these organisms lack the glyoxylate cycle enzymes isocitrate lyase and malate synthase. These enzymes convert acetyl-CoA to malate, the citric acid cycle intermediate. (Refer to Figure 9.17). Malate is then converted to oxaloacetate, a substrate for gluconeogenesis.

16.62 The appetite center in the brain is integrated with taste receptor cells. If sweet taste receptors indicate that sugar has been consumed and the glucostat in the hypothalamus does not detect a rise in blood glucose ; the brain over-compensates for the discrepancy by stimulating appetite for carbohydrae-containing foods. Hence, the use of artificial sweeteners leads to weight gain.

16.64 Effects of starvation include the breakdown of muscle protein and the use of amino acids to generate energy. Large amounts of amino nitrogen are excreted as urea.

16.65 Leptin is a satiety–promoting adipokine that is synthesized in adipocytes in proportion to adipose tissue mass. Leptin's actions within the appetite center in the arcuate nucleus of the hypothalamus include stimulation of appetite-depressing POMC neurons and inhibition of the appetite-enhancing AgRP/NPY neurons. Weight control is promoted when leptin is produced in appropriate amounts and leptin signaling pathway in the brain is working properly. If leptin is not produced in adequate amounts in relation to fat stores, or if there is leptin resistance, weight control is dysfunctional.

16.67 Several processes may trigger the same response. For example, glucagon is released into the blood when blood glucose levels are low between meals, and epinephrine, the hormone released under threatening circumstances, triggers the quick release of glucose into blood. The overlap in function of these and other sets of hormones allows an effective response to what may be subtle differences in physiological conditions.

16.68 IRS1 is one of the numerous proteins in the signal transduction mechanisms triggered by insulin in target tissues. If IRS1 is defective on in low concentration, the expression of several genes will be altered. A defective IRS1 mechanism, combined with obesity, itself a risk factor for insulin resistance, results in dysfunctional insulin signaling pathway and abnormal insulin receptor processing. Type II diabetes is the foreseen consequence of these anomalies.

16.70 Blood glucose is the immediate source of energy for a large number of tissues (e.g. brain and muscle). As blood glucose levels fall the liver releases glucose derived from glycogen (17.2 kJ/g) or gluconeogenesis. The body's long-term energy is stored in triacylglycerol molecules (38.9 kJ/g), which are stored in significantly higher amounts than those of glycogen. The fat reserves of an average person may last for an extended period of time. The amino acids from muscle protein, and triacylglycerol molecules are mobilized more slowly than glucose. In the average body, there is sufficient glucose available from muscle and liver glycogen for short term needs.

16.71 Under the stated conditions, the fat stores will last 70.5 days.

16.73 Since the brain relies exclusively on glucose for energy, excessively high doses of insulin will result in the death of brain cells.

16.74 Fructose is degraded in the liver when its metabolism in the glycolytic pathway is essentially unregulated. As a result, fructose is rapidly converted to phosphorylated trioses causing increased AMP levels which triggers uric acid synthesis. High blood uric acid levels have been experimentally and clinically linked to damaged blood vessels leading to hypertension.

17 Nucleic Acids

Brief Outline of Key Terms and Concepts

OVERVIEW

GENETICS; MOLECULAR BIOLOGY; CENTRAL DOGMA
REPLICATION; GENE; GENOME
TRANSCRIPTION; TRANSCRIPT; TRANSCRIPTOME
TRANSLATION; PROTEIN; PROTEOME
GENE EXPRESSION; METABOLOME

17.1 DNA

STRUCTURAL FEATURES OF DNA

Two antiparallel polynucleotide strands, held together by hydrogen bonding between complementary base pairs, wind around each other to form a relatively stable right-handed double helix, with nitrogenous bases facing toward the interior and a negatively charged sugar-phosphate backbone.

DNA STRUCTURE: THE NATURE OF MUTATION

DNA is vulnerable to certain types of disruptive force that can result in mutations, permanent changes in its base sequence.

POINT MUTATIONS: TRANSITION MUTATION; TRANSVERSION MUTATION;
MUTAGENIC FACTORS
TAUTOMERIC SHIFTS; DEPURINATION; DEAMINATION
THYMINE DIMERS
XENOBIOTICS: BASE ANALOGUES, ALKYLATING AGENTS, NONALKYLATING AGENTS, INTERCALATING AGENTS

DNA STRUCTURE: THE GENETIC MATERIAL: FROM MENDEL'S GARDEN TO WATSON AND CRICK

The model of DNA structure proposed by James Watson and Francis Crick in 1953 was based on information derived from the efforts of many individuals.

PHAGE (BACTERIOPHAGE); CHARGAFF'S RULES

DNA STRUCTURE: VARIATIONS ON A THEME

B-DNA; A-DNA (LESS HYDRATED); Z-DNA (ZIGZAG, LEFT-HANDED HELIX AND SLIMMER)

Higher order structures: **CRUCIFORMS; INVERTED REPEATS** (palindromes); **PROTOCRUCIFORMS; SUPERCOILING**

DNA SUPERCOILING

NEGATIVE SUPERCOILING; POSITIVE SUPERCOILING; NICK; TOPOISOMERASES; DNA GYRASE

CHROMOSOMES AND CHROMATIN

Prokaryotic chromosome: a supercoiled circular DNA molecule complexed to a protein core.

Eukaryotic chromosome: a single linear DNA molecule complexed with histones; **NUCLEOHISTONE; NUCLEOSOME; ORGANELLE DNA**

GENOME STRUCTURE

Each organism's **GENOME** organizes and stores the information required to direct living processes. Genomes from different types of organisms differ in their sizes and levels of complexity.

TANDEM REPEATS

17.2 RNA

RNA is involved in various aspects of protein synthesis and in the regulation of gene expression. The most abundant types of RNA are **TRANSFER RNA**, **RIBOSOMAL RNA**, and **MESSENGER RNA**.

TRANSFER RNA (tRNA) molecules have specific amino acids attached to them by specific enzymes and transport these to the ribosome, where they are properly aligned during protein synthesis into newly synthesized protein.

RIBOSOMAL RNAS (rRNA) are components of ribosomes, the sites of catalytic activity.

MESSENGER RNA (mRNA) contains within its nucleotide sequence the coding instructions for synthesizing a specific polypeptide.

NONCODING RNA

Several classes of **NONCODING RNAS** have diverse roles in genome regulation and protection. Important **NONCODING RNAS** include **miRNAs (micro RNAs)**, **siRNAs (small interfering RNAs)**, **snoRNAs**, and **snRNAs**.

17.3 VIRUSES

THE STRUCTURE OF VIRUSES

Viruses are composed of nucleic acid enclosed in a protective coat. The nucleic acid may be a single- or double-stranded DNA or RNA. In simple viruses the protective coat, called a **CAPSID**, is composed of protein. In more complex viruses the **NUCLEOCAPSID**, composed of nucleic acid and protein, is surrounded by a membranous envelope derived from host cell membrane.

RETROVIRUS; REVERSE TRANSCRIPTASE

BACTERIOPHAGE T4: A VIRAL LIFESTYLE

The lifestyle of the T4 bacteriophage involves lysogeny and the lytic cycle

OVERVIEW

DNA contains all of an organism's genetic information[1] encoded as a series of nitrogenous bases (purines and pyrimidines).

GENETICS is the scientific investigation of inheritance.

MOLECULAR BIOLOGY is the study of gene structure and genetic information processing.

THE CENTRAL DOGMA: REPLICATION, TRANSCRIPTION, AND TRANSLATION

The central dogma describes the flow of genetic information from DNA to RNA to proteins.

DNA → DNA	REPLICATION	GENE	GENOME
DNA → RNA	TRANSCRIPTION	TRANSCRIPT	TRANSCRIPTOME
RNA → PROTEINS	TRANSLATION	PROTEIN	PROTEOME

Regulation of these processes controls GENE EXPRESSION. METABOLOME

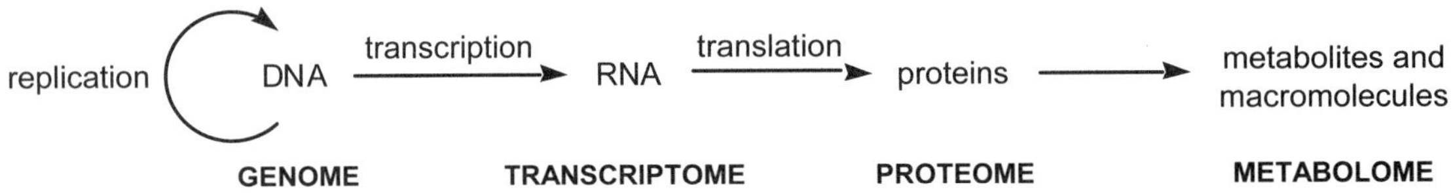

REVERSE TRANSCRIPTASE (in some viruses) is an important exception to the central dogma.

REVIEW: NUCLEOTIDES AND DINUCLEOTIDES

EXAMPLE OF A NUCLEOTIDE[2]

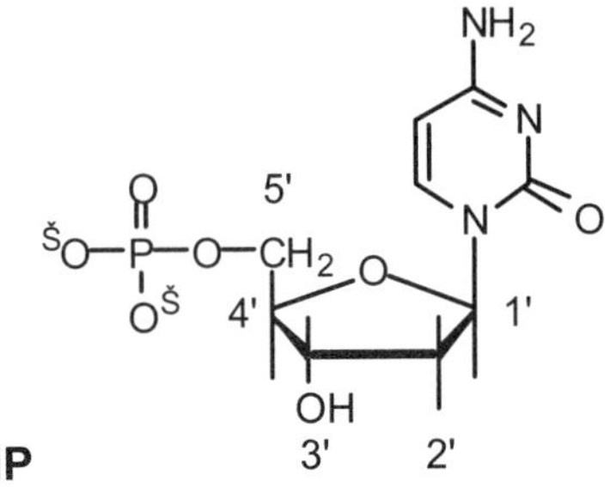

dCMP

Each 2′-deoxyribose has a nitrogenous base at its 1′ carbon and a phosphoryl group at its 5′ carbon. Nucleotides are connected by 3′,5′-phosphodiester bonds in DNA or RNA.

Compare the following functional groups:

ESTER **PHOSPHOESTER** **PHOSPHODIESTER**

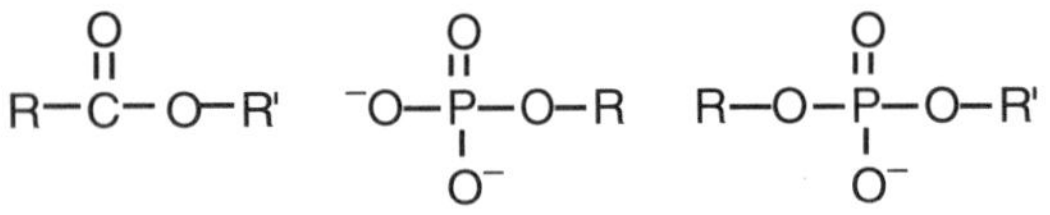

EXAMPLE OF A DINUCLEOTIDE

Note the phosphodiester bond between the 3′-OH on C and the 5′-phosphoryl group on A.

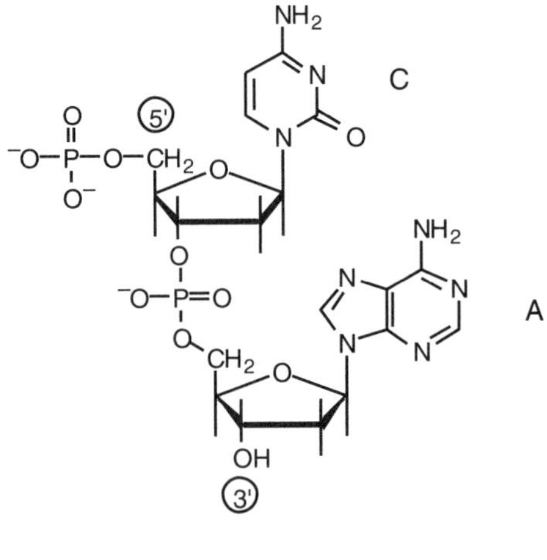

5'-CA-3'

---

[1] Some viruses use RNA rather than DNA for their genetic information.

[2] Remember the difference between a nucleoside and a nucleotide: the -Side is the Sugar + base, and the -Tide is the Total Package, with the Phosphate. Review the structures of purine and pyrimidine bases, nucleosides, and nucleotides in Chapter 14.

When writing a DNA sequence, the order of nucleotides must be direction specific.

DNA function is based upon the nucleotides being in a precise order, proceeding from the 5′ end to the 3′ end. (*5′ end* and *3′ end* refer to the open positions in the sugars at each end of the chain.)

The direction of one DNA strand can be noted as either 5′ → 3′ or 3′ → 5′. If the direction is not specified, it is assumed to be 5′ → 3′.

Nucleic acid sequences are always 5′ → 3′ unless noted otherwise.

For example, the strand on the right can be written as any of the following:

5′-CATG-3′ CATG 3′-GTAC-5′

5′-CATG-3′

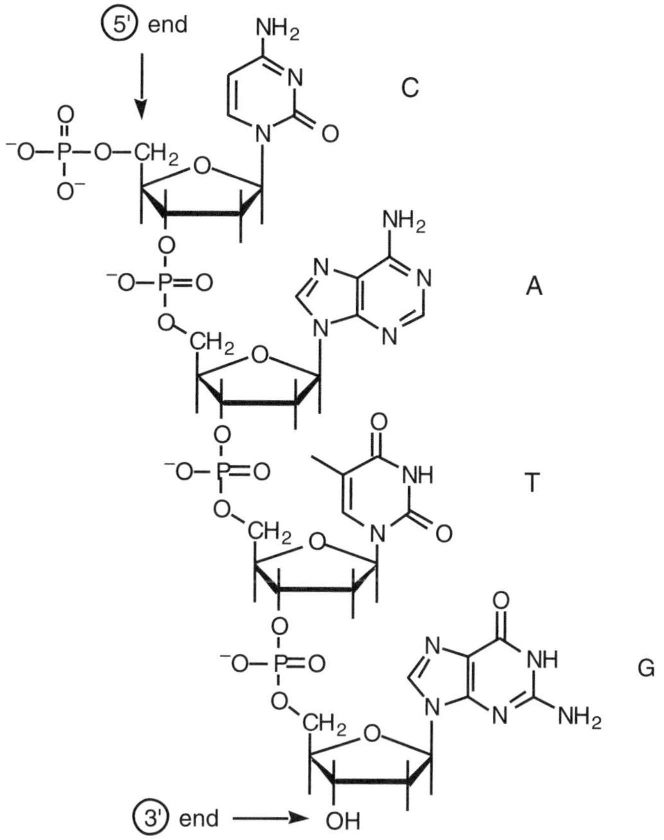

17.1 DNA: A RIGHT-HANDED, ANTIPARALLEL DOUBLE HELIX

Two **ANTIPARALLEL** polynucleotide strands, held together by hydrogen bonding between complementary base pairs, wind around each other to form a relatively stable right-handed **DOUBLE HELIX**.

Nitrogenous bases face toward the interior of the helix. The negatively charged **SUGAR-PHOSPHATE BACKBONE** faces outward.

Antiparallel chains run in opposite directions. One strand is 5′ → 3′ and the other strand is 3′ → 5′.

BASE-PAIRING ALLOWS DNA TO TRANSFER INFORMATION.
Base-pairing brings two DNA strands together, and the chains are held together by hydrogen bonds between the bases.

BASE PAIR (bp)—two bases that are on opposite strands of DNA and have hydrogen bonds between them

Possible DNA base pairs: A:::T (two hydrogen bonds, noted as AT)
C⁝⁝⁝G (three hydrogen bonds, noted as GC)

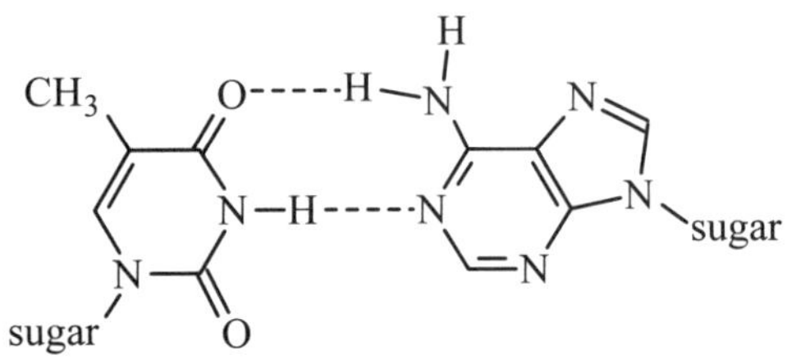

Thymine (T) Adenine (A)

AT base pair

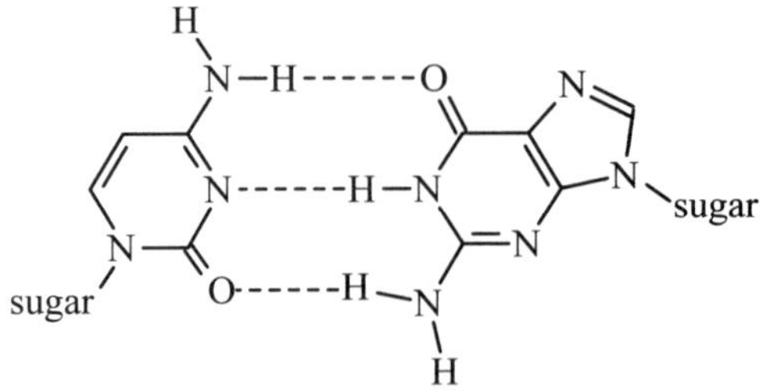

Cytosine (C) Guanine (G)

GC base pair

DNA STRANDS ARE COMPLEMENTARY

Because base-pairing is specific, the strands are complementary and can serve as templates for each other. So, if the sequence of one strand is known, then the sequence of the other strand can be determined.

Complementary copies can be made from a single strand. One strand serves as the template for synthesis of the other strand.

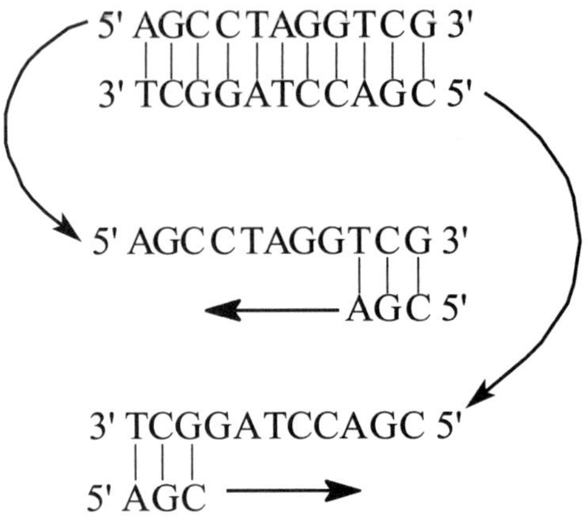

FORCES THAT STABILIZE THE DOUBLE HELIX

1. **HYDROPHOBIC INTERACTIONS:** The double helix structure allows the hydrophobic purine and pyrimidine rings to reside in the interior, away from water. Recall that hydrophobic interactions are driven by the significant increase in the entropy of the surrounding water, which increases the total entropy of the system.
2. **HYDROGEN BONDS:** Hydrogen bonding between bases (i.e., base-pairing), is *very* specific. A only base-pairs with T, and C only base-pairs with G. A purine-purine would be too bulky, and a pyrimidine-pyrimidine base pair would create a gap. Also, the combinations AT and CG have the most effective hydrogen bonding.
   T cannot effectively hydrogen bond with G, and the same holds true for C and A. Try this for yourself. Draw the structures of T and G, and try to form hydrogen bonds between them.
3. **BASE-STACKING:** The base pairs are parallel to each other and therefore allow van der Waal contacts between them. With just a few bases, the force is *very* weak. But when a DNA molecule has ≈10,000 bases, the cumulative effect is great.
4. **HYDRATION:** DNA binds a significant amount of water at its phosphoryl groups, 3′- and 5′-oxygens, and electronegative atoms in the nucleotide bases. B-DNA can bind about 18-19 H_2O molecules per nucleotide (measured under laboratory conditions).
5. **ELECTROSTATIC INTERACTIONS:** The pK_a of a phosphodiester is about 2, so at pH ≈7, DNA is a polyanion with many negative charges along its sugar-phosphate backbone. To relieve some of the repulsion between adjacent negative charges, the charged backbone is stabilized, or shielded, by cations such as Mg^{2+}, polyamines, and cationic proteins (proteins with basic amino acids such as arginine or lysine; e.g., histones) and H_2O.

DNA STRUCTURE: THE NATURE OF MUTATION

MUTATIONS are permanent changes in the base sequence of DNA molecules.

POINT MUTATIONS involve a single base pair. Types of point mutations include:

TRANSITION MUTATION: a pyrimidine substitutes for another pyrimidine or a purine substitutes for another purine.

TRANSVERSION MUTATION: a pyrimidine substitutes for a purine or vice versa.

MUTAGENIC FACTORS: WHAT CAUSES MUTATIONS?

SPONTANEOUS CHEMICAL REACTIONS

TAUTOMERIC SHIFTS that occur during DNA replication can cause base-mispairing and transition mutation. (Remember that tautomeric shifts are the spontaneous interconversion of amino and imino groups or of keto and enol groups.)

DEPURINATION REACTIONS: spontaneous hydrolysis that cleaves a purine from its sugar. One cause is the protonation of guanine's N-3 and N-7.

DEAMINATION REACTIONS: For example, C can change to U via deamination and a tautomeric shift. The resulting U would base-pair with A (rather than the G that was supposed to base-pair with C).

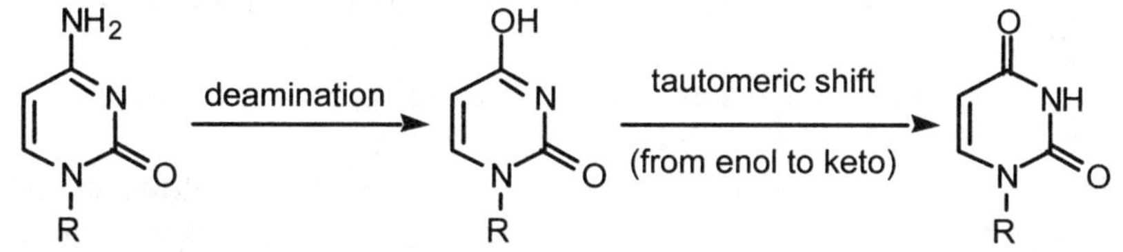

C U

IONIZING RADIATION SUCH AS UV, X-RAYS, AND γ-RAYS

Free-radical mechanisms cause strand breaks, DNA-protein crosslinking, ring openings, base modifications (thymine glycol, 5-hydroxymethyl uracil, 8-hydroxyguanine), and the formation of pyrimidine dimers. **THYMINE DIMERS** are the most common and are caused by UV energy absorption by double bonds. Thymine dimers distort the helix and stall DNA synthesis.

XENOBIOTICS:

BASE ANALOGUES have structures that are similar to normal nucleotide bases and can be inadvertently incorporated into DNA.

ALKYLATING AGENTS add carbon-containing alkyl groups that may result in a transversion mutation (methylguanine) or may promote tautomer formation. Examples include dimethylsulfate, dimethylnitrosamine, and mitomycin C.

NONALKYLATING AGENTS modify DNA structure in other ways and typically prevent base-pairing. Examples include the following: HNO_2 deaminates bases and benzo[a]pyrene is metabolized to a highly reactive intermediate that forms adducts to bases.

INTERCALATING AGENTS can insert themselves between stacked base pairs, distorting the DNA double helix and causing deletion or insertion of base pairs (resulting in frame-shift mutations).

DNA Structure: The Genetic Material: From Mendel's Garden to Watson and Crick

Historical developments that led to the discovery of DNA structure:

1865	The scientific revolution that eventually necessitated the determination of DNA structure began when Gregor Mendel discovered the basic rules of inheritance.
1869	Friedrich Miescher discovered "nuclein," later renamed nucleic acid.
1882-1897	The chemical composition of DNA was determined largely by Albrecht Kossel.
1928	Fred Griffith proposed the concept of transmission of genetic information between bacterial cells caused by a "transforming agent."
1944	Avery and McCarty demonstrated that the digestion of DNA by deoxyribonuclease inactivated the transforming agent, concluding that genetic information was carried by DNA.
1952	By using T2 bacteriophage, Hershey and Chase demonstrated the separate functions of viral nucleic acid and protein. These experiments reconfirmed DNA as the genetic material.

Information Used by Watson and Crick to Determine DNA Structure

1. The chemical structures and dimensions of the nucleotides had been elucidated.
2. Adenine:thymine and guanine:cytosine exist as 1:1 ratios in DNA. (**Chargaff's Rules**)
3. X-ray diffraction studies performed by Rosalind Franklin indicated that DNA was a symmetrical molecule, in all likelihood a helix.
4. The diameter and pitch of the helix were estimated by Wilkins and Stokes.
5. Linus Pauling had recently shown that protein, another biopolymer, could exist as a helix.

DNA Structure: Variations on a Theme

B-DNA

Right-handed helix; base pairs are at right angles to the backbone axis.

2.37 nm	diameter (this distance allows *only* purine-pyrimidine base pairing)
10.5 bp	base pairs per turn of the helix (may vary slightly with pH and salt concentration)
3.32 nm	length per turn of the helix
0.34 nm	distance between adjacent base pairs

A-DNA

Right-handed helix; base pairs tilt 20° away from the horizontal. A-DNA occurs when DNA becomes partially dehydrated (about 13-14 H_2O/nucleotides, as opposed to B-DNA's 18-19 H_2O/nucleotides), as when extracted with solvents such as ethanol.

Z-DNA

Left-handed helix; base pairs are in a zigzag conformation. Base pairs with alternating pyrimidine-purine bases are most likely to be in Z-DNA configuration.

B-DNA	A-DNA	Z-DNA	
2.37 nm	2.55 nm	1.84 nm	Helix diameter
10.5 bp	11 bp	12 bp	Base pairs per turn of the helix
3.32 nm	2.4 nm	4.56 nm	Helix rise (length per turn of the helix)

HIGHER-ORDER STRUCTURES:

CRUCIFORMS—crosslike structures; occurence is most likely when a DNA sequence contains a palindrome—an inverted repeat. A protocruciform is a small bubble that eventually forms a cruciform.

DNA SUPERCOILING

SUPERCOILING packages large DNA into a compact form. One strand is nicked, the double helix is overwound or underwound, and then the nicked strand is resealed.

NEGATIVE SUPERCOILING:

Underwound DNA twists to the right to relieve strain.
DNA winds around itself to form an interwound supercoil.
Negative supercoiling stores potential energy in the form of torque, which facilitates strand separation during replication and transcription.

POSITIVE SUPERCOILING:

Overwound DNA twists to the left to relieve strain.
DNA winds around a protein core to form a toroidal[3] supercoil.
Positive supercoiling interferes with replication and must be removed by topoisomerases such as DNA gyrase.

CHROMOSOMES AND CHROMATIN

PROKARYOTES:

CHROMOSOME: circular DNA that is extensively looped and coiled. A **NUCLEOID** is supercoiled DNA complexed with a protein core. Polyamines (e.g., spermidine and spermine) are polycations at cellular pH and bind to DNA's sugar-phosphate backbone to overcome the charge repulsion between adjacent DNA coils.

EUKARYOTES:

CHROMOSOME: one linear DNA molecule complexed with histones; structural units are the nucleosomes. (Humans possess 23 pairs of chromosomes.)

HISTONES—a group of small basic proteins found in all eukaryotes; positively charged side chains interact with DNA's negatively charged sugar-phosphate backbone. Five classes of histones: H1, H2A, H2B, H3, and H4. Nucleohistones[4] are histones that form complexes with DNA.

NUCLEOSOME—a complex of histones and DNA (about 145 bp):
—a positively supercoiled DNA segment (about 145 bp) that forms a toroidal coil around an octameric histone core.
—histone core: 2 H2A•H2B dimers and a $H3_2$•$H4_2$ tetramer; each histone contains a **HISTONE FOLD** (3 α-helices separated by 2 unstructured segments)
—H1, an additional histone that attaches to the coil in two places to hold the wrapping in place around the core.

A 60-bp segment of linker DNA connects adjacent nucleosomes.

Nucleosomes are coiled into **30-nm FIBERS**, which are further coiled to form higher order structures.

---

[3] A toroid is the shape of a donut.

[4] "Nucleohistone" is an older term that was used to describe the DNA-protein complexes that resulted from laboratory separations of cellular components. The term "nucleosome" reflects recent research results regarding DNA-histone structure.

CHROMATIN—partially decondensed form of chromosomes; occurs when cell is not dividing (interphase). Chromatin compacts to form chromosomes.

HETEROCHROMATIN; EUCHROMATIN

ORGANELLE DNA—in mitochondria and chloroplasts

GENOME STRUCTURE

GENOME—a complete set of genetic information, encoded in the nucleotide base sequence of DNA. Genes are contained within genomes.

PROKARYOTIC VS. EUKARYOTIC GENOMES

FEATURE:	PROKARYOTIC GENOMES	EUKARYOTIC GENOMES
GENOME SIZE	Relatively small; fewer genes	Relatively large (but size does not imply complexity of an organism)
CODING CAPACITY; CODING CONTINUITY	Genes are compact and continuous Little to no noncoding DNA	Most DNA is noncoding Most genes are discontinuous Intergenic sequences ("junk DNA") that do not code for gene products
GENE EXPRESSION MECHANISMS	Operon—set of functionally related genes, the transcription of which is regulated as a unit	Introns are interspersed between exons. Introns—noncoding sequences Exons—expressed sequences that encode for a gene product
OTHER UNIQUE FEATURES	PLASMIDS—additional DNA, typically circular, that codes for biomolecules that provide a growth or survival advantage	Chromosomes contain repetitive sequences: tandem repeats, interspersed genome-wide repeats

TYPES OF REPETITIVE SEQUENCES IN EUKARYOTIC GENOMES

TANDEM REPEATS (SATELLITE DNA)

—multiple copies arranged next to each other

—certain types play structural roles in centromeres[5] and telomeres[6]

TYPES OF RELATIVELY SMALL TANDEM REPEATS

MINISATELLITES—10- to 100-bp repeat sequences; total = 10^2-10^5 bp

MICROSATELLITES—**SINGLE-SEQUENCE REPEATS (SSR)**—1- to 4-bp repeated 10-100 times; vary with individuals; used as markers in genetic disease diagnosis, forensics investigations; also called **SHORT TANDEM REPEATS (STR)**

INTERSPERSED GENOME-WIDE REPEATS

—repetitive sequences scattered around the genome

—most result from transposition.

In transposition, transposons (transposable DNA elements) excise themselves and insert at another site (or involve RNA transposons or retrotransposons).

---

[5] Centromeres are structures that contain kinetochores and attach chromosomes to the mitotic spindle during mitosis and meiosis.

[6] Telomeres are structures at the end of chromosomes that buffer the loss of critical coding sequences after a round of DNA replication.

LINEs and SINEs: LONG and SHORT INTERSPERSED NUCLEAR ELEMENTS

LINEs are retrotransposons longer than 5000 bp. About 1 of every 1200 mutations is due to LINE insertions.

SINEs have less than 500 bp and need a functional LINE sequence to undergo transposition. About 11% of the sequences in the human genome are SINEs.

ALU ELEMENTS—the only SINE insertions linked to human disease;
—mediate chromosome rearrangements, insertions, deletions, and recombinations
—cause mutations that result in genetic diseases such as hemophilia B, Lesch-Nyhan disease, and Tay-Sachs disease

17.2 RNA: STRUCTURE, FUNCTION, AND SYNTHESIS (TRANSCRIPTION)

DIFFERENCES BETWEEN RNA AND DNA:

1. The sugar of RNA is ribose; so the 2′ –OH group makes RNA more reactive than DNA.
2. Uracil replaces thymine. So, the base pairs in RNA are AU and GC.
3. RNA is a single strand (not a double helix), so it can coil back on itself and form unique and complex 3D structures. RNA does not follow Chargaff's rules.
4. RNA's 3D structure can form binding pockets, which gives some RNAs catalytic properties. Mg^{2+} is usually needed in RNA catalysis to stabilize transition states. Ribozymes are catalytic RNA molecules; most catalyze cleavage of itself or of other RNAs.

TYPES OF RNA:

Of the total RNA of the cell, ≈80% is rRNA, ≈15% is tRNA, and ≈5% is mRNA.

1. **TRANSFER (tRNA)**—carries individual amino acids to ribosomes to be properly aligned and assembled into proteins.
2. **RIBOSOMAL (rRNA)**—an integral part of ribosomes.
3. **MESSENGER (mRNA)**—carries the genetic information for protein synthesis from the nucleus to the ribosomes. mRNA molecules are copies of DNA "genes" and contain the genetic information needed to synthesize specific polypeptides.
4. **NONCODING RNA (ncRNA)**—does not code for proteins or RNA.

TRANSFER RNA (tRNA): ADAPTER MOLECULES IN PROTEIN SYNTHESIS

tRNA are single-stranded polynucleotide chains about 75 nucleotides long that convert a nucleotide sequence into a specific amino acid.

AMINOACYL-tRNA SYNTHETASES link each amino acid to the **3′-TERMINUS** on its specific tRNA.

tRNA carries the amino acid to the ribosomes to be aligned, ordered, and assembled into proteins.

tRNA contains modified bases that help to stabilize and protect tRNA from degradation. These modifications can also occur in rRNA.

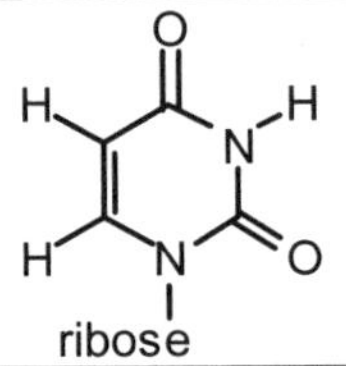	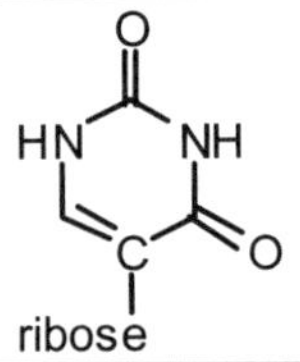	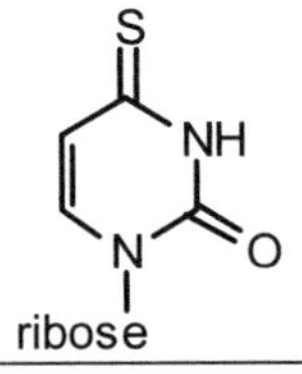	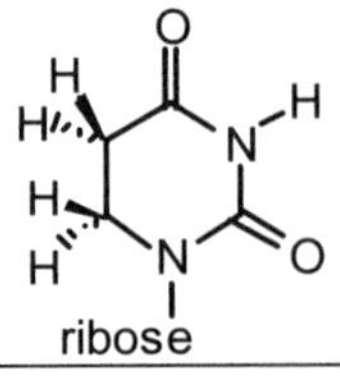	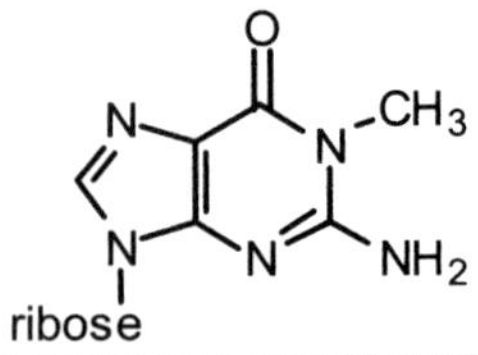
uridine	pseudouridine (ψ)	4-thiouridine	dihydrouridine	1-methyl guanosine
(shown for comparison)	Attachment to ribose is via a carbon atom.	S replaces a carbonyl O.		

THE SHAPE OF tRNA RESEMBLES A WARPED CLOVERLEAF WITH THE FOLLOWING FEATURES:

SPECIFIC—bonds to a specific amino acid

ANTICODON LOOP—contains a 3-bp sequence that is complementary to the DNA triplet code for a specific amino acid.

D LOOP—contains dihydrouridine

TψC LOOP—contains thymine, pseudouridine (ψ), and cytosine

VARIABLE LOOP—length varies between 4-5 and 20.

RIBOSOMAL RNA (rRNA) IS AN INTEGRAL COMPONENT OF RIBOSOMES

Ribosomes are cytoplasmic ribonucleoprotein structures that synthesize proteins and consist of 60-65% rRNA (the remainder is protein). Ribosomes consist of two subunits of unequal size, with several different kinds of rRNA and protein in each subunit.

MESSENGER RNA (mRNA) CARRIES DNA'S GENETIC INFO FOR PROTEIN SYNTHESIS

mRNA specifies the order of amino acids in a protein.

CISTRON—a DNA sequence that contains coding information for a polypeptide plus signals needed for ribosome function

PROKARYOTIC VS. EUKARYOTIC mRNA:

POLYCISTRONIC (prokaryotes) vs. **MONOCISTRONIC** (eukaryotes): contain information for many (poly) or one (mono) polypeptide chain(s).

Prokaryotic mRNA is not processed further and can be translated immediately. Eukaryotic mRNA is extensively modified, including

- Capping the 5′ end with a 7-methylguanosine
- Splicing (removing introns)
- Attaching a poly (A) tail (polyadenylate) to the 3′ end

NONCODING RNA (ncRNA)

miRNA	micro RNA
siRNA	small interfering RNA
snoRNA	small nucleolar RNA
snRNA	small nuclear RNA—involved in splicing and other RNA processing; combine with proteins to form small nuclear ribonucleoprotein particles (snRNPs or snurps)

SPLICEOSOMES are molecular machines made of snRNPs and other proteins. Spliceosomes splice by excising introns from pre-mRNA and join the exons together.

17.3 VIRUSES

VIRUSES: OBLIGATE, INTRACELLULAR PARASITES OR MOBILE GENETIC ELEMENTS?

Each virus contains a piece of nucleic acid within a protective coat. When a virus infects a host cell, its nucleic acid hijacks the cell's nucleic acid and protein-synthesizing machinery, directing the synthesis of complete new viral particles that are released from the host cell (often rupturing the cell in the process). Or the viral nucleic acid may insert itself into a host chromosome, resulting in cell transformation.

Viral infection disrupts cell function. By suppressing some cellular genes and activating others, the viral genome directs the host cell to produce new virus in a process that often results in cell death.

WHAT PROPERTIES OF VIRUSES MAKE THEM USEFUL RESEARCH TOOLS?

Because a virus subverts normal cell function to produce more viruses, a viral infection can provide unique insight into cellular metabolism. Several eukaryotic genetic mechanisms have been elucidated with the aid of viruses and/or viral enzymes. Viral research has also provided substantial information concerning genome structure and carcinogenesis. Viruses have also been invaluable in the development of recombinant DNA technology.

THE STRUCTURE OF VIRUSES

VIRIONS—complete viral particles = nucleic acid + capsid (+ membrane envelope in more complex viruses)

- **CAPSID**—a protein coat made of **CAPSOMERES** (interlocking proteins)
- **NUCLEOCAPSID** = nucleic acid + capsid

TYPES OF VIRAL GENOMES

Most common: double-stranded DNA (**dsDNA**), single-stranded RNA (**ssRNA**)

- Positive-sense RNA genome [(+)-ssRNA]
- Negative-sense RNA genome [(–)-ssRNA]: Viruses with (–)-ssRNA genomes must provide reverse transcriptase.

Also observed: single-stranded DNA (**ssDNA**), double-stranded RNA (**dsRNA**)

AFTER MANY YEARS OF RESEARCH AND EXPENDING BILLIONS OF DOLLARS, WHY IS AIDS STILL CONSIDERED AN INCURABLE DISEASE?

HIV is a retrovirus that causes AIDS. Retroviruses are a class of RNA viruses that possess a reverse transcriptase activity that converts their RNA genome to a DNA molecule. This DNA is then inserted into the host cell genome, causing a permanent infection.

Because the viral genome mutates frequently (i.e., its surface antigens become altered), a vaccine for the HIV virus has been difficult to develop. In the case of HIV, mutations occur because the reverse transcriptase doesn't have any proofreading capabilities. So, now and then it makes mistakes that are related to the surface antigens, which are continuously altered.

Similarly, that's why we need to get a flu shot every year. Flu viruses are very active and mutate constantly because of continuous replication. Their enzymes make mistakes that are beneficial for the survival of the virus.

CHAPTER 17: SOLUTIONS TO REVIEW QUESTIONS

17.1 a. molecular biology – the science devoted to elucidating the structure and function of genomes

b. genetics – the scientific investigation of inheritance

c. replication – the process in which an exact copy of parental DNA is synthesized using the polynucleotide strands of the parental DNA as templates

d. transcription – the process in which single–stranded RNA with a base sequence complementary to the template strand of DNA is synthesized

e. transcriptome – the complete set of RNA molecules that are produced within a cell

17.2 a. transcript – an RNA molecule that is produced by the transcription of a DNA sequence

b. proteome – the complete set of proteins produced within the cell

c. metabolome – the complete set of organic metabolites that are produced within as cell under the direction of the genome

d. double helix – a term used to describe DNA structure where two antiparallel strands of DNA intertwine via base pairing to form a helical structure.

e. base stacking – the parallel stacking of base pairs in the double helix of DNA

17.4 a. single nucleotide polymorphism – SNPs – single nucleotide variations or point mutations that occur in at least 1% of the population

b. nonsense mutation – a point mutation that converts the code for an amino acid into a premature stop signal.

c. indel – insertion and deletion mutations; occurs when from one to thousands of bases are either inserted or removed from a DNA sequence

d. inversion – a mutation in which a deleted DNA fragment is reinserted into its original position, but in the opposite direction

e. translocation – a eukaryotic abnormality in which a DNA fragment from one chromosome inserts into a different position on the same chromosome or into a different chromosome

17.5 a. alkylating agent – an electrophile that attacks molecules that possess an unshared pair of electrons, adding alkyl groups

b. base analogue – a molecule with a structure similar to that of a normal nucleotide base; can be incorporated into DNA

c. nonalkylating agent – a molecule that can modify DNA structure; for example HNO_2 derived from nitrosamines or $NaNO_2$, deaminates adenine, guanine and cystosine

d. intercalating agent – a planar polycyclic aromatic molecule that distorts DNA structure by inserting between stacked base pairs

e. ethidium bromide – a mutation causing intercalating agent; used as a fluorescent tag in DNA staining techniques

17.7 a. A-DNA - a short compact DNA structure in which the base pairs are not at right angles to the helical axis; occurs when DNA becomes partially dehydrated

b. B-DNA – the commonly found form of DNA, as the sodium salt under highly humid conditions

c. Z-DNA – form of DNA that is twisted into a left–handed spiral; named for the conformation, which is slimmer than that of B-DNA

d. cruciform – cross-like structure that forms when a DNA sequence contains a palindrome

e. palindrome - a sequence that provides the same information whether it is read forward or backward

17.8 a. positive supercoiling – the right–handed DNA helix twisted in the right– handed direction twists to the left to relieve tension

b. negative supercoiling – the right-handed DNA helix twisted in a left handed direction, twists to the right to relieve strain; stores potential energy in the form of torque

c. polyamines – polycationic molecules that assist in chromosomal compression by binding to the negatively charged DNA backbone

d. chromatin – the complex of DNA and histones found in the nucleus of eukaryotic cells

e. nucleosome – a repeating structural element in eukaryotic chromosomes composed of a core of eight histone molecules around which about 140 base pairs of DNA are wrapped; an additional 60 base pairs connect adjacent nucleosomes

17.10 a. satellite DNA – DNA sequences that are highly repetitive; when genomic DNA is digested and centrifuged, a satellite band forms

b. transposition – a mechanism whereby certain DNA sequences, referred to as mobile genetic elements, can be duplicated and enabled to move within the genome

c. transposon - a transposable DNA element; a DNA segment that carries the genes required for transposition and moves about the chromosome

d. retrotransposons – RNA transposon; a mobile genetic element that requires an RNA transcript intermediate

e. endogenous retrovirus – an LTR (long terminal repeat) retrotransposon that is believed to be a decayed virus that is embedded in a eukaryotic genome

17.11 a. epigenetics – alterations in DNA methylation and histone covalent modifications that alter gene expression via interconversion of facultative heterchromatin and euchromatin

b. CpG island – a region of a genome where CpGs constitute more than 50% of the bases

c. epimutation – an abnormal epigenetic change in a genome

d. DNA methylation – methylation of carbon-5 of cytosine residues in CpGs

e. histone acetylation – the acetylation of specific lysine residues in histone tails; promotes transcription by facilitating access of DNA to transcription factors

17.13 a. hypochromic effect – the decrease in the absorption of UV light (260nm) that occurs when purine and pyrimidine bases are incorporated into base pairs in polynucleotide sequences

b. DNA denaturation – the disruption of the forces that hold the complementary strands of DNA together; promoted by heat, low salt concentration, and extremes in pH

c. restriction endonucleases – enzymes isolated from various bacteria that cut DNA at specific sequences

d. DNA hybridization – a laboratory technique in which fragments of single– stranded DNA from different sources anneal; the rate at which a DNA hybrid forms is a measure of the similarity of the two strands

e. Southern blotting – a laboratory technique in which radioactively labeled DNA or RNA profiles are used to locate a complementary sequence in a DNA digest

17.14 a. cistron – a DNA sequence that contains the coding information for a polypeptide and the signals required for ribosome function

b. operon – a set of linked genes that are regulated as a unit

c. miRNA – microRNAs – a type of noncoding RNA between 22- and 26-nt in length; involved in gene expression and regulation

d. siRNA – small interfering RNAs – a type of noncoding RNAs between 21–

to 23-nt in length; play a crucial role in RNA interference

e. snoRNAs – small nuclear RNAs – a type of noncoding RNAs , 70- to 300-nt in length, facilitate chemical modifications of rRNA within the nucleolus

17.16 Biological processes that are facilitated by supercoiling include packaging of DNA into compact forms (i.e., chromosomes), DNA replication and transcription.

17.17 Nucleosomes are the structural units of chromatin. Each nucleosome consists of a left-handed supercoiled DNA segment wound around eight histone molecules

17.19 Promoters are DNA sequences immediately before a gene that is recognized by RNA polymerase. Enhancers are DNA sequences that when bound to an activator transcription factor stimulates the activity of an RNA polymerase. Silencers are DNA sequences that inhibit the transcription of a nearby gene when bound to a repressor protein. An insulator DNA sequence bound to an insulator binding protein blocks the interaction between enhancers and promoters of neighboring genes.

17.20 The transition from B-DNA to Z-DNA can occur when the nucleotide base sequence is composed of alternating purine and pyrimidines (e.g., CGCGCG). Because alternate nucleotides can assume different conformations (syn or anti), these DNA segments form a left-handed helix. The phosphate groups in the backbone of this DNA conformation zigzag hence the name Z-DNA.

17.21 There are approximately 6 million base pairs in a single human cell. Assuming that there are 10^{14} body cells, the total length of the DNA in the human body is approximately 2×10^{11} km. This estimated length is about 1000 time greater than the distance for the earth to the sun. (Note that 1 nm is 10^{-9} m.)

17.22 According to Chargaff's rules, if a DNA sample contains 21% adenine then it also contains 21% thymine. If the A-T content is 42%, then the G-C content is 58%. Consequently the guanine and cytosine percentages are both 29% in the DNA sample.

17.23 Guanine – cytosine pairs contain three hydrogen bonds, while adenine – thymine contain only two. The greater the number of hydrogen bonds holding the DNA strands together, the higher the melting point will be.

17.25 The complementary DNA strand (written in the standard 5' to 3' direction is 5'-AACGATAACGGCCCCT-3'. The RNA strand is 5'-AACGAUAACGGCCCCU-3'

17.26 RNA forms three-dimensional structures by coiling back on itself, allowing its single strand to form double-stranded regions with stabilizing features that are analogous to DNA, namely, hydrogen bonding between complementary base pairs (A-U, G-U, and G-C) and base stacking. These double-stranded regions in RNA are primarily intrachain, but may occur between single strands from adjacent molecules. Additional stabilizing interactions may occur between double-stranded regions and free loops of RNA.

17.28 A DNA replication error will continue to be propagated as the replicated strands themselves undergo replication, and so on, for all subsequent "generations" of DNA molecules. A transcription error will result in only a small number of defective gene products, as opposed to all of the products of a mutated DNA sequence. Considering that a replication error, as opposed to a transcription error, is potentially permanent, a replication error would be expected to cause more cellular damage.

17.29 When an aromatic hydrocarbon intercalates between two stacked base pairs, either adjacent base pairs are deleted or new base pairs are inserted. Intercalating agents may also cause chromosomes to break.

17.31 Chargaff's rules do not apply to RNA because RNA is single-stranded.

17.32 The nucleophilic attack of the 3'-hydroxyl oxygen on the phosphorous of the α-phosphate group of a deoxyribonucleotide yields a proton.

17.34 DNA structure is stabilized by several types of noncovalent interactions. These include hydrophobic interactions between stacked bases, hydrogen bonds between the bases in each base pair, hydration of phosphate groups, weak van der Waals forces between parallel stacked bases, and the 3'- and 5'- oxygen atoms of ribose, and electrostatic interactions between phosphate groups and substances that minimize repulsive forces (water molecules, magnesium ions and polycationic molecules such as polyamines and histones).

17.35 Polyamines are polycationic molecules that bind to the negatively charged phosphate groups along the DNA backbone. Positively charged polyamines help to stabilize the highly compressed structure of a chromosome by relieving the electrostatic repulsion between the negatively charged, tightly coiled DNA molecules. Examples of polyamines are spermine and spermidine.

17.37 The information used to construct the Watson-Crick DNA model includes (1) the structures and molecular dimensions of deoxyribose, the nitrogenous bases, and phosphate; (2) Chargaff's rules – the 1:1 ratios of adenine and thymine and guanine and cytosine; (3) the x-ray diffraction images produced by Rosalind Franklin that indicated that DNA has a helical structure; (4) the helix diameter and pitch measurements provided by Maurice Wilkins; and (5) the precedent of the polypeptide helix, discovered by Linus Pauling, which indicated that helical structure was a biological phenomenon.

17.38 The genomes of prokaryotes are small when compared with those of eukaryotes. For example the genomes of *E. coli* and humans are 4.6 Mb and 3200 Mb (haploid), respectively. In contrast to prokaryotes which have modest coding capacities (i.e., genes are compact and continuous with little or no noncoding DNA), eukaryotes have enormous coding capacity. For example, it has been estimated that no more than 1.4% of the human genome codes for proteins. Prokaryotes and eukaryotes also differ in coding continuity. In contrast to prokaryotic genes, most eukaryotic genes are discontinuous. Noncoding sequences called introns are interspersed between coding sequences called exons.

17.40 Alu elements are the only SINE (sort interspersed nuclear elements) linked to human disease. By inserting into human chromosomes the retrotransposons Alu elements can cause mutations via chromosome rearrangements, indels, and DNA sequence recombinations.

Chapter 17: Solutions to Fill-in-the-Blank Questions

17.41 Metabolome

17.43 Silent

17.44 Nonsense mutation

17.46 Telomeres

17.47 Introns

17.49 Epimutations

17.50 Restriction enzymes

Chapter 17: Solutions to Short-Answer Questions

17.52 The change in coat color distribution resulted from varying degrees of hypomethylation of the CpG sites within the inserted genetic element near the AgRP gene.

17.53 The results of this experiment suggest that the methyl donors overrode the hypomethylation effects of BPA. In other words, a nutritious diet can provide some degree of protection against the effects of environmental toxins.

17.55 DNA stability is enhanced when the polycationic polyamine molecules associate with the negatively charged DNA stand backbone. Because polyamine binding reduces the number of DNA strand negative charges, the DNA molecule can more easily form supercoils.

Chapter 17: Solutions to Thought Questions

17.56 The principal structural difference between DNA and RNA is the 2′-OH group of ribose in RNA molecules. In DNA, which lacks the 2′-OH group in the deoxyribose sugar, hydrogen-bonded complementary strands can easily adopt the B-form double helix. In contrast, double-stranded regions of RNA molecules cannot adopt this conformation because of steric hindrance. Instead, they adopt the less compact A-helical form in which there are 11 bp per turn and the base pairs tilt 20° away from the horizontal.

17.58 RNA can coil back on itself to form complex three-dimensional structures.

17.59 Relaxed circular DNA with a nicked strand is less compact than supercoiled circular DNA and so has a lower effective density. As a result it will not migrate as far in the centrifuge tube as the supercoiled DNA.

17.61 Because nucleotide base sequences and amino acid sequences are such completely different "languages," a complex mechanism is required for the "translation" of one type of information into another. In the absence of any evidence to the contrary it does not appear likely that information expressed in proteins can be utilized to direct the synthesis of nucleic acids.

17.62 Nuclei and mitochondria are separately obtained from source tissue by means of cell homogenization followed by density gradient centrifugation. The nucleic acids in each organellar fraction are then extracted with the aid of detergents, solvents and proteases (to remove proteins). RNA is removed by treating each sample with RNase. The DNA from each type of organelle is then further purified by centrifugation.

17.64 Each cell is constantly receiving information from its environment. Cells adapt to changing conditions as information in the form of nutrients, hormones, growth factors, and other types of molecules triggers changes in the molecular mechanisms that ultimately cause changes in gene expression. The transcriptome, the set of mRNA molecules produced under specified conditions, is a measure of the current status of gene expression.

17.65 Each base in the nucleotides of DNA and RNA has a specific shape that has unique information content. The enormous number of possible sequences of the nucleotides make possible a very large coding capacity that living organisms use to specify the molecular structure of all of their biomolecules

17.67 The protein components of the electron transport system are located in the inner membrane of mitochondria. If these components are damaged they need to be replaced immediately. If the genes coding for these molecules are in the mitochondria then they can be easily synthesized as needed. If the genes were in the cell nucleus a time-consuming signal mechanism would be required to initiate gene expression. The newly synthesized proteins would then have to be delivered back to the mitochondrion and then transported across the outer mitochondrial membrane and inserted into the inner membrane.

17.68 Viruses frequently infect more than one host. Some of the nucleic acid sequences obtained from one or more hosts provide a virus particle with a selective advantage (e.g., making it easier to attack a host cell). For example, human influenza viruses frequently become more virulent after they have infected chickens and pigs.

17.70 When DNA molecules become alkylated, the chains break or misreading is facilitated, both of which can result in mutations. If such genetic changes cause the genes that control cell growth to be turned off, uncontrolled cell division will occur. Abnormal methylation of DNA sequences can result in epimutations.

17.71 During each brother's life, his epigenome will change in response to interactions with the environment. Examples include exposure to agents that change methylation of DNA bases. The older the brothers, the greater the differences will be. The two individuals could be distinguished by examining the epigenomic differences (e.g., by staining chromosomes to reveal DNA methylation patterns). This is not true of DNA base sequences which would remain the same in both individuals

17.73 The enolization of the fluorouracil will promote the wrong hydrogen bonding pattern (i.e. it will hydrogen bond with guanine instead of adenine) and a transition mutation will result. As these mutations accumulate rapidly dividing cells (the majority of which are cancer cells) are fatally damaged.

18 Genetic Information

Brief Outline of Key Terms and Concepts

18.1 GENETIC INFORMATION: REPLICATION, REPAIR, AND RECOMBINATION

DNA REPLICATION is DNA synthesis. **SEMICONSERVATIVE REPLICATION MECHANISM**—each strand of DNA serves as a template to synthesize a new strand. The **LEADING STRAND** is synthesized continuously; the **LAGGING STRAND** is synthesized in pieces which are then covalently linked. Both syntheses are in the **5′ → 3′ DIRECTION**. **REPLICATION FACTORIES; REPLISOMES**

DNA SYNTHESIS IN PROKARYOTES

Replication requires enzymatic activities in DNA unwinding (**TOPOISOMERASES, HELICASES, SSB; REPLICON, REPLICATION FORK, REPLICATION EYE**), **PRIMER** synthesis (**PRIMASE, PRIMOSOME**), **ELONGATION** (**DNA POLYMERASES, PROCESSIVITY, β-2-CLAMP, CLAMP LOADER**), supercoiling control and ligation (**DNA LIGASE**). Also **3′ → 5′ EXONUCLEASE ACTIVITY**; **5′ → 3′ EXONUCLEASE ACTIVITY**

DNA SYNTHESIS IN EUKARYOTES

Significant differences between DNA replication in eukaryotes versus prokaryotes include replication time and rate, replication origin numbers, **OKAZAKI FRAGMENT** size, and replication machinery structure. **PREINITIATION REPLICATION COMPLEX (preRC), ORIGIN OF REPLICATION COMPLEX (ORC), MCM COMPLEX (MINICHROMOSOME MAINTENANCE COMPLEX);**

REPLICATION LICENSING FACTORS (RLFs), REPLICATION PROTEIN A (RPA), REPLICATION FACTOR C (RFC);

TELOMERASE, TELOMERE END-BINDING PROTEINS (TEBPs), TELOMERE REPEAT-BINDING PROTEINS (TRFs)

DNA REPAIR

Each organism's survival depends on its capacity to repair DNA structural damage. Types of DNA repair mechanisms are direct single-stranded repairs such as **BASE EXCISION REPAIR, NUCLEOTIDE EXCISION REPAIR, PHOTOREACTIVATION (LIGHT-INDUCED) REPAIR,** transcription-coupled repair, and mismatch repair, and double-strand break repair such as nonhomologous end joining (NHEJ). The double-strand break repair model and the synthesis-dependent strand annealing model are forms of **RECOMBINATIONAL REPAIR.**

DNA RECOMBINATION

GENERAL RECOMBINATION—exchange of DNA sequences in homologous chromosomes

SITE-SPECIFIC RECOMBINATION, the exchange of DNA sequences, requires only short homologous sequences. DNA-protein interactions are responsible for the exchange of largely nonhomologous sequences.

DNA GLYCOSYLASE

TRANSFORMATION, CONJUGATION, TRANSDUCTION

TRANSPOSITION, the movement of genetic elements (**TRANSPOSONS**) from one place to another within a genome, can cause genetic changes such as insertions, deletions, and translocations. The movement of **RETROTRANSPOSONS**, found in large numbers in eukaryotic genomes, can cause disease or can provide opportunities for genetic diversity.

RETROELEMENT, RETROPOSON, TRANSPOSABLE ELEMENT

18.2 TRANSCRIPTION (SYNTHESIS OF RNA)

During transcription, an RNA molecule is synthesized from a DNA template (**CODING STRAND**).

TRANSCRIPTION IN PROKARYOTES

In prokaryotes transcription involves a single RNA polymerase activity. Transcription is initiated when the RNA polymerase complex binds to a specific DNA sequence called a **PROMOTER**. **CONSENSUS SEQUENCE**

TRANSCRIPTION IN EUKARYOTES

Transcription is significantly more complex. In addition to chromatin-remodeling and RNA-processing reactions, gene transcription requires the binding of unique sets of **TRANSCRIPTION FACTORS** to **PROMOTER SEQUENCES**. Eukaryotic RNA transcripts undergo several processing reactions.

Regulation of transcription differs significantly between prokaryotes and eukaryotes. Transcription processes observed only in eukaryotes include RNA processing, such as **CAPPING, POLY(A) TAIL** synthesis, and **RNA SPLICING. SPLICEOSOME; ACCEPTOR SITE; SUPRASPLICEOSOME**

18.3 GENE EXPRESSION

CONSTITUTIVE GENES are routinely transcribed, whereas **INDUCIBLE GENES** are transcribed only under appropriate circumstances.

GENE EXPRESSION IN PROKARYOTES

In prokaryotes, inducible genes and their regulatory sequences are grouped into **OPERONS**.

RIBOSWITCHES are metabolite-sensing domains in mRNAs that regulate the transport or synthesis of certain types of molecules.

GENE EXPRESSION IN EUKARYOTES

Mechanisms that control gene expression include DNA methylation, histone covalent modification, **CHROMATIN REMODELING**, RNA-processing reactions such as alternative splicing, **RNA EDITING**, RNA transport, and translational controls.

OVERVIEW

Molecular machinery, consisting primarily of DNA-binding proteins, bends, twists, unwinds, and unzips DNA in replication and transcription.

DNA-BINDING PROTEINS tend to have:
—a twofold axis of symmetry and often form dimers.
—very specific DNA binding via noncovalent interactions between amino acids and the edges of nucleotide base pairs along the major groove.
—one of the following supersecondary structures: helix-turn-helix, helix-loop-helix, leucine zipper, or zinc finger.

18.1 GENETIC INFORMATION: REPLICATION, REPAIR, AND RECOMBINATION

DNA REPLICATION: DNA POLYMERASE MAKES COPIES OF DNA

GENERAL FEATURES OF DNA REPLICATION:

1. The two strands of DNA are separated by unwinding to form a **REPLICATION FORK.**
2. **SEMICONSERVATIVE REPLICATION**: Each strand serves as a template for the synthesis of a complementary strand, so each new DNA molecule has one old strand and one new strand. (This was proven in the Meselson-Stahl experiment.) An existing DNA template is essential to provide the proper sequence of bases.
3. Deoxyribonucleotides are added to the 3′-OH of a pre-existing nucleic acid chain (either DNA or RNA) by forming new phosphodiester bonds.
4. Deoxyribonucleotide triphosphates (dNTPs) such as dATP, dGTP, dCTP, and dTTP are the substrates for **DNA POLYMERASE** (with Mg^{2+} as a cofactor).

$$(DNA)_n + dNTP \rightarrow (DNA)_{n+1} + PP_i$$

5. **ELONGATION** of the new complementary strand proceeds ***only*** in the 5′ → 3′ direction. Because of this, and because elongation can only occur at the 3′-OH of a pre-existing chain, one of the new strands (the **LEADING STRAND**) can be synthesized continuously, but the other strand (the **LAGGING STRAND**) must be synthesized in short segments (**OKAZAKI FRAGMENTS**), with new RNA primer being synthesized as the replication fork opens.

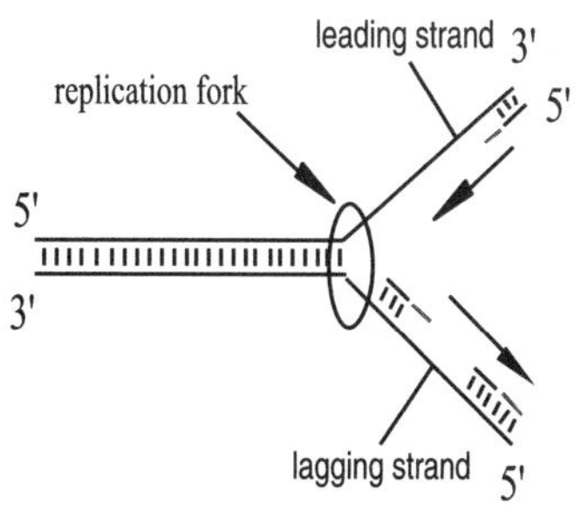

LEADING STRAND

The 5′ → 3′ direction is *toward* the replication fork. dNTPs can react to add to the continuous, growing, complementary strand as the replication fork opens.

LAGGING STRAND

The 5′ → 3′ direction proceeds *away from* the fork. DNA can only be synthesized in Okazaki fragments as the replication fork opens.

REPLICATION FACTORIES are specific nuclear compartments where DNA replication occurs. Replication begins when there are enough copies of **DnaA** (a DNA-binding protein) and when there is a high ATP/ADP ratio.

DNA SYNTHESIS IN PROKARYOTES (*E. COLI*)

1. **DNA UNWINDING UTILIZES THREE TYPES OF PROTEINS:**
   TOPOISOMERASES work ahead of the replication fork to relieve torque. (See supercoiling control, No. 5, below.)
   HELICASES unwind the DNA at the fork (and require ATP).
   SSB (SINGLE-STRANDED BINDING PROTEIN) stabilizes and protects single-stranded DNA segments.
   Prokaryotic replication begins at oriC (initiation site on *E. coli*) and proceeds in two directions, forming a **REPLICATION EYE**. **REPLICON**—a DNA molecule or segment that contains an initiation site and regulatory sequences.
2. **PRIMERS**—short RNA segments needed to initiate replication—are synthesized by **PRIMASE.** An RNA **PRIMER** provides a 3′-OH group, to which DNA polymerase can add its first nucleotide. The leading strand only needs one primer per replication fork, but the lagging strand needs a primer for each Okazaki fragment.
   PRIMOSOME: a multienzyme complex containing primase (an RNA polymerase) and several auxiliary proteins.
3. **DNA SYNTHESIS IS CATALYZED BY DNA POLYMERASES**
   REPLISOME—the prokaryotic DNA replicating machine—DNA unwinding proteins, the primosome, and two copies of the pol III holoenzyme.

DNA POLYMERASES are large multienzyme complexes that catalyze the formation of phosphodiester bonds between dNTPs in the 5′ → 3′ direction.

DNA POLYMERASE III (POL III) (in prokaryotes) has at least 10 subunits.

- **CORE POLYMERASE**—three subunits: α, ε, and θ: α subunit forms the phosphodiester bonds (5′ → 3′ polymerase); ε subunit has 3′ → 5′ exonuclease activity (see below); θ subunit function is unknown.
- **SUBUNIT τ** forms a dimer of 2 core polymerase.
- **β_2-CLAMP (β-PROTEIN** or **SLIDING CLAMP PROTEIN**; two subunits)—forms a ring around the template DNA strand; promotes **PROCESSIVITY** by preventing the dissociation of the polymerase and the template DNA during replication.
- **γ COMPLEX** (5 subunits)—transfers the β_2-clamp to the core polymerase.

The RNA primers used in DNA replication need to be removed and replaced by DNA. This is where the 5′ → 3′ exonuclease activity of **DNA POLYMERASE I (POL I)** comes into play. Pol I degrades RNA primers ahead of it as the enzyme synthesizes DNA. RNA primer sequences are thereby replaced with DNA sequences.

OTHER DNA POLYMERASES:

DNA POLYMERASE I (pol I, Kornberg enzyme)—DNA repair enzyme; also removes RNA primer.

DNA POLYMERASE II (pol II)—can reinitiate DNA synthesis beyond gaps caused by damaged segments.

EXONUCLEASE REMOVES NUCLEOTIDE(S) FROM AN END OF A STRAND

3′ → 5′ EXONUCLEASE ACTIVITY functions as a "proofreader." If pol III adds an incorrect base to the newly synthesized strand, the 3′ → 5′ exonuclease activity removes the mistake and lets the polymerase try again. All three DNA polymerases have 3′ → 5′ exonuclease activity.

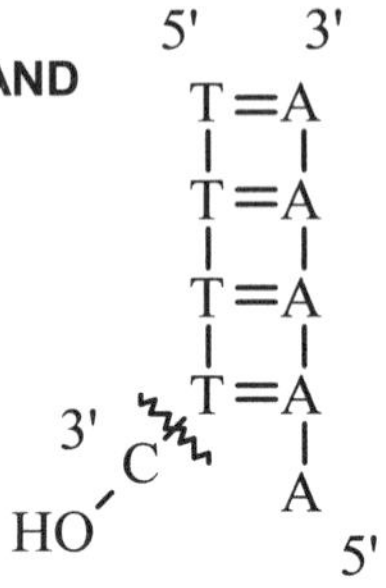

5′ → 3′ EXONUCLEASE ACTIVITY (POL I) removes primers or repairs damaged DNA.

Replication ends when replication forks meet at the **TER REGION** (termination site).

4. **DNA LIGASE JOINS DNA FRAGMENTS** by catalyzing phosphodiester bond formation. Discontinuous DNA synthesis of the lagging strand requires DNA ligase to join the newly synthesized fragments.
5. **SUPERCOILING CONTROL: DNA TOPOISOMERASES** prevent the DNA strands from tangling, which would prevent further unwinding of the double helix.
   DNA topoisomerases change the supercoiling of DNA by breaking one or both strands, passing the DNA through the break, and rejoining the strands. This process functions to relieve torque.
   Controlled supercoiling can facilitate DNA unzipping.
   TYPE I TOPOISOMERASES: make transient single-strand breaks in DNA
   TYPE II TOPOISOMERASES: make transient double-strand breaks
   DNA GYRASE: a prokaryotic type II topoisomerase that helps to separate the replication products and to create the negative supercoils needed for genome packaging.

DIFFERENCES BETWEEN PROKARYOTIC AND EUKARYOTIC DNA REPLICATION . . .

. . . appear to be related to the size and complexity of eukaryotic genomes.

TIMING OF REPLICATION: Eukaryotic cells only replicate DNA during a specific period of time during their cell cycle (the S phase). (The eukaryotic cell cycle includes times of rest—G0, G1, and G2—in addition to the S phase and the cell division phase.) In contrast, prokaryotes replicate DNA throughout most of their cell cycle.

REPLICATION RATE: Because of the complex structure of chromatin, DNA replication is significantly slower in eukaryotes (about 50 nucleotides/second per replication fork) than in prokaryotes (about 500 nucleotides/second per replication fork).

MULTIPLE REPLICONS are used by eukaryotes to compress the replication of their large genomes into short time periods. Instead of replisomes, eukaryotes have **REPLICATION FACTORIES**—immobilized sites that contain a large number of replication complexes. DNA is threaded through these complexes as it is synthesized.

OKAZAKI FRAGMENTS: Eukaryotic Okazaki fragments are significantly shorter (100-200 nucleotides) than prokaryotic fragments (1000-2000 nucleotides).

EUKARYOTIC REPLICATION ENZYMES

- Five types of eukaryotic DNA polymerase: α, β, δ, ε, and γ
- **DNA POLYMERASE α:** initiates synthesis of both leading and lagging strands
- **DNA POLYMERASE δ:** two complexes: one synthesizes the leading strand and one synthesizes the lagging strand; binds to PCNA, a sliding clamp protein that acts like β-protein in *E. coli*
- **REPLICATION PROTEIN A (RPA)** (acts like SSB): keeps DNA strands separated during DNA synthesis
- **FEN1 (MF1):** removes RNA primer from each Okazaki fragment; activity is associated with the δ complex.
- **DNA LIGASE:** joins Okazaki fragments
- **DNA POLYMERASE β:** involved in DNA repair
- **DNA POLYMERASE ε:** 3′ → 5′ exonuclease activity; other functions unknown
- **DNA POLYMERASE γ:** catalyzes mitochondrial genome replication
- **EUKARYOTIC TYPE II TOPOISOMERASES:** catalyze only the removal of superhelical tension (refer to the description of prokaryotic DNA gyrase, above).

DNA Repair Mechanisms[1]

DNA LIGASE can repair breaks in the phosphodiester linkages.

DNA repair can be classified according to whether single strands or double strands require repair. Examples of single-strand repairs include base excision repair, nucleotide excision repair, transcription-coupled repair and mismatch repair. Photoreactivation repair can also be considered a form of single-strand repair. Double-strand break repair is accomplished by nonhomologous end joining and recombinational repair.

BASE EXCISION REPAIR: Incorrect bases are removed (excised) and replaced with the correct ones. Excision repair involves a series of enzymes.

IN NUCLEOTIDE EXCISION REPAIR a **REPAIR ENDONUCLEASE** (also, **EXCISION NUCLEASE** or **EXCINUCLEASE**) detects a distorted DNA segment and then cuts the damaged DNA and removes a single-stranded sequence about 12 nucleotides[2] in length. The enzyme pol I replaces the nucleotides in the gap left by the excised segment, and DNA ligase seals the break in the phosphodiesterase backbone.

TRANSCRIPTION-COUPLED REPAIR occurs only during transcription by the transcribing enzyme.

MISMATCH REPAIR corrects helix-distorting base mispairings that are the result of replication proofreading errors or the result of replication slippage.

PHOTOACTIVATION REPAIR (also called **LIGHT-INDUCED REPAIR**) of pyrimidine dimers: DNA photolyase[3] has flavin and pterin chromophores that use energy from visible light to cleave pyrimidine dimers, restoring them to two separate pyrimidines and leaving the phosphodiester bonds intact.

DOUBLE-STRAND REPAIRS are accomplished either by nonhomologous end joining or by recombinational repair.

Nonhomologous end joining repair ligates two DNA end together. In mammals nonhomologous end joining is initiiated by ATM and ATR (see p. 676).

RECOMBINATIONAL REPAIR: Postreplication repair of DNA. Damaged DNA interrupts replication; the replication complex detaches from the DNA and reinitiates after the damaged site. This results in a gap in the daughter strand. This gap is repaired by an exchange of the corresponding segment of the homologous DNA (this process is called recombination). After recombination, DNA polymerase and DNA ligase complete the repair process. There are two models that explain the mechanisms of double-strand repair: the double strand repair model and the synthesis-dependent strand annealing model.

DNA RECOMBINATION PRODUCES NEW COMBINATIONS OF GENES (AND FRAGMENTS): VARIATIONS THAT MAKE EVOLUTION POSSIBLE

DNA recombination is the rearrangement of DNA sequences by exchanging segments from different molecules. **GENERAL RECOMBINATION, SITE-SPECIFIC RECOMBINATION**

GENERAL RECOMBINATION (requires precise pairing of homologous DNA molecules)

1. Pairing of two homologous DNA molecules.
2. Nicking: Two of the DNA strands (one in each molecule) are cleaved.
3. Crossover: The two strand segments cross over, forming a Holliday intermediate.
4. Sealing nicks: DNA ligase seals the cut ends.

---

[1] (At this point, it might be helpful to review the types of DNA mutations that were discussed in Chapter 17. Remember what causes pyrimidine dimers and which bonds needs to be cleaved to separate them)

[2] 27-29 in eukaryotes.

[3] DNA photolyase is not present in humans.

5. Branch migration caused by base-pairing exchange leads to the transfer of a segment of DNA from one homologue to the other.
6. A second series of DNA strand cuts occurs.
7. DNA polymerase fills any gaps, and DNA ligase seals the cut strands.

FORMS OF INTERMICROBIAL DNA TRANSFER (IN BACTERIA) THAT INVOLVE GENERAL RECOMBINATION:

1. **TRANSFORMATION**—naked DNA fragments enter through an opening in the cell wall and are introduced into the bacterial genome
2. **TRANSDUCTION**—bacteriophages inadvertently carry bacterial DNA to a recipient cell; after recombination, the cell uses the transduced DNA
3. **CONJUGATION**—an unconventional sexual mating: a donor cell synthesizes a sex pilus (via a specialized plasmid) that attaches to the surface of the recipient cell. The pilus transfers a fragment of the donor's DNA, which can undergo recombination or exist in plasmid form.

SITE-SPECIFIC RECOMBINATION

Site-specific recombination requires only short segments of homologous DNA [**ATTACHMENT (ATT) SITES** or **INSERTIONAL (IS) ELEMENTS**] of DNA homology and depends more on protein-DNA interactions than on sequence homology. *Example*: Integration of bacteriophage λ into the *E. coli* chromosome (See Figure 18.25, p. 686 of your text.)

TRANSPOSITION—a variation of site-specific recombination—transposable elements (certain DNA sequences) are moved from one chromosome or chromosomal region to another; differs from site-specific recombination in that a specific protein-DNA interaction occurs on only one of the two recombining sequences. The recombination of the second DNA sequence is nonspecific.

TRANSPOSONS—transposable elements (**JUMPING GENES**) that can jump between bacterial chromosomes, plasmids, and viral genomes

IS ELEMENTS (INSERTION ELEMENTS)—bacterial transposons that consist only of a gene that codes for a transposase (transposition enzyme) flanked by inverted repeats (short palindromes)

COMPOSITE TRANSPOSONS—bacterial transposons that contain additional genes, several of which may code for antibiotic resistance

1. **REPLICATIVE TRANSPOSITION**—transposon inserts a replicated copy, stays in its original site
2. **NONREPLICATIVE TRANSPOSITION**—transposon is cut out of its original (donor) site and inserted into the target site; the donor site must be repaired

EUKARYOTIC TRANSPOSONS: many contain **LTR** (**LONG TERMINAL REPEATS** or **DELTA REPEATS**); many mechanisms involve an RNA intermediate and resemble the replicative phase of a retrovirus

18.2 TRANSCRIPTION: RNA SYNTHESIS

CATALYZED BY RNA POLYMERASE: $NTP + (NMP)n \rightarrow (NMP)n + 1 + PP_i$.

NONTEMPLATE STRAND = plus (+) strand; also called the coding strand because it has the same base sequence as the RNA transcription product (except U substitutes for T)

TEMPLATE STRAND = minus (–) strand

The direction of the gene = the direction of the coding strand. So, polymerization proceeds from the 5′-end to the 3′-end of both the coding strand and the gene.

TRANSCRIPTION IN PROKARYOTES

RNA POLYMERASE IN E. COLI: core enzyme catalyzes RNA synthesis; σ-factor binds transiently to the core enzyme and allows it to bind both the correct template strand and the proper site to initiate transcription.

Pribnow box—region that occurs 10 nucleotides before the transcription initiation site.

STAGES OF TRANSCRIPTION IN *E. COLI*:

1. **INITIATION**: RNA polymerase binds to a promoter (a specific DNA sequence).
   A short DNA segment near the Pribnow box unwinds.
   The first nucleoside triphosphate binds to the RNA polymerase complex, beginning transcription, and then attacks the second NTP to form the first phosphodiester bond.
   After the transcribed sequence is about 10 nucleotides long, the conformation of RNA polymerase complex changes.
   The σ-factor detaches, RNA polymerase affinity for the promoter site decreases, and the initiation phase ends.
2. **ELONGATION PHASE:** Core RNA polymerase converts to an active transcription complex and binds several accessory proteins.
   DNA unwinds ahead of the transcription bubble. (Topoisomerases resolve positive and negative supercoiling ahead of and behind the bubble.)
   Elongation continues until a termination sequence is reached.
3. **TERMINATION**: Termination sequences contain palindromes, and their RNA transcripts form a stable hairpin turn.

PRODUCTS OF TRANSCRIPTION:

mRNA is used immediately for protein synthesis.
tRNA and rRNA require posttranscriptional processing.

TRANSCRIPTION IN EUKARYOTES: SIGNIFICANTLY MORE COMPLEX THAN IN PROKARYOTES

UNIQUE FEATURES OF EUKARYOTIC TRANSCRIPTION:

1. RNA polymerase activity: requires three nuclear RNA polymerases:
   - RNA polymerase I: transcribes large rRNAs
   - RNA polymerase II: transcribes precursors of mRNA and most snRNAs
   - RNA polymerase III: transcribes precursors of tRNAs and 5S rRNA

   Eukaryotic RNA polymerases cannot initiate transcription. Various transcription factors must be bound at the promoter before transcription can begin.

2. **PROMOTERS**: larger, more complicated, and more variable than prokaryotic promoters. Many promoters for RNA polymerase II contains consensus sequences (**TATA BOX**) about 25-30 bp upstream from the initiation site.
   CAAT BOX AND GC BOX are examples of sequences upstream that bind transcription factors and affect the frequency of transcription initiation.
   ENHANCERS are regulatory sequences that may be thousands of base pairs away from the gene, but affect activity of promoters.

3. **POSTTRANSCRIPTIONAL PROCESSING:** Eukaryotic mRNAs are extensively processed (unlike prokaryotic mRNAs, which typically have little to no posttranscriptional processing).

   WHY MODIFY MRNA? —to help stabilize mRNA
   —to help mRNA transport out of the nucleus

EUKARYOTIC MRNA MODIFICATIONS:

CAPPING the 5′-phosphate end with a 7-methylguanosine protects the 5′-end from exonucleases and promotes mRNA translation. (The first two nucleotides of the transcript are methylated at the 2′–OH.)

ATTACHING A POLY A TAIL (a polyadenylate with 100-250 As) to the 3′-end protects from the action of 3′,5′-exonucleases and promotes mRNA export to the cytoplasm.

SPLICING removes **INTRONS** (DNA sequences that intervene and occur within a particular gene); each intron is excised as a **LARIAT** (a configuration that resembles a loop); then the exons are joined.

EXONS are coding sequences (DNA sequences that designate the amino acid sequence).

Not all of the DNA sequences in eukaryotes code for amino acids. Some DNA sequences serve regulatory or structural roles. So, the removal of introns is necessary to synthesize an mRNA that will be translated into a continuous (and correct) amino acid sequence.

SPLICEOSOME are multicomponent structure with several snRNAs, and several proteins occurs.

RIBOZYME is a catalytic RNA that exhibits self-splicing; found in several organisms.

18.3 GENE EXPRESSION

. . . How cells produce the genes they need, when they need them.

CONSTITUTIVE or **HOUSEKEEPING GENES** are transcribed routinely because their products are needed for cell function.

INDUCIBLE GENES are expressed (turned on) only under certain circumstances.

OPERONS control inducible genes. Operons are groups of structural and regulatory genes that are linked.

Most gene expression mechanisms involve DNA-protein binding.

DEREPRESSION inhibits a repressor to activate a gene. (It is like taking your foot off the brake in a stopped car that is pointed downhill. It might not be the same as hitting the gas, but the car still moves forward!)

GENE EXPRESSION IN PROKARYOTES

REGULATION OF LACTOSE[4] METABOLISM IN *E. COLI* CELLS BY THE LAC OPERON

SUMMARY: When lactose is present, β-galactosidase converts a few molecules to allolactose, which turns the lac operon on. The lac operon codes for lactose metabolism enzymes until the lactose supply is consumed. Then, the repressor protein can once again bind to the operator site, and the lac operon turns off.

lac operon	structural genes Z, Y, and A that code for lactose enzymes; plus a control element—a promoter site that contains the CAP site and overlaps the operator site
CAP site	where the CAP protein binds
operator site	DNA sequence that binds a repressor protein and helps to regulate adjacent genes
repressor gene i	codes for the lac repressor protein when lactose is absent; directly adjacent to lac operon

[4] Lactose is the disaccharide galactose β(1,4) glucose.

lac repressor — protein that binds to the operator and prevents binding of RNA polymerase to the promoter (that is, it turns the lac operon off)

allolactose — β-1,6-isomer of lactose, and an inducer. When allolactose binds to the lac repressor, its conformation changes and it dissociates from the operator. When the operator is free of the repressor, the lac operon is turned on, and transcription of the structural genes can begin. In the absence of inducer, the lac operon remains off.

Glucose is a preferred carbon and energy source for *E. coli*. An organism that has both glucose and lactose will use glucose first. Only after the glucose is gone will the lac operon enzymes be synthesized. This is how it happens:

When glucose is depleted (and the cell needs energy), cAMP levels in the cell rise.[5] cAMP binds to CAP (catabolite gene activator protein), which binds to the lac promoter. CAP-promoter binding increases the affinity of RNA polymerase for the lac promoter, thus promoting transcription and activating lactose metabolism.

GENE EXPRESSION IN EUKARYOTES: HOW ARE EUKARYOTIC GENES REGULATED?

GENOMIC CONTROL

MOST COMMON REGULATORY CHANGES:

DNA METHYLATION: *Example*: methylation of cytosines in certain 5′-CG-3′ sequences turns off genes.

HISTONE ACETYLATION: acetylation of lysine residues in H3 and H4 reduces their affinity for DNA; histone acetylation promotes genes expression.

TRANSCRIPTIONAL CONTROL IS HEAVILY INFLUENCED BY:

CHROMATIN STRUCTURE: heterochromatin (too condensed to do transcription) vs. euchromatin (less condensed; varying levels of transcription activity)

GENE REGULATORY PROTEINS: bind to DNA to activate or repress genes.

Mechanisms of gene regulatory proteins:

—competitive DNA binding of transcription factor proteins
—masking the activation surface
—direct interaction with (binding to) transcription factors

GENE REARRANGEMENTS regulate certain genes and may be involved in cell differentiation. *Example*: rearrangement of antibody genes in B lymphocytes.

GENE AMPLIFICATION increases the number of copies of a gene by repeated rounds of replication within the amplified region. This occurs when the need for specific gene products is unusually high. For example, during the early developmental stages of fertilized eggs, the huge demand for protein synthesis requires amplification of rRNA genes.

BIOCHEMISTRY IN PERSPECTIVE: CARCINOGENESIS: TERMS: PROTOONCOGENE; GTPASE-ACTIVATING PROTEIN, GUANINE NUCLEOTIDE EXCHANGE FACTOR; MITOGEN; ONCOGENE; TUMOR PROMOTER

ALTERNATIVE RNA PROCESSING TO CONTROL GENE EXPRESSION:

ALTERNATIVE SPLICING joins different combinations of exons ultimately resulting in different proteins. Example: tissue-specific forms of α-tropomyosin, a structural protein produced in various tissues.

ALTERNATIVE SITES TO ATTACH POLY A TAILS

LENGTH OF POLY A TAILS: Longer tails give more stability to mRNA, resulting in increased opportunity for translation.

---

[5] cAMP again!! One way to tie a number of chapters together is to list the regulatory effects of cAMP in various types of cells and pathways.

RNA EDITING: certain bases are chemically modified, deleted, or added. *Example*: Deaminating cytosine to produce uracil changes a CAA codon for glutamine into a UAA codon, which is a stop signal, producing a shorter version of the protein.

mRNA TRANSPORT CONTROL THROUGH NUCLEAR PORE COMPLEXES

NUCLEAR EXPORT SIGNALS: include capping, and association with or presence of specific proteins, such as CBP (cap-binding protein)

TRANSLATIONAL CONTROL: ALTERING PROTEIN SYNTHESIS

allows eukaryotic cells to respond to various stimuli (e.g., heat shock, viral infections, and cell cycle phase changes).

COVALENT MODIFICATION of translation factors (nonribosomal proteins that aid translation) alters translation rate or enhances translation of specific mRNAs.

SIGNAL TRANSDUCTION: RESPONDING TO ENVIRONMENTAL SIGNALS BY ALTERING GENE EXPRESSION; INVOLVES SIGNAL MOLECULES

Gene expression changes are initiated by ligand binding to a receptor (cell surface or intracellular).

Complicating features of intracellular signal transduction mechanisms include the following:

1. Each type of signal may activate one or more pathways.
2. Signal transduction pathways may converge or diverge.

Checkpoints in cell cycle phases prevent the cell from entering the next phase until conditions are optimal and specific signals are received. The mechanism of progression: alternating synthesis and degradation of cyclins, a group of regulatory proteins that bind to and activate cyclin-dependent protein kinases (Cdks). Cdks phosphorylate a variety of proteins that signal the cell past a checkpoint to the next phase of mitosis.

REGULATION OF CELL DIVISION:

POSITIVE CONTROL: Growth factors bind to specialized cell surface receptors, initiating a cascade of reactions that induces two classes of genes:

EARLY RESPONSE GENES are rapidly activated. *Example*: protooncogenes—normal genes that, if mutated, can promote carcinogenesis

DELAYED RESPONSE GENES are induced by activities of transcription factors and other proteins that were produced or activated during the early response phase. Examples of delayed response gene products include Cdks, cyclins, and other components needed for cell division.

NEGATIVE CONTROL: tumor suppressor genes (*examples*: Rb gene and p53 gene); apoptosis (programmed cell death) occurs if too much DNA damage has occurred and/or if DNA repair mechanisms are incomplete.

BIOCHEMISTRY IN THE LAB: GENOMICS KEY TERMS: GENOMICS; FUNCTIONAL GENOMICS, RECOMBINANT DNA TECHNOLOGY, COSMID, ELECTROPORATION, TRANSFECTION, TRANSGENIC ANIMAL VECTOR, BACTERIAL ARTIFICIAL CHROMOSOME, YEAST ARTIFICIAL CHROMOSOME, COLONY HYBRIDIZATION TECHNIQUE, MARKER GENE, POLYMERASE CHAIN REACTION CDNA LIBRARY, SHOTGUN CLONING, CHROMOSOMAL JUMPING, CONTIG, DNA MICROARRAY, ANNOTATION, BIOINFORMATICS

CHAPTER 18: SOLUTIONS TO REVIEW QUESTIONS

18.1 a. replication – the process in which an exact copy of parental DNA is synthesized using the polynucleotide strands of the parent DNA as templates

b. semiconservative – DNA synthesis in which each polynucleotide strand serves as a template for the synthesis of a new strand

c. replication factory – a specific nuclear compartment (or nucleoid) in which DNA replication occurs

d. primosome – a multienzyme complex involved in the synthesis of RNA primers at various intervals along the DNA template strand during *E .coli*

DNA replication

e. clamp loader – the γ-complex that recognizes single DNA strands with primer and transfer β_2-clamp dimer to the core polymerase

18.2 a. processivity – the prevention of frequent dissociation of a polymerse from the DNA template

b. replisome – the large complex of polypeptides, including the primosome, that replicates DNA in *E. coli*

c. exonuclease – an enzyme that removes nucleotides from the end of polynucleotide strands

d. DNA ligase – an enzyme that catalyzes the formation of a covalent phosphodiester bond between the 3' OH end of one DNA segment with the

5'-phosphate end of another segment during replication or DNA repair processes

e. replication fork – the Y shaped region of a DNA molecule that undergoes replication; results from separation of two DNA strands

18.4 a. ORC - origin of replication complex – a protein complex that binds to the DNA replication origin during the initiation phase of DNA synthesis; contains analogs of the protein DnaA

b. licensing factors – replication licensing factors – proteins that bind to the ORC and complete the structure of preRC

c. RPA – replication protein A – a protein that stabilizes the separated DNA strands during replication

d. TEBP – telomerase end binding protein – a protein that binds to and stabilizes GT-rich telomeric sequences

e. TRF – telomerase repeat-binding factors – a protein that binds to and secures the 3'-overhang sequence of a telomere

18.5 a. RFC – replication factor C – a eukaryotic clamp loader protein in DNA replication

b. DNA glycosylase – a DNA repair enzyme that cleaves N-glycosidic linkages between a damaged base and the deoxyribose component of the nucleotide

c. apurinic site – a nucleotide residue in a DNA strand from which a purine base has been removed.

d. apyrimidinic site – a nucleotide residue in a DNA strand from which a pyrimidine base has been removed

e. mismatch repair – a single–strand DNA repair mechanism that corrects helix – distorting base mispairings

18.7 a. nonreplicative transposition – transposition of a DNA segment occurs in cut and paste mechanism; a DNA sequence is removed from a donor site and then spliced into a target site without duplication of the transposed sequence

b. replicative transposition – one strand of donor DNA is transferred to the target site; replication followed by site–specific recombination results in duplication of the transposed sequences

c. composite transposon – a transposable element composed of two transposons linked by a DNA sequence between them

d. retrotransposon – one of a subclass of transposons that use an RNA intermediate

e. insertional element – a short DNA sequence involved in site specific recombination; also called an IS element or att site

18.8 a. transfection – a mechanism by which bacteriophage inadvertently transfer bacterial chromosome or plasmid sequences to a new host cell

b. cosmid – a cloning vehicle that contains the γ-bacteriophage COS sites incorporated into plasmid DNA sequences with one or more selectable markers

c. electroporation – a method of introducing a cloning vector into a host cell that involves treatment with an electrical current

d. transgenic animal – animals created by the microinjection of recombinant DNA into fertilized ova

e. colony hybridization technique – bacteria are screened for the presence of recombinant DNA by using a radioactively labeled nucleic acid probe, an RNA molecule or a single-stranded DNA molecule with a sequence complementary to that of a specific sequence within the recombinant DNA

18.10 a. promoter – the sequence of nucleotides immediately before a gene that is recognized by RNA polymerase and signals the start point and direction of transcription

b. consensus sequence – the average of several similar sequences; for example the consensus sequence of the – 10 box of *E. coli* promoter of TATAAT.

c. operon – a set of linked genes that are regulated as a unit

d. chromatin–remodeling complex – a multisubunit complex that facilitates the release of the histones from nucleosomal DNA during transcription

e. general transcription factors – a set of six transcription factors, which are the minimum number of proteins that are necessary for eukaryotic transcription

18.11 a. RNA splicing – the process in which introns are cut out and the exons are linked together to form a functional RNA product

b. spliceosome - a multicomponent complex containing protein and RNA; used in the splicing phase of mRNA processing

c. operator – a prokaryotic DNA regulatory sequence that binds to specific repressor or activator proteins that modulate gene expression

d. riboswitch – a bacterial RNA-based control mechanism that usually represses gene expression

e. lac operon – a bacterial control element and structural genes that code for the enzymes of lactose metabolism; repressed in the absence of lactose and availability of glucose

18.13 a. cell transformation – the process in which an apparently normal cell is converted into a malignant cell

b. oncogene – an abnormal version of a protooncogene; mediates cancerous transformatons

c. apoptosis – the genetically programmed series of events that lead to cell death.

d. early response gene – one of a set of genes coding for transcription factors that are rapidly activated after growth factors bind to cell surface receptors

e. delayed response gene – one of a set of cell cycle regulatory genes that are induced by the activities of early response genes

18.14 a. cohesin – a ring-shaped DNA-binding protein with multiple functions, including transcription regulation in association with promoter, enhancer, and insulator elements

b. trancription factory – an active transcription unit composed of several clustered RNA polymerases that is located in a discrete site within a eukaryotic nucleus

c. P bodies – cytoplasmic structures that consist of enzymes involved in several mRNA degradation or silencing processes

d. CTCF – a transcriptional repressor protein involved in many cellular processes including transcription regulation; interacts with cohesin at insulator elements to prevent enhancer interaction with certain promoters

e. ENCODE – *e*ncyclopedia of *DN*A *e*lements; a follow-up project to the human genome project with a goal of identifying all functional elements in the human genome

18.16 Briefly, prokaryotic DNA replication consists of DNA unwinding, RNA primer formation, DNA synthesis catalyzed by DNA polymerase and the joining of Okazaki fragments by DNA ligase. Prokaryotic DNA replication differs from the eukaryotic process in that prokaryotic replication is faster, and in prokaryotes the Okazaki fragments are longer

18.17 a. Helicase is an enzyme activity that relives torque generated by supercoiling ahead of the replication machinery.

b. Primase is an enzymatic activity that catalyzes the synthesis of several RNA primers.

c. DNA polymerase is an enzymatic activity that catalyzes nucleotide polymerization during DNA replication

d. DNA ligase forms phosphodiester linkages between newly synthesized DNA fragments.

e. Topoisomerase is an enzymatic activity that prevents the tangling of DNA strands during DNA replication.

f. DNA gyrase facilitates the separation of DNA strands during prokaryotic replication

18.19 a. ROS may cause single and double-stranded breaks, pyrimidine dimer formation and the loss of purine and pyrimidine bases.

b. Intercalating agents cause deletion or insertion mutations.

c. Small alkylating agents attach to the nitrogen atoms of the purines and pyrimidines, destabilizing glycosidic linkages (leading to depurination), interfering with hydrogen bonding, and promoting both transversion and transition mutations.

d. Large alkylating agents have the same effect as small alkylating agents, but in addition they behave similarly to intercalating agents, leading to frameshift mutations and breakage of the DNA chain.

e. Nitrous acid deaminates bases. For example, cytosine is converted to uracil.

18.20 Viruses can cause mutations that affect the expression of protooncogenes by inserting their genomes into host cell regulatory sequences, thereby inactivating them.

18.22 Genetic recombination promotes species diversity. General recombination, a process in which segments of homologous DNA molecules are exchanged, is most commonly observed during meiosis. In site-specific recombination, protein-DNA interactions promote the recombination of nonhomologous DNA. Transposition is an example of site-specific recombination.

18.23 Most mutations are silent. Of those that do affect the functioning of an organism, most are deleterious because of the complex nature of living processes. Change in the properties of any of the thousands of different gene products is potentially disruptive. Only on rare occasions does a mutation improve the viability of an individual organism.

18.25 In replicative transposition, a replicated copy of a transposable element is inserted into a new chromosome location in a process that involves the formation of an intermediate called a cointegrate. In nonreplicative transposition, sequence replication does not occur, that is, the transposable element is spliced out of its donor site and inserted into the target site. The donor site must be repaired.

18.26 In DNA, if cytosine is converted to uracil, which forms a base pair with adenine, an AT base pair is substituted for a GC base pair. Such a change in RNA is not as important because RNA molecules are short lived and disposable. In contrast, because DNA molecules are the cell's permanent repository of genetic information, any change in base sequence may affect an organism's viability.

18.28 In both cases DNA copies are produced. However, in DNA replication usually only one copy of each DNA molecule is synthesized and several proofreading mechanisms ensure accurate copying. PCR technology is designed to produce multiple copies of a DNA molecule and proofreading is limited to the DNA polymerase that is employed.

18.29 Marker genes are useful in recombinant DNA technology because their function is known and their presence, which indicates that a successful recombinant event has occurred, is easily detected. For example, an antibiotic resistance gene, which codes for the synthesis of a substance that provides protection for a bacterium from the effects of an antibiotic, allows the growth of recombinant cells in a medium containing that antibiotic. Cells that do not contain the marker gene, that is, those in which the recombinant DNA is not present, do not survive

18.31 The processing steps that prepare a typical eukaryotic mRNA for its functional role include capping (the linkage of a 7-methylguanosine to the 5'-end), cleavage of mRNA and addition of a poly (A) tail to the 3'- end, and splicing (the removal of introns).

18.32 In relatively simple genomes, such as those in bacteria, operons provide a convenient mechanism for regulating genes. Proteins required in the same metabolic pathway or functional process are synthesized together because their genes are controlled by the same promoter.

18.34 Mechanisms to regulate gene expression differently in different types of cells include: genomic control (including gene rearrangements and selective gene amplification), transcription initiation control, RNA processing, RNA editing, RNA transport control, translational control, and signal transduction-triggered gene expression. See Solution 18.35 for examples of these mechanisms.

18.35 Examples of genomic control includes gene rearrangements of antibody genes in B lymphocytes and the selective gene amplification of rRNA genes in oocytes. An example of RNA processing is the alternative splicing of the vertebrate tropomyosin mRNA gene, in which different combinations of its 13-15 exons result in protein isoforms that serve the needs of different cells types, namely, striated muscle, smooth muscle, fibroblast, and brain cells. Examples of RNA editing are found in intestinal cells and some brain neurons. Intestinal cell apolipoprotein B-100 mRNA undergoes a C→U conversion, resulting in apolipoprotein B-48, a truncated version produced for chylomicron particles.

18.37 The purpose of gene amplification is to rapidly produce multiple copies of specific gene products that are required in greater quantities during certain stages in a cell's development.

18.38 Gene amplification occurs via repeated rounds of replication within the amplified region.

18.40 RNA molecules are more reactive than DNA because of the presence of the 2′-OH group of ribose. In addition the complexity of its three-dimensional structures that result from single-stranded RNA coiling back on itself provides more opportunities for more diverse intermolecular interactions and binding than does DNA.

18.41 Line element (L1) transposition occurs by a 'cut and paste" process in which the L1 is first transcribed to form L1 RNA, which then exits the nucleus to be translated. Retrotransposon proteins (transposase and reverse transcriptase) and the L1 RNA form a complex, reenter the nucleus, and bind to a target DNA sequence. Transposase makes a staggered cut in the DNA sequence, reverse transcriptase synthesizes a DNA copy from the L1 RNA, and a second DNA strand is synthesized. DNA ligase completes the insertion by joining the ends of the target DNA and the newly synthesized L1 DNA element.

18.43 The function of telomere repeat-binding proteins is to secure the 3′ overhang as part of the process that sequesters and stabilizes telomeres.

18.44 Mismatch repair is a single strand repair mechanism that corrects helix distorting base mispairings resulting from replication proofreading errors or replication slippage. Mutations in MMr proteins cause microsatellite instability, which has been linked to several types of human cancers.

18.46 In eukaryotic transcription-coupled nucleotide excision the stalled RNA polymerase serves as the damage recognition signal.

CHAPTER 18: SOLUTIONS TO FILL-IN-THE-BLANK QUESTIONS

18.47 Recombination

18.49 DNA ligase

18.50 1000

18.52 Site-specific

18.53 Homologous

18.55 Topoisomerases

18.56 Consensus sequence

CHAPTER 18: SOLUTIONS TO SHORT-ANSWER QUESTIONS

18.58 Riboswitches are regulatory devices usually located in the 5′-UTRs of bacterial mRNAs. They are composed of two structural elements, an aptamer that binds a specific metabolite and a gene expression regulator called the expression platform. Several genes involved in lysine synthesis would be expected to have a riboswitch. As each of these genes is transcribed, the aptamer sequence begins to emerge from the transcription complex and into the cytoplasm, where it may or may not encounter and bind a lysine molecule. If a molecule of lysine does bind to the aptamer (indicating that the cell's lysine level is sufficient), an allosteric rearrangement of the riboswitch's gene expression regulator prevents the translation the mRNA.

18.59 The tumor suppressor protein pRB inhibits cell division in the presence of DNA damage by inhibiting the transcription of proteins that are vital for cell cycle progression. It does so by suppressing transcription of proteins that would allow entry into the S phase by binding to and inhibiting the transcription activator E2F/DP and by promoting tighter chromatin coiling with the deacetylation of histone acetylated lysine residues.

18.61 Premature stop codons are detected during translation when there is a stop codon upstream of an exon junction complex bound to an exon-exon junction. Premature stop codons are caused by RNA splicing errors, random mutations, and faulty DNA rearrangements.

CHAPTER 18: SOLUTIONS TO THOUGHT QUESTIONS

18.62 DNA replication time is calculated as follows:

$$\frac{150{,}000{,}000 \text{ base pairs}}{50 \text{ bases/s}} = 3 \times 10^6 \text{ s} = 34.5 \text{ days}$$

Consequently, approximately one month is required for this DNA replication. Eukaryotic DNA synthesis is significantly faster than expected because each chromosome contains multiple replication units (replicons).

18.64 Mustard gas cross links the strands in DNA with permanent covalent bonds.

18.65 The increase in pathogenicity of the streptococcus organism may result from the incorporation of a segment of the virus genome into the streptococcus genome. The genome of the current virulent streptococcus and the DNA of preserved specimens can be compared with the use of a probe for the viral toxin gene. The presence or absence of the viral sequence in modern streptococcus can be determined.

18.67 Recall that phorbol esters mimic the action of DAG, the normal cell metabolite that activates protein kinase C (PKC). PKC initiates a phosphorylation cascade that results in the activation of numerous molecules involved in cell growth and division, including jun and fos, which then combine to form AP-1. AP-1 is a transcription factor whose presence promotes cell division. Its formation causes an affected cell to have a growth advantage over nearby cells. Because phorbol esters are tumor promoters any exposure to them increases the risk that initiated cells may progress toward a cancerous state.

18.68 Antibiotic resistance arises because the overuse of antibiotics acts as a selection pressure, i.e., they provide a growth advantage for disease-causing organisms that possess resistant genes. So-called superbugs are organisms that are resistant to several types of antibiotics because they possess plasmids containing several resistant genes. If the circumstances that cause antibiotic resistance continue, antibiotics may eventually become ineffective against most infectious diseases.

18.70 Errors that occur during DNA replication have the potential to become permanent if repair processes fail. Errors made during transcription affect only a few RNA molecules and are temporary.

18.71 Gene amplification, the selective duplication of certain genes, can occur via a reverse transcriptase–mediated event. The creation of one or more cDNAs from an mRNA is followed by insertion of these sequences into the genome.

18.73 The DNA polymerase enzyme complex is a sophisticated machine that is composed of a large number of fragile components. By keeping the complex stationary, cells protect it from mechanical damage that would occur if it moved through the crowded nucleoplasm.

18.74 The riboswitch must have a binding site for the metabolite. This binding usually triggers a conformational change that blocks translation. So the riboswitch must also have a sequence that changes shape.

19 Protein Synthesis

Brief Outline of Key Terms and Concepts

PROTEIN SYNTHESIS

TRANSLATION of nucleotide base sequences into the amino acid sequence of polypeptides.

POSTTRANSLATIONAL MODIFICATION

TARGETING

19.1 THE GENETIC CODE

THE GENETIC CODE

—a mechanism by which ribosomes translate nucleotide base sequences into the primary sequence of polypeptides.

—consists of 64 codons: 61 codons that specify the amino acids and 3 STOP CODONS.

CODON; READING FRAME; OPEN READING FRAME

CODON-ANTICODON INTERACTIONS

During TRANSLATION, the genetic code is translated through base-pairing interactions between mRNA CODONS and tRNA ANTICODONS. The WOBBLE HYPOTHESIS explains why cells usually have fewer tRNAs than expected.

THE AMINOACYL-tRNA SYNTHETASE REACTION

—links a tRNA to its amino acid, using 2 ATP equivalents. ACTIVATION; MIXED ANHYDRIDE; tRNA LINKAGE; COGNATE tRNA; PROOFREADING SITE

19.2 PROTEIN SYNTHESIS

TRANSLATION consists of three phases: INITIATION, ELONGATION, and TERMINATION. Each phase requires several types of PROTEIN FACTOR.

INITIATION: INITIATOR tRNA; POLYSOME

ELONGATION: TRANSPEPTIDATION, TRANSLOCATION

TERMINATION: RELEASING FACTOR

POSTTRANSLATIONAL MODIFICATIONS

TARGETING

PROKARYOTIC PROTEIN SYNTHESIS is a rapid process involving several protein factors.
Most prokaryotic gene expression appears to be regulated by transcription initiation.

INITIATION

INITIATION COMPLEX ; SHINE-DALGARNO SEQUENCE

ELONGATION

GUANINE NUCLEOTIDE EXCHANGE FACTOR (GEF); EF-TU PEPTIDYL TRANSFERASE CENTER, PRETRANSLOCATION STATE; POSTTRANSLOCATION STATE

TERMINATION

RELEASING FACTORS, RIBOSOME RECYCLING FACTOR

POSTTRANSLATIONAL MODIFICATIONS prepare the polypeptide for its function, assist in folding, or target it to a specific destination and include PROTEOLYTIC PROCESSING, CONJUGATION, METHYLATION, PHOSPHORYLATION, and insertion of cofactors.

SIGNAL PEPTIDES

TRANSLATIONAL CONTROL MECHANISMS

Prokaryotes and eukaryotes differ in their usage of translational control mechanisms. Prokaryotes:

- vary the SHINE-DALGARNO sequences; and
- use NEGATIVE TRANSLATIONAL CONTROL (i.e., the translation of a POLYCISTRONIC mRNA is repressed by one of its products).

In contrast, eukaryotes exhibit a wide range of eukaryotic translational control mechanisms, from

- global controls in which thc translation rate of a large number of mRNAs is altered to
- specific controls that alter the translation of a specific mRNA or small group of mRNAs.

EUKARYOTIC PROTEIN SYNTHESIS

COMPARED TO PROKARYOTIC PROTEIN SYNTHESIS, EUKARYOTIC PROTEIN SYNTHESIS IS SLOWER, MORE COMPLEX

- is slower, due in part to mRNA SECONDARY STRUCTURE, mRNA SCANNING.
- uses more protein factors (translation factors) at each step.
- has a more complex initiation mechanism;
  - 43S PREINITIATION COMPLEX; CAP-BINDING COMPLEX (CBC); GUANINE NUCLEOTIDE ACTIVATION FACTOR; POLY(A) BINDING PROTEIN (PABP) includes a cap-binding protein.
- forms CIRCULAR mRNA POLYSOMES.
- KINETIC PROOFREADING
- has posttranslational processing and targeting mechanisms that are much more complicated.

POSTTRANSLATIONAL MODIFICATIONS (EUKARYOTES)

PROTEOLYTIC CLEAVAGE (PROPROTEIN; PREPROPROTEIN); DISULFIDE BOND FORMATION, DISULFIDE EXCHANGE; PROTEIN SPLICING (INTEIN, EXTEIN); GLYCOSYLATION; LIPOPHILIC MODIFICATIONS; HYDROXYLATION; METHYLATION; CARBOXYLATION; PHOSPHORYLATION

TARGETING (EUKARYOTES)

TRANSCRIPT LOCALIZATION

SIGNAL HYPOTHESIS, SIGNAL RECOGNITION PARTICLE

DOCKING PROTEIN

COTRANSLATIONAL TRANSFER, TRANSLOCON,

OVERVIEW

Translation of the nucleotide-base-sequence code results in a sequence of amino acids that form a polypeptide. Protein synthesis also includes posttranslational modification and targeting.

19.1 THE GENETIC CODE

GENETIC CODE a coding dictionary that specifies a meaning for each base sequence

CODON a three-base-sequence (triplet) in mRNA that codes for each amino acid. Example: "CCC" codes for proline.

Of the 64 triplets that are possible, 61 code for amino acids and 3 are **STOP SIGNALS** that terminate the growing polypeptide chain. (Stop codons: UAA, UGA, UAG)

PROPERTIES OF THE GENETIC CODE:

1. **SPECIFIC:** Each codon signals a specific amino acid. *Exception*: AUG codes for both Met and a start signal (initiating codon). Also, many of the codons that code for a specific amino acid have similar sequences. *Example*: UC? codes for Ser, whether the third base is U, C, A, or G.
2. **DEGENERATE:** many codons have the same meaning; that is, most amino acids are specified by more than one codon. (Only Met and Trp have one codon.)
3. **NONOVERLAPPING AND WITHOUT PUNCTUATION:** The code reads from the initiating codon (AUG) straight through to a stop codon, without repeating or skipping any bases.

   READING FRAME: set of side-by-side codons in mRNA

   OPEN READING FRAME: series of codons without a stop codon

   Why are reading frames important? Since the codons are sequential and nonoverlapping, the actual amino acid sequence depends on the starting point for the first triplet. Each different starting point defines a unique potential protein. Changing the starting point changes the reading frame, which changes the final product—the amino acid sequence. Check out the effects of changing the reading frame for the following sequence:

   5′- A G G C A G A A C U A A C C A G G U C U A -3′

Frame 1:	AGG	CAG	AAC	UAA	CCA	GGU	CUA	
	Arg	Gln	Asn	Stop				
Frame 2:	A	GGC	AGA	ACU	AAC	CAG	GUC	UA
		Gly	Arg	Thr	Ile	Gln	Val	
Frame 3:	AG	GCA	GAA	CUA	ACC	AGG	UCU	A
		Ala	Glu	Leu	Thr	Arg	Ser	

4. **ALMOST UNIVERSAL:** Most codons have the same meaning in different species. (Exceptions: mitochondrial DNA, protozoa, and yeast have some deviations from the general genetic code.)

CODON-ANTICODON INTERACTIONS

The genetic code is translated through base-pairing interactions between **mRNA CODONS** and **tRNA ANTICODONS**. The **WOBBLE HYPOTHESIS** explains why cells usually have fewer tRNAs than expected.

ANTICODON—a three-base sequence on tRNA that base-pairs with its complementary codon on mRNA.

Codon-anticodon base-pairings are **ANTIPARALLEL.**

tRNA has an anticodon on one side and a specific amino acid on the other.

EXAMPLE: What is the anticodon for the mRNA codon AGG?

mRNA codon 5′- A G G -3′
base-pairs with
tRNA anticodon 3′- U C C -5′

Remember that base sequences are always written in the 5′ → 3′ direction, so the final answer is the tRNA anticodon is 5′-CCU-3′.

Careful (exam alert!):
Again, codon-anticodon base-pairings are antiparallel, but the base sequences are always written in the 5′ → 3′ direction *(the incorrect 3′ → 5′ order will be one of the options in a multiple-choice question*!)

THE WOBBLE HYPOTHESIS

One tRNA may recognize more than one codon (i.e., a tRNA can have multiple codon-anticodon interactions).

In a codon-anticodon interaction,

- the first two base-pairings confer most of the specificity needed for protein synthesis.
- interactions between the third codon and anticodon nucleotides are not as strict, and nontraditional base pairs can occur.

That third anticodon nucleotide (in the 5′-position) corresponding to the third codon nucleotide in the mRNA sequence (in the 3′-position) is the "wobble" position.

The wobble hypothesis explains why cells usually have fewer tRNAs than expected. As few as 32 tRNAs (31 + a tRNA for initiating protein synthesis) are needed to translate all 61 codons.

AMINOACYL-TRNA SYNTHETASE REACTION:

amino acid + ATP + tRNA → aminoacyl-tRNA + AMP + PP_i

Aminoacyl-tRNA synthetases catalyze attachment of amino acids to tRNAs in two steps: **AMINO ACID ACTIVATION** and **tRNA LINKAGE**. This reaction is irreversible due to the hydrolysis of PP_i.

At least one aminoacyl-tRNA synthetase exists for each amino acid.

Many synthetases have a proofreading site to correct mistakes. Example: If isoleucyl-$tRNA^{Ile}$ synthetase makes Val–tRNA, its proofreading site will fit the Val end (but not Ile) and hydrolyze the incorrect bond.

This reaction consumes ***two ATP equivalents***, even though only one ATP reacts. This is because AMP (not ADP) is produced. AMP needs ***two*** phosphoryl groups to turn it back into an ATP. [Recall that AMP + ATP → 2 ADP]

LINKING AN AMINO ACID TO THE 3′-TERMINUS OF THE CORRECT tRNA:

1. **ACTIVATION OF THE AMINO ACID:** Amino acid + ATP → Aminoacyl-AMP + PP_i

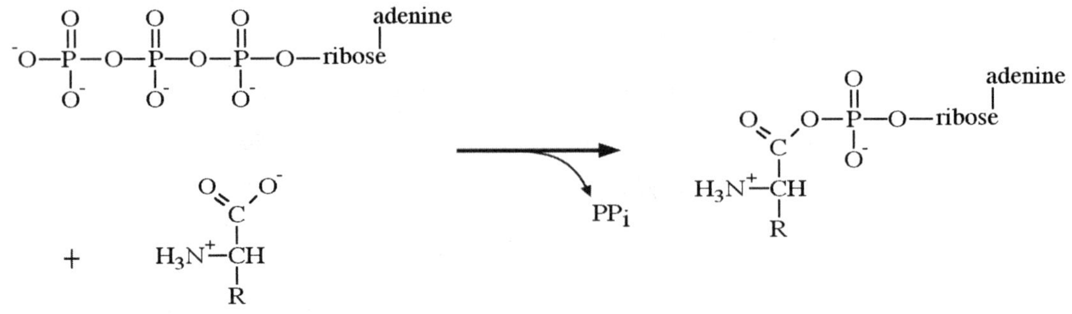

2. **tRNA LINKAGE:** Aminoacyl-AMP + tRNA → Aminoacyl-tRNA + AMP

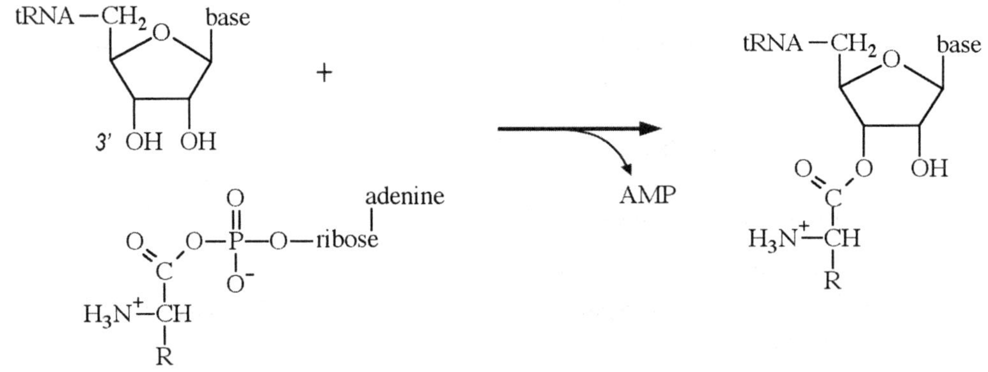

Linkage always happens at the 3′ position on the ribose. The aminoacyl group can move between the 2′ and 3′ positions, but only the 3′-aminoacyl esters are used in protein synthesis.

19.2 PROTEIN SYNTHESIS: TRANSLATION

POLYSOME—an mRNA with several ribosomes bound to it. One mRNA can be read at the same time by several ribosomes.

P-SITE, A-SITE—two ribosomal sites for codon-anticodon interactions
(P = peptidyl) (A = acyl)

Polypeptide synthesis
—proceeds from the N-terminal to the C-terminal because the code is read in the 5′ → 3′ direction.
—requires GTP (as an energy source) plus protein factors.

PHASES OF TRANSLATION: INITIATION, ELONGATION, AND TERMINATION

INITIATION:

1. The **SMALL RIBOSOMAL SUBUNIT** binds an mRNA.

2. **INITIATOR tRNA** base-pairs with the initiation codon AUG on the mRNA.
3. The **LARGE RIBOSOMAL SUBUNIT** combines with the small subunit; the initiator tRNA is bound to the P-site.

ELONGATION:

1. A second aminoacyl-tRNA base-pairs to the A-site.
2. Peptidyl transferase catalyzes peptide bond formation (transpeptidation)—α-amino group of the A-site amino acid attacks the C=O of the P-site amino acid. Now both amino acids are on the A-site tRNA.
3. The P-site tRNA leaves.
4. **TRANSLOCATION**—the ribosome moves along the mRNA. The tRNA with the growing peptide chain moves to the P-site, and the next codon enters the A-site.

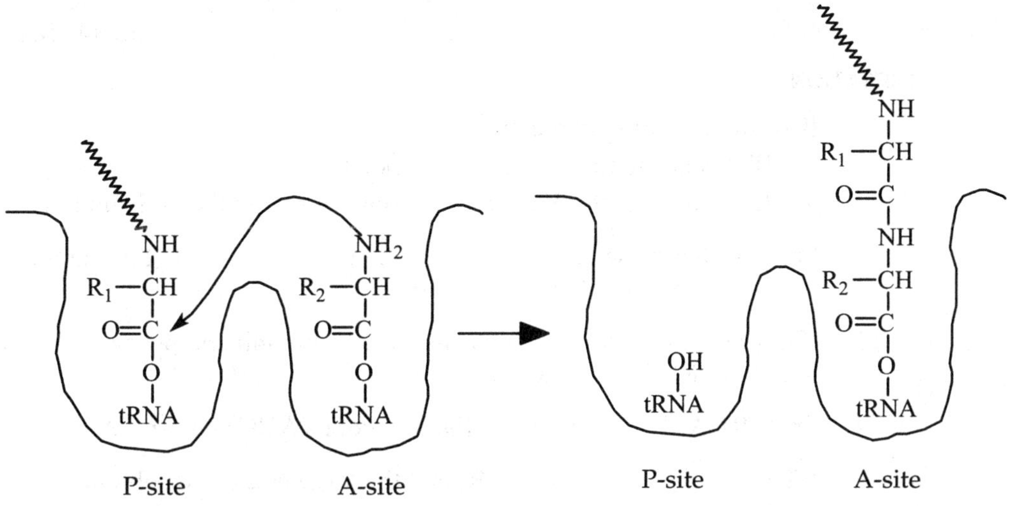

TERMINATION:

1. Stop codon enters the A-site.
2. **PROTEIN RELEASING FACTOR** binds to the A-site.
3. **PEPTIDYL TRANSFERASE** hydrolyzes the polypeptide-tRNA ester bond, releasing the polypeptide chain.
4. The ribosome releases mRNA and dissociates into its subunits.

POSTTRANSLATIONAL MODIFICATIONS—PURPOSES:

—to prepare a polypeptide for its specific function
—**TARGETING**—to direct a polypeptide to a specific location

PROKARYOTIC PROTEIN SYNTHESIS

PROKARYOTIC RIBOSOMES are 70S in size and composed of:

50S (large) subunit (that contains catalytic site for peptide bond formation)
30S (small) subunit (that serves as a guide for regulatory translation factors)

IF = INITIATION FACTOR **EF = ELONGATION FACTOR** **RF = RELEASE FACTOR**

The **SHINE-DALGARNO SEQUENCE** on the mRNA is located upstream from the initiation codon AUG. It binds to a complementary sequence on the small ribosomal subunit and distinguishes between AUG as a start codon and AUG as a methionine codon. Each gene has its own start codon and its own Shine-Dalgarno sequence.

The initiating tRNA is *N*-formylmethionine-tRNA or fmet-$tRNA^{fmet}$. The formyl group also helps to distinguish the start amino acid from an internal methionine.

N-formylmethionine is synthesized on the charged tRNA.

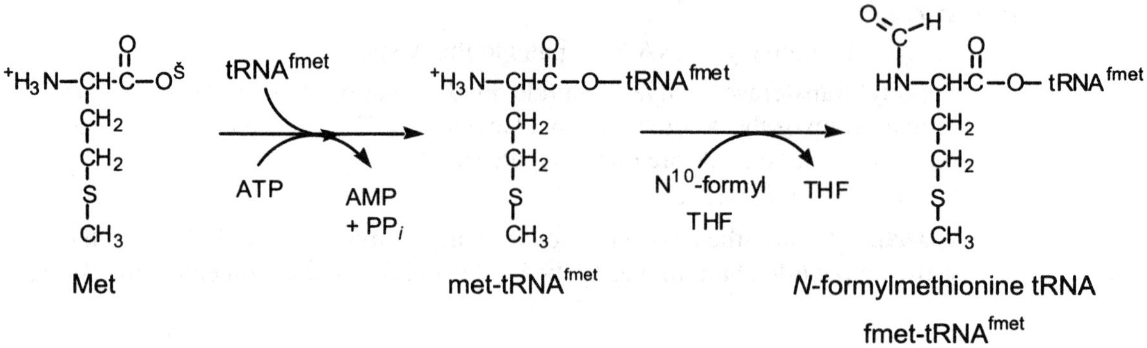

INITIATION

1. IF-1 and IF-3 bind to the 30S subunit
   IF-1 binds to the A-site and blocks it.
   IF-3 prevents the 30S subunit from binding to the 50S subunit too soon.
2. Shine-Dalgarno sequence base-pairs to the 30S subunit, guiding the mRNA, which also binds to the 30S subunit.
3. IF-2 (with a bound GTP) binds to the 30S subunit and promotes binding of the initiating tRNA, fmet-$tRNA^{fmet}$.
4. fmet-$tRNA^{fmet}$ binds to the initiating codon (AUG) on mRNA.
5. GTP hydrolyzes to GDP + PP_i, and the large and small subunits join.

ELONGATION

1. The aminoacyl-tRNA needs to enter the empty A-site of the ribosome complex. An aminoacyl-tRNA binds *first* to EF-Tu-GTP and *then* to the A-site; GTP hydrolysis releases EF-Tu-GDP (and a P_i) from the ribosome.

   EF-Ts regenerates EF-Tu-GTP:

   EF-Ts + EF-Tu-GDP → EF-Ts/EF-Tu-GDP → EF-Ts/EF-Tu + GDP

   EF-Ts/EF-Tu + GTP → EF-Ts/EF-Tu-GTP → EF-Ts + EF-Tu-GTP
2. **PEPTIDYL TRANSFERASE** catalyzes peptide bond formation. Now both amino acids are on the A-site tRNA, and the tRNA on the P-site leaves.
3. **TRANSLOCATION**: The polypeptide-tRNA moves from the A-site to the P-site, and the ribosome moves so that the next codon on mRNA is in the A-site. This stage requires EF-G and GTP hydrolysis.

The elongation process repeats until the ribosome reaches a stop codon.

TERMINATION REQUIRES RELEASING FACTORS RF-1, RF-2, AND RF-3

There are no tRNAs for stop codons (UAA, UGA, UAG). Instead, **RELEASING FACTORS** (RFs) are used. RF-1 recognizes UAA and UAG, and RF-2 recognizes UAA and UGA. (RF-3 may promote RF-1 and RF-2 binding.) A GTP is hydrolyzed, and peptidyl transferase hydrolyzes the polypeptide-tRNA bond, releasing the polypeptide. The mRNA and tRNA dissociates from the ribosome, which separates into its subunits.

POSTTRANSLATIONAL MODIFICATIONS

1. **PROTEOLYTIC PROCESSING:** Cleavage reactions include the removal of formylmethionine and signal peptide sequences. **SIGNAL PEPTIDES** (leader peptides) are short peptides, typically near the N-terminus, that determine a polypeptide's destination.
2. **CONJUGATION** reactions typically form lipoproteins. Some instances of glycosylation (covalently linking carbohydrates to proteins) are also known.
3. **METHYLATION:** Protein methyltransferases use SAM to add methyl groups (reversibly) to amino acid residues of proteins. (Protein methylation plays a role in signal transduction.)
4. **PHOSPHORYLATION/DEPHOSPHORYLATION** are catalyzed by protein kinases and phosphatases; used in chemotaxis and the regulation of nitrogen metabolism.

TRANSLATIONAL CONTROL MECHANISMS IN PROKARYOTES

- Control of the rate of ***transcription*** initiation is the predominant method of translational control.
- Differences in Shine-Dalgarno sequences may cause different translation rates.
- Negative translational regulation (e.g., some ribosomal proteins inhibit the translation of their own or related operons).

EUKARYOTIC PROTEIN SYNTHESIS

EUKARYOTIC RIBOSOMES are 80S in size, with a 40S small subunit and a 60S large subunit. The number and size of rRNAs and ribosomal proteins also differ.

RATE OF SYNTHESIS = ~50 amino acids per minute = *much* slower than prokaryotes (prokaryotic translation rate = ~1,200 amino acids per minute)

eIF = **e**ukaryotic **I**nitiating **F**actor *(at least 12)*
eEF = **e**ukaryotic **E**longation **F**actor
eRF = **e**ukaryotic **R**eleasing **F**actor

Eukaryotic initiation is more complex than prokaryotic initiation, in part because of:

1. **mRNA SECONDARY STRUCTURE:** Eukaryotic mRNA has a cap and a poly(A) tail, and its introns have been removed. Also, mRNA is made inside the nucleus, but the ribosome is outside the nucleus; mRNA might interact with cellular proteins before it gets there. (**RIBONUCLEOPROTEIN PARTICLE** = mRNA + protein)
2. **mRNA SCANNING:** Ribosomes bind to the capped 5′-end and move toward the 3′-end, searching for a translation start site (because there are no Shine-Dalgarno sequences in eukaryotic mRNA).

Eukaryotic elongation and termination are similar to those in prokaryotes.

FEATURES OF EUKARYOTIC TRANSLATION

INITIATION (refer to Figure 19.11, p.747 of your text.)

1. Assembly of the **43S preinitiation complex**:

    small (40S) subunit | eIF-3 | eIF-1A | eIF-2-GTP | met-$tRNA_i$

eIF-3 helps to prevent the 40S subunit from binding to the large (60S) subunit during this phase of initiation.

eIF-2 is a GTP-binding protein[1] that mediates met-tRNA$_i$ binding to the 40S subunit.

met-tRNA$_i$ is an initiating species of methionyl-tRNA^{met}.

2. At the same time, the **CAP-BINDING COMPLEX (CBC)** binds to the mRNA 5′-cap structure.

CBC			
eIF-4A (helicase)	eIF-4E (translation initiation factor)	eIF-4G (scaffold protein)	mRNA 5′-cap

3. The helicase eIF-4B removes any secondary structures in the 5′-UTR that could interfere with initiation.
4. The 43S preinitiation complex binds to the mRNA (at the 5′-end).
5. eIF-G and a **poly(A)-binding protein (PABP)** interact and bring the 3′-poly(A) tail of the mRNA close to the 5′-capped end, causing mRNA to become circular.
6. The initiation complex begins to scan for the start codon, 5′-AUG-3′.
7. When the start codon is recognized, the initiation complex binds the 60S subunit (now dissociated from eIF-6), which triggers the release of the initiation factors eIF-2, eIF-3, eIF-4B, and eIF-1A. An eIF-5 subunit acts as a **GUANINE NUCLEOTIDE ACTIVATING PROTEIN (GNAP)** to hydrolyze the GTP bound to eIF-2; this hydrolysis is involved in the formation of the complete ribosome (Figure 19.12 in your text), the **80S COMPLEX**.

ELONGATION is very similar to prokaryotic elongation.

eEF-1α is analogous to EF-Tu. eEF-1β and eEF-1γ work together to regenerate eEF-1α-GTP from eEF-1α-GDP. (This is similar to the action of EF-Ts). In translocation, eEF-2 is analogous to EF-G.

KINETIC PROOFREADING: if incorrect pairing occurs between the A-site and the eEF-1α-GTP-aminoacyl-tRNA, the complex leaves the A-site.

TERMINATION is also very similar to prokaryotic termination.

GTP binds to eRF-3, which then forms a complex with eRF-1. This complex binds in the A-site when a stop codon (UAG, UGA, or UAA) enters.

GTP hydrolysis promotes dissociation of the eRFs from the ribosome; and peptidyl transferase hydrolyzes the polypeptide-tRNA bond, releasing the polypeptide. The mRNA and tRNA dissociates from the ribosome, which separates into its subunits.

EUKARYOTIC POSTTRANSLATIONAL MODIFICATIONS INCREASE THE STRUCTURAL AND FUNCTIONAL DIVERSITY OF PROTEINS.

- **PROTEOLYTIC CLEAVAGE**: hydrolysis of specific peptide bonds by proteases

  Purpose: a common regulatory mechanism

  Examples: removing N-terminal Met and signal peptides;

[1] eIF-2B, a GEF, regenerates eIF-2-GTP from eIF-2-GDP when it releases its GDP.

conversion of inactive **PROPROTEINS** to their active forms (proinsulin → insulin; zymogens or proenzymes → enzymes) (**PREPROPROTEIN** = proprotein + signal peptide)

- **GLYCOSYLATION** adds sugar groups (secreted proteins have complex oligosaccharides; ER membrane proteins have high mannose species).
- **HYDROXYLATION** of the amino acids proline and lysine by prolyl-4-hydroxylase, prolyl-3-hydroxylase, and lysyl hydroxylase in the **RER**; substrate requirements are highly specific. For example, hydroxylated proline and lysine are required for the structural integrity of collagen and elastin.
- **PHOSPHORYLATION** for metabolic control and signal transduction
- **LIPOPHILIC MODIFICATIONS** covalently attach lipid moieties to proteins to enhance membrane binding capacity and/or certain protein-protein interactions. Most common are acylation (attaches a fatty acid) and prenylation.
- **METHYLATION** by methyltransferases may alter the cellular roles of certain proteins. Methylation of altered Asp residues promotes either the repair or the degradation of damaged proteins. Other examples of amino acid methylation are Lys, His, and Arg.
- **CARBOXYLATION** of glutamyl residues forms γ-carboxyglutamyl residues, which increases a protein's sensitivity to Ca^{2+}-dependent modulation. This carboxylation requires vitamin K and NADPH.
- **DISULFIDE BOND (S–S) FORMATION**: in secreted proteins and certain membrane proteins, because reducing agents (like GSH) in the cytoplasm will reduce –S–S– bonds to –SH + HS–.

  DISULFIDE EXCHANGE—disulfide bonds rapidly migrate from one position to another until the most stable structure is achieved. (S–S bonds are not necessarily formed sequentially.)
- **PROTEIN SPLICING**: An **INTEIN** (internal section of a protein) is cut out of a protein, and the two flanking sections—the **EXTEINS**—are spliced together. Protein splicing is self-catalyzed (i.e., it requires no other enzymes, cofactors, or energy sources).

TARGETING DIRECTS THE PROTEIN TO ITS PROPER DESTINATION

TARGETING MECHANISMS

TRANSCRIPT LOCALIZATION: A specific mRNA binds to receptors in certain cytoplasmic locations. Translation of this localized mRNA results in a cytoplasmic protein gradient, that is, an asymmetrical distribution of this protein in the cytoplasm.

SIGNAL PEPTIDES are sorting signals that target polypeptides to their proper location (for secretion or for use in the plasma membrane or any of the membranous organelles). Signal peptides help to insert the polypeptide that contains it into an appropriate membrane.

THE SIGNAL HYPOTHESIS

—was proposed to explain how polypeptides translocate across RER membrane.

SIGNIFICANCE OF THE SIGNAL HYPOTHESIS: It helps to explain the ability of proteins to be specifically targeted to their proper location. The fate of a targeted polypeptide depends on the location of the **SIGNAL PEPTIDE** and other signal sequences.

SIGNAL RECOGNITION PARTICLE (SRP) binds to a ribosome and interrupts translation. The SRP then mediates binding of the ribosome to RER via a **DOCKING PROTEIN** (**SRP RECEPTOR PROTEIN**). Translation restarts, the growing polypeptide inserts into the membrane, and SRP is released.

TRANSLOCON—an integral membrane protein complex believed to mediate polypeptide translocation after SRP is released.

COTRANSLATIONAL TRANSFER—simultaneous translocation of a polypeptide during ongoing protein synthesis

POSTTRANSLATIONAL TRANSLOCATION—previously synthesized polypeptides are pulled across the RER membrane by an ATP-binding peripheral translocon-associated protein (hsp70).

After a protein is in the RER, it typically undergoes initial posttranslational modifications and is transferred to the Golgi complex via transport vesicles that bud off from the ER and fuse with the *cis* face of the Golgi membrane. Further modifications are made inside the Golgi complex. Transport vesicles exit from the *trans* face of the Golgi and move to target locations.

TRANSLATION CONTROL MECHANISMS IN EUKARYOTES

mRNA EXPORT

The spatial separation of transcription and translation that is afforded by the nuclear membrane appears to provide eukaryotes with the opportunity to control translation. Export through the nuclear pore complex is a carefully controlled, energy-driven process whose minimum requirements include a 5′-cap and a 3′-poly(A) tail.

mRNA STABILITY = HOW WELL mRNA CAN AVOID DEGRADATION BY NUCLEASES

In general, the translation rate of any mRNA species is related to its abundance, which in turn is dependent on both its rates of synthesis and degradation. The length of the poly(A) tail is significant and affects its stability.

NEGATIVE TRANSLATIONAL CONTROL

The translation of some mRNAs is controlled by the binding of repressor proteins to the 5′-ends of the mRNA. This effectively blocks ribosome binding and scanning.

INITIATION FACTOR PHOSPHORYLATION

The phosphorylation of eIF-2 in response to certain stimuli (e.g., heat shock, viral infections, and growth factor deprivation) has been observed to decrease protein synthesis. However, the translation of certain mRNA increases (e.g., hsp synthesis in response to heat shock).

TRANSLATIONAL FRAME-SHIFTING

This process, often observed in retroviruses-infected cells, allows the synthesis of more than one polypeptide from a single mRNA.

BIOCHEMISTRY IN PERSPECTIVE:

TRAPPED RIBOSOMES: RNA TO THE RESCUE

BIOCHMISTRY IN PERSPECTIVE:

CONTEXT-DEPENDENT CODING REASSIGNMENT

SELENOCYSTEINE; SECIS ELEMENT; PYRROLYSINE

BIOCHEMISTRY IN THE LAB: PROTEOMICS

GOALS OF PROTEOMICS:

- to study the global changes in the expression of cellular proteins over time.
- to determine the identity and the functions of all proteins produced by organisms.

TOOLS OF PROTEOMICS include **TWO-DIMENSIONAL GEL ELECTROPHORESIS** and **MASS SPECTROMETRY.**

CHAPTER 19: SOLUTIONS TO REVIEW QUESTIONS

19.1 a. codon – an mRNA triplet base sequence that specifies the incorporation of a specific amino acid into a growing polypeptide chain during translation or acts as a start or stop signal

b. anticodon – a tRNA base triplet that is complementary to a codon on an mRNA, thereby specifying the insertion of the correct amino acid during protein synthesis

c. genetic code – the set of nucleotide base triplets (codons) that code for amino acid in proteins as well as start and stop signals.

d. open reading frame – a series of triplet base sequences in mRNA that do not contain a stop codon

e. codon usage bias – the preference of an organism for specific synonymous codons in polypeptide synthesis

19.2 a. wobble hypothesis – the explanation of the observation that cells often have fewer tRNAs than expected; freedom in the pairing of the third base of the codon to the first base of the anticodon allows some tRNAs to pair with several codons.

b. cognate tRNA – the tRNA that associates with a specific amino acid

c. AUG sequence – the initiation codon in an mRNA

d. Shine-Dalgarno sequence – a purine rich sequence on an mRNA close to

AUG (the initiation codon) that binds to a complementary sequence on the 30S ribosome subunits thereby promoting the formation of the correct preinitiation complex

e. amino acyl-tRNA synthetase – an enzyme that catalyses the attachment of an amino acid to its cognate tRNA

19.4 a. signal peptide – a short sequence typically near the amino terminal of the polypeptide that determines its insertion into a membrane of an organelle

b. preinitiation complex – the multisubunit small ribosomal subunit protein complex, formed during the initiation phase of eukaryotic protein synthesis, that is able to bind to an mRNA

c. initiation complex – the ribosome-protein complex required to initiate the first step in the translation of mRNA

d. PABP – poly(A) binding protein - a eukaryotic scaffold protein that interacts with eIF-G to bring the 3'-poly(A)-tail of the mRNA close to the 5'capped end, forming a circular mRNA molecule that increases the overall efficiency of translation; PARP is a component of cap-binding complex

e. cap-binding complex – eTF-4F(CBC) – a complex that binds to the mRNA 5'–cap structure prior to mRNA binding to the 43S preinitiation complex; the CBC consists of eIF-4A (a helicase), eIF-4E (a translation initiation factor), and eIF-G (a scaffold protein)

19.5 a. mRNA scanning – the mechanism used by eukaryotic ribosomes to migrate along an mRNA in order to identify the initiating AUG sequence

b. transcript localization – the creation of cellular protein gradients by the binding of mRNA transcripts to receptors in certain cytoplasmic locations

c. glycosylation – a posttranslational mechanism whereby carbohydrate groups are covalently attached to polypeptides

d. targeting – a series of mechanisms that directs newly synthesized polypeptides to their correct cellular locations

e. lipophilic modifications – covalent attachment of lipid moieties to proteins to improve membrane–binding capacity and/or certain protein–protein interactions

19.7 a. SRP – signal recognition particle – a multisubunit complex that recognizes RER–directing signal sequences and mediates binding of the ribosome to the ER

b. translocon – a pore-containing protein complex that facilitates the translocation of polypeptides across the RER membrane and their subsequent processing

c. docking protein – SRP recognition protein – a GTPase heterodimer that mediates the binding of the ribosome to the ER

d. SRP receptor protein – docking protein - a GTPase heterodimer that mediates the binding of the ribosome to the ER
e. signal peptidase – the enzyme that removes signal peptides from nascent polypeptides

19.8 a. cotranslational transfer – the simultaneous synthesis and transmembrane transfer of a polypeptide

b. posttranslational transfer – previously synthesized polypeptides are transported across the RER membrane

c. TOM receptor protein – translocase of the mitochondrial outer membrane – a protein complex containing a protein–conducting channel on the outer mitochondrial membrane

d. TIM complex – translocase of the mitochondrial inner membrane – a transmembrane channel that transports proteins across the inner mitochondrial membrane driven by the electrochemical proton gradient

e. mthsp70 – molecular chaperone that facilitates the transport of proteins targeted to the mitochondrion

19.10 a. nascent – newly synthesized

b. signal hypothesis – a mechanism that explains how secreted or membrane proteins are synthesized on ribosomes bound to the

rough endoplasmic reticulum; a sequence of amino acid residues on the nascent polypeptide chain that mediates the insertion of the polypeptide into the RER membrane.

c. posttranslational modification – one of a set of reactions that alter the structure of newly synthesized polypeptides

d. context–dependent codon reassignment – variation in the genetic code in which a specific codon codes for an amino acid (or a stop signal) different from that which typically occurs; an example is the stop codon UAG that in some methane–producing Archaea codes for the nonstandard amino acid pyrolysine

e. dipthamide – a unique modification of a histidine residue; occurs at a specific location in eEF-2

19.11 a. proteomics – the investigation of protein synthesis patterns and protein–protein interactions

b. protein chip – protein microarray – a slide to which are attached specific molecules that are used to identify protein molecules extracted from living organisms

c. yeast two–hybrid screening – a technique used to identify functional protein-protein interactions

d. mass spectroscopy – an analytical technique in which molecules are vaporized and then bombarded by a high-energy electron beam, causing them to fragment as cations

e. mixed anhydride – an acid anhydride with two different R groups

19.13 The observations upon which the wobble hypothesis is based are: (1) the first two base pairings in a codon-anticodon interaction confer most of the specificity required during translation, and (2) the interactions between the third codon and anticodon nucleotides are less stringent. Because of the "wobble rules," only a minimum of 31 tRNAs are required for the translation of all 61 codons.

19.14 The sequential reactions that occur within the active site of aminoacyl-tRNA synthesis are (1) the formation of aminoacyl-AMP, which contains a high-energy mixed anhydride bond, and (2) linkage of the aminoacyl group to its specific tRNA.

19.16 In prokaryotes most of translation control occurs at the level of transcription initiation because transcription and translation are temporally linked. Other central mechanisms include the short half lives of mRNAs, which limits their use, and variations in Shine-Dalgarno sequences, which affects the rate of translation initiation.

19.17 During the elongation phase of protein synthesis, the second aminoacyl-RNA becomes bound to the ribosome in the A site. Peptide bond formation is then catalyzed by peptidyl transferase. Subsequently, the ribosome is moved along the mRNA by a mechanism referred to as translocation.

19.19 A signal recognition particle (SRP) is a large complex composed of protein and RNA that binds to a ribosome that has begun translating a polypeptide possessing a signal peptide component. Once the SRP has bound to the ribosome translation is temporarily arrested. The SRP then mediates ribosomal binding to docking proteins on the surface of a membrane (e.g. RER membrane). Translation subsequently recommences, and the growing polypeptide inserts into the membrane.

19.20 Cotranslational transfer is a process in which nascent polypeptides are inserted through an intracellular membrane during ongoing protein synthesis. An integral membrane protein complex referred to as the translocon, mediates the transfer of polypeptides (each of which contain some hydrophilic residues) across the hydrophobic core of the membrane.

19.22 Synthesis of a secretory glycoprotein begins on a ribosome. An appropriate signal peptide mediates the translocation of the polypeptide into the ER lumen. The core N-linked oligosaccharides are then covalently linked to appropriate asparagine residues in the polypeptide in a reaction catalyzed by glucosyl transferase. Subsequently, the molecule is transferred in transport vesicles to the Golgi complex where additional glycosylation reactions occur. Eventually, the glycoprotein is incorporated into secretory vesicles that migrate to the plasma membrane. Secretion of the glycoprotein then occurs via exocytosis.

19.23 tRNAs are adaptor molecules because they bind to specific amino acid molecules and then position those molecules in the ribosome according to the base sequence of an mRNA. In other words, they bridge the gap between the base code of the nucleic acids and the amino acid sequence of polypeptides.

19.25 The proteins involved in the initiation of prokaryotic protein synthesis are: IF-1 (binds to the A site of the 30S subunit, blocking it during initiation), IF-2 (binds to the 30S subunit and promotes the binding of the initiating tRNA to the initiation codon of mRNA), and IF-3 (prevents the 30S subunit from binding prematurely to the 5OS subunit).

19.26 The prokaryotic 50S subunit contains 23S and 5S rRNAs and 34 proteins. The smaller prokaryotic subunit (30S) consists of a 16S rRNA and 21 proteins. The eukaryotic 60S ribosomal subunit is composed of 28S, 5S and 5.8S rRNAs and 47 proteins. The smaller 40S subunit is composed of an 18S rRNA and 32 proteins.

19.28 Answer d is the process of translation.

19.29 The major classes of eukaryotic posttranslational modification are proteolytic cleavage (the hydrolysis of specific peptide bonds), glycosylation (attachment of sugar residues to specific amino acid residues in the protein), hydroxylation (the adding of OH groups to proline and lysine residues), phosphorylation (the addition of phosphate groups to specific amino acid residues on a protein), lipophilic modification (covalent attachment of lipid groups to a protein), methylation (attachment of methyl groups), and disulfide bond formation (formation of –S-S- bonds between cysteine residues).

19.31 The aminoacyl-tRNA synthetases correctly attach each amino acid to its cognate tRNA and proofread the product. This process increases the accuracy of protein synthesis.

19.32 Since there are four serine codons and assuming that there is no codon usage bias the total number of different mRNA sequences is 4^{10} or 1048576.

19.34 In eukaryotes polypeptides are targeted to their final destination via transcript localization (transport of mRNAs and their subsequent binding specific cellular receptors) and signal peptides, polypeptide segments that allow binding to specific membrane-translocating complexes.

19.35 Eukaryotic translation is controlled by (1) the metabolic status of a cell (primarily via the mTORC1 signaling pathway); (2) mRNA export through nuclear pores; (3) mRNA stability (e.g., presence or absence of nuclease resistant sequences such as palindromes and poly (A) tails; and (3) negative translation controls in which in certain circumstances repressor proteins bind to sequences near the mRNA's 5'-end.

19.37 The phases of protein synthesis during which each process occurs are as follows: a. initiation, b. elongation, c. elongation, and d. termination.

19.38 The total amount of nucleotide bond energy that is required to synthesize Lys-Ala-Ser-Val is equivalent to the hydrolysis of either 16 or 17 phosphoryl groups from ATP or GTP, determined as follows:

1 GTP to form the complete ribosome and initiate translation

8 ATP to create 4 aminoacyl-tRNAs; 2 ATP equivalents per amino acid

6 GTP to create 3 peptide bonds; 1 GTP for elongation and 1 GTP for translocation for each peptide bond formed

1 or 2 GTP to terminate translation; 1 GTP in eukaryotes, 2 GTP in prokaryotes

19.40 There are many correct answers to this question. The three-letter sequences listed below each amino acid in the first table below are the possible mRNA sequences that code for that specific amino acid. An asterisk "*" is used to designate any of the four possible nucleotides. For example, GC* is either GCU, GCC, GCA, or GCG. Choose one 3-letter sequence from each column to build an mRNA sequence that will code for this peptide

19.41 The deletion of a single DNA base will alter the amino acid sequence in the polypeptide produced from mRNA, and may truncate the polypeptide, depending upon the base in the Ser codon that was deleted. As noted in Solutions 19.39 and 19.40, many possible base sequences will code for this polypeptide. Using the solution from Solution 19.39:

5′-	Ala	Ser	Phe	Tyr	Ser	Lys	Lys	Leu	Ala	Asp	Val	Ile	-3′
DNA: 3′-	CGA	AGA	AAA	ATA	AGA	TTT	TTT	AAT	CGA	CTA	CAA	TAA	-5′

Removing the first base, A, from the second Ser residue results in the following altered DNA and mRNA base sequences and a new polypeptide that corresponds to this sequence.

DNA: 3′-	CGA	AGA	AAA	ATA	GAT	TTT	TTA	ATC	GAC	TAC	AAT	AA	-5′
mRNA: 5′-	GCU	UCU	UUU	UAU	CUA	AAA	AAU	UAG	CUG	AUG	UUA	UU	-3′
Amino Acid	Ala	Ser	Phe	Tyr	Leu	Lys	Asn	Stop					

19.43 One possible codon sequence for the peptide sequence is GGUAGUUGUAGAGCU. The number of possible codons for the amino acids in this peptide sequence is as follows: glycine (4), serine (6), cysteine (2), arginine (6), and alanine (4). The total number of possible codon sequences for this peptide sequence is therefore 1152

19.44 The nucleotide GTP is the source of the energy required to drive conformational changes in GTP-binding translation factors that facilitate the translation mechanism.

CHAPTER 19: SOLUTIONS TO FILL-IN-THE-BLANK QUESTIONS

19.46 Termination

19.47 To direct a polypeptide to a specific site

19.49 Preproteins

19.50 Posttranslational modifications

19.52 Phosphodiester bond

19.53 Wobble hypothesis

CHAPTER 19: SOLUTIONS TO SHORT-ANSWER QUESTIONS

19.55 Mitochondrial proteins, synthesized as preproteins, bind to a multichaperone complex composed of the ATPases hsp70 and hsp90. The preprotein-chaperone complex then docks in an ATP-dependent process with receptors of the TOM complex, which is embedded in the outer mitochondrial membrane. As soon as the N-terminal signal sequence emerges from the TOM complex entering the intermembrane space, it binds to a receptor of the TIM complex. Transport of the preprotein through the TIM complex in the inner membrane is driven by the proton gradient. After the signal sequence is removed, the newly arrived matrix protein folds into its native conformation in a process assisted by ATP-dependent chaperones.

19.56 The signal hypothesis explains how polypeptides synthesized by cytoplasmic ribosomes are translocated across the RER membrane. As soon as the first 70 amino acid residues of a polypeptide emerge from a ribosome, its signal sequence binds to the signal recognition particle (SRP). The subsequent binding of the SRP-signal sequence to the SRP receptor in the RER membrane, driven by GTP hydrolysis, results in the binding of the ribosome to the translocon complex and the dissociation of the SRP from its receptor. Polypeptide synthesis then resumes and continues until translation is completed. The signal peptide is removed by signal peptidase in the RER lumen.

19.58 Each type of GEF is a protein that activates monomeric GTPases by stimulating the release of GDP to allow the binding of GTP. For example, in prokaryotic protein synthesis the elongation factor EF-Ts is a GEF that displaces GDP from EF-Tu, the elongation factor that positions aminoacyl-tRNAs in the A site within a ribosome. When GTP subsequently binds to EF-Tu, the newly formed EF-Tu-GTP can then bind another aminoacyl-tRNA.

19.59 The two types of anhydride bonds are distinguished as follows. In an anhydride bond, two carbonyl groups are linked through an oxygen atom. A mixed anhydride bond is formed from two different acids, in this case a carboxylic acid group of the amino acid and phosphoric acid group of AMP. The formation of an amino acyl-AMP is driven to completion because of the hydrolysis of pyrophosphate.

CHAPTER 19: SOLUTIONS TO THOUGHT QUESTIONS

19.61 Despite considerable species differences in the amino acid and nucleotide sequences of ribosomal proteins and RNA, respectively, the overall three-dimensional structures of these molecules are remarkably similar. This similarity is presumably due to high selection pressure. In other words, ribosomal function is such an important factor in species viability that evolution has conserved their tertiary structure.

19.62 Because the accuracy of protein synthesis depends directly on codon-anticodon interactions, the specificity with which t-RNAs are linked to amino acids is critically important. The process in which the amino acid-tRNA synthetases catalyze the covalent binding of each of the t-RNAs with its correct amino acid has, therefore, been referred to as the second genetic code.

19.64 Although they differ in structures and complexity, translation factors in prokaryotes and eukaryotes have the same basic functions. They facilitate the formation of the ribosomal initiation complex, the positioning of aminoacyl-tRNA in the A site, translocation, and translation termination.

19.65 Four high-energy phosphate bonds are required to incorporate each amino acid into a polypeptide (i.e., 2 GTP and 2 ATP). The polymerization of 200 amino acids requires 400 GTP and 400 ATP.

19.67 Each Shine-Dalgarno sequence in a prokaryotic mRNA occurs near a start codon (AUG). The Shine-Dalgarno sequence provides a mechanism for promoting the correct alignment of the start codon on the ribosome (as opposed to a methionine codon) because it binds to a nearby complementary sequence in the 16S rRNA component of the 30S ribosome. Eukaryotic ribosomes identify the initiating AUG codon by binding to the capped 5′ end of the mRNA and scanning the molecules for a translation start site.

19.68 While you can go directly and predictably from a nucleotide sequence to one and only one amino acid sequence, the reverse is not true because of the degeneracy of the genetic code.

19.70 Preproteins contain signal sequences that direct them to the ER for translocation and Golgi for modification. The cleavage of an inactive protein and other posttranslational modification processes ensures that the protein is active only when it has been targeted to its site of function.

19.71 The principal factors that ensure accuracy in protein synthesis are codon-anticodon base pairing and the mechanism by which amino acids are linked to their cognate tRNAs. The level of accuracy of protein synthesis, while quite high, is still less than that achieved during replication or transcription.

19.73 The process would be similar to the pyrolysine insertion outlined

in the text. Assuming that the pyrovaline is available from a metabolic pathway, the following circumstances must occur:

1. A codon is assigned to pyrovaline.

2. A tRNA with the requisite anticodon is available.

3. An aminoacyl tRNA synthetase binds pyrovaline to its cognate tRNA.

4. A stem-loop or similar structure upstream of the newly assigned codon in the mRNA promotes the codon reassignment.

5. The tRNA bound to pyrovaline would then enter the ribosome, where its amino acid is incorporated into the protein.